In the Sun King's Cosmos

Rethinking the Early Modern

Series Editors
Marcus Keller
Ellen McClure
Feisal Mohamed

In the Sun King's Cosmos

Comets and the Cultural Imagination of Seventeenth-Century France

Claire Goldstein

Northwestern University Press
Evanston, Illinois

Northwestern University Press
www.nupress.northwestern.edu

Printed in the United States of America

10 9 8 7 6 5 4 3 2 1

ISBN 978-0-8101-4812-3 (cloth)
ISBN 978-0-8101-4813-0 (ebook)

Cataloging-in-Publication Data are available from the Library of Congress.

CONTENTS

ILLUSTRATIONS

ACKNOWLEDGMENTS

This is not a very long manuscript, but I have worked on it for many years, supported by a great number of friends, colleagues, family members, and doctors. I cannot fully express how grateful I am to be able to offer some of them a small acknowledgment for their tremendous and sustaining presence in my life and the many contributions they made to this undertaking.

It is my great joy to thank my wonderful doctor, the brilliant and caring Dr. Jonathan Wesley Riess, as well as the legion clinicians, researchers, and clinical trial participants who have made the past six-plus years and millions of beautiful experiences possible post–stage 4 cancer diagnosis. I am grateful also to the worldwide community of ALK Positive, who prop me up with information and empathy. I might never have had the audacity to complete this manuscript without a challenge from the late great Gina Hollenbeck to get back to work on something hard and meaningful. Stars every time, always for you, dear Gina.

I have had the good fortune to find intellectual and professional homes in two wonderful departments of French and Italian, at Miami University (Oxford, OH) and the University of California, Davis. Thanks to all my dear colleagues, with special thanks to Jonathan Strauss and Jeanne De Groote Strauss, Elisabeth Hodges, Anna Klosowska, the late Jim Creech, Paul Sandro, and Noah Guynn. My research was supported at Miami by the L. P. Irvin Fund and university research grants, and at UC Davis by the Academic Senate Small Grants for Research and a quarter in residence at the Davis Humanities Institute. Funds from the UC Davis Dean of Letters and Science and the Office of Research helped support the publication of this book. I have been supported in this work by librarians at UC Davis, the BNF, the Bibliothèque Municipale de Lyon, and the College of Physicians in Philadelphia.

This project germinated in conversation with the kind, funny, discerning, and field-changing editor Jerry Singerman, to whose confidence in me I have always felt I owe this career I have so loved. How grateful I am to have found a new home for my comets at Northwestern University Press, and such a talented editorial collaborator in the magnificent Faith Wilson Stein, whose incisive suggestions have made this book better. This project was enriched by Maia Rigas's sharp eye and deep reserve of patience. Madeline Schultz helped these comets shine in the world. I am grateful to the series editors, Marcus Keller, Feisal Mohamed, and especially my colleague and interlocutor Ellen McClure, for offering my book a place in such good company. Two extraordinary readers offered insightful engagement that strengthened the finished product in myriad ways.

My thinking about comets and other spectacular things was greatly enriched by invitations to share my work in various stages of progress at Harvard, the University of Kentucky, the University of Minnesota TEMS workshop, UNC Chapel Hill, Indiana University, Bowdoin College, the University of Pittsburg, and the University of Sydney Powers Institute. Thank you to colleagues Sylvaine Guyot, Jeffrey N. Peters, Juliette Cherbuliez, Ellen Welch, Allison Calhoun, Hall Bjørnstad, Katherine Dauge-Roth, Chloe Hogg, Mark De Vitis, and Mark Ledbury for these opportunities. What good fortune that so many years ago I picked a field full of generous and inspiring colleagues, also friends, who have challenged and nurtured me and my work. In addition to those I have just mentioned, I am overwhelmingly grateful for Faith Beasley, Katherine Ibbett, and Allison Stedman and the wider SE17 community, which is a collegial and stimulating intellectual home.

This project was nurtured in its earliest stages by conversations and draft exchanges with my brilliant friend Cindy Klestinec and Chloe Hogg, precious interlocutor of the past thirty years; feedback from the Miami University Early Modern Collective; and the generously intense engagement of the Miami University Department of French and Italian Works in Progress Series. When I moved to California, Seeta Chaganti welcomed me into a writing group with Gina Bloom, Mark Jerng, and Margaret Ronda. My colleagues in UC Davis Early Modern Studies, especially Daniel Stolzenberg, Tiffany Worth, and Margie Ferguson, shared generously of their knowledge and insights. Douglas Boyce agreed to consult on a lute piece about comets, and instead composed a partita and enlisted the talented Dan Lippel to travel to California with him to perform "La Comète" on my home campus. It was a marvelous and surreal experience to share this work with friends and students in this celebratory setting. Fran Dolan started as an idol and became a true friend. Her radiant spirit keeps me afloat on the bleakest days and the best ones. I simply could not have persevered through these years without the kindness, humor, and constant friendship of my poddies Fran and Tiffany and the Three Ladies lunch and writing group: the original Claire (Waters) and Anna Uhlig.

I am lucky for superlative friends near and far. Michal Kurlaender and Bryce, Noa, and Oren Vinokurov welcomed me and my family into their circle of warmth. Sareena Malhi propelled me forward with the sheer force of friendship. Anna, Stefan, and Solomon Uhlig became family. Chloe Paul asked about my progress every Friday in the crosstown French carpool. Thank you for all the love Karen Popernik Aitken; Jessica Butzel; Mark De Vitis; Jeff Hurok; Rayna Kalas; Prithvi Mruthyunjaya; Rachel Posner; Martha Schoolman and Caitlin Wood; Laurie Stillman; my CBH community; my family, with a special thanks to my lovely cousin Aaron Macks who patiently shared insight about historical dance; also the Jansen-van Tulder family; Håkan Palm; my beloved and always supportive brother Josh; and my indominable mother and first and forever model of an engaged intellectual (who reads every footnote), Athene Schiffmann Goldstein.

This book is dedicated with overflowing love and gratitude to my most precious family and best friend, without whom neither this book nor this beautiful life would be possible:

My kind and generous children, Asher and Noam: being your parent has been the absolute joy of my life. I love you so much. I am so very proud of the magnificent people you are and full of gratitude for every day together.

SER, you read every word, fill my fountain pens with lovely colors, serenade me, love me, hold me up when I am falling, and make me better in every way. I love you eternally.

And Juliette Cherbuliez: I decided we would be friends on my graduate school visit. Thank you for sharing every single thing these last three decades. I don't know what or who I would be without you. It's all sweeter, better, more possible because we have tackled it together.

In the Sun King's Cosmos

INTRODUCTION

From 1680 to 1683 Venetian cosmographer Vincenzo Coronelli worked on the fabrication and decoration of a pair of massive globes, one terrestrial, one celestial, commissioned by the cardinal César d'Estrées as a present for King Louis XIV. Constructed of layers of plaster and canvas over intricately crafted wood supports, the spheres measure close to four meters in diameter and weigh 3.2 tons each.[1] Alongside landmasses and oceans, the brilliant, polychrome terrestrial globe features minute inscriptions and detailed depictions of mythological and classical iconography; navigation routes; naval battles; coffee, tea, and sugar cane production; flora and fauna; and the customs and habits of various cultures.[2] On the celestial globe, rendered exclusively in shades of blue, brass studs of various sizes mark the positions and relative brightness of the stars and planets on September 5, 1638, at 11:15 in the morning. The event memorialized on Coronelli's celestial globe was a royal birth: after twenty-three years of marriage and four stillbirths, a son had been born to King Louis XIII of France and his queen, Anne of Austria.[3] Louis Dieudonné, God-given Louis, was crowned king Louis XIV in 1643 at age four.

The globes designed as a present for the now-adult king make concrete the proposition that his reign was, and had always been, destined to extend over land, ocean, and even the heavens.[4] On the celestial globe, however, one anachronism intrudes on the frozen temporality of the king's birthday sky. While all of the planets and stars record their exact positions at the moment of the royal birth, Coronelli also painted a scattering of golden comets from other historical moments, including the 1607 comet in a position observed by Johannes Kepler, Isaac Newton's placement of the comet in the winter of 1664–65, and the comet seen in the winter of 1680–81 as mapped by French royal astronomer Giovanni Domenico Cassini. A final comet—observed and charted during the three-year period in which Coronelli was at work on the globes—would later come to be known as Halley's comet. This comet, and the Newtonian mathematics that would allow Edmond Halley to describe its path, would eventually bring to an end a fascinating period in which comets, the last celestial bodies whose orbit seemed to lie beyond the grasp of mathematics and observation, troubled the

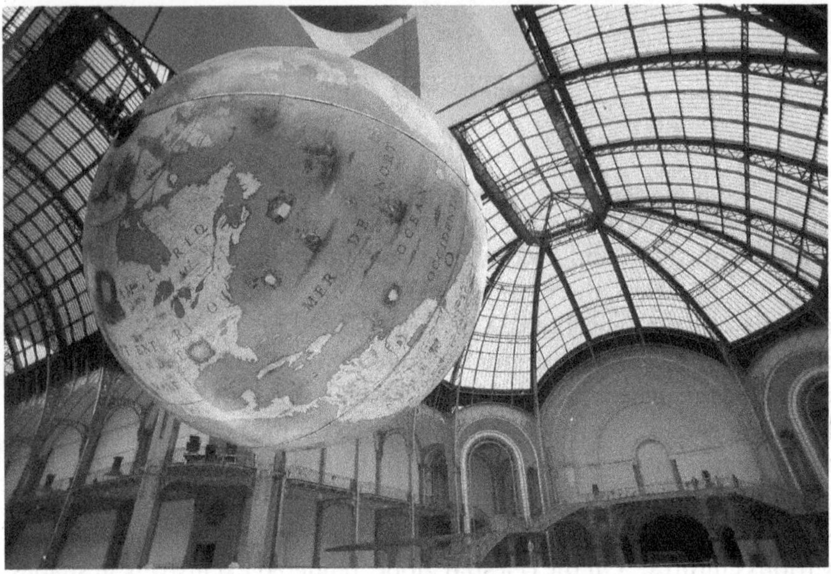

Fig. 1. Vincenzo Coronelli, Terrestrial Globe (1680–83). Installation view, Grand Palais. Bibliothèque nationale de France.

orderly ballet of the heavens. During the period this book examines, however, and during the period in which Coronelli was at work on Louis XIV's globes, comets still seemed utterly unpredictable. They eluded all attempts to discern or forecast regularity in their movements; they seemed to defy cosmic order.[5]

The cultural effects of this disruptive element are the subject of this book. *In the Sun King's Cosmos* is a study of how unusually bright comets visible in 1664–65 and 1680–81 appeared not only in the sky but also in ballets and theater, letters and early journalism, architecture and institutions, theology and literary style.[6] I trace how comets—considered at the time to be without discernable form or pattern—nonetheless organized curiosity, scrutiny, and doubt regarding the epistemological status of observation, as well as also how they crystalized alternative—nonofficial, sometimes contestatory—networks in which information, bodies, and texts circulated against the current of a centralizing force that sought to dictate the shape, availability, and dissemination of knowledge.

Comets galvanized public attention on all levels of society, inspiring what astronomer, priest, and mathematician Alexandre-Gui Pingré (1711–96) called *cométomanie*, a comet mania that drew the learned and the illiterate, the fashionable, the seedy, and the serious

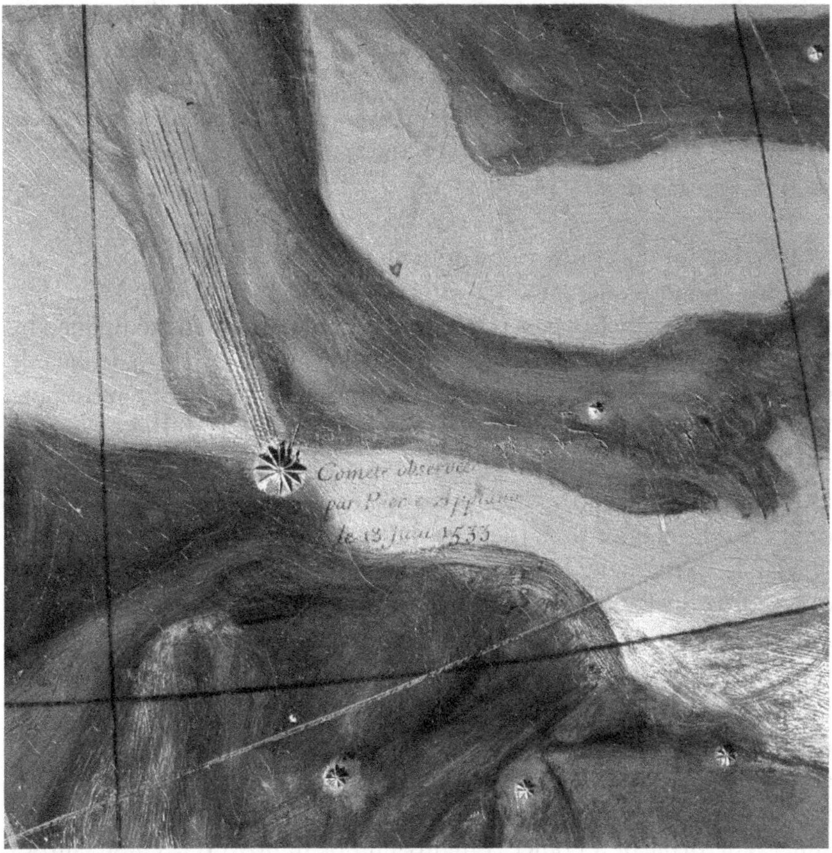

Fig. 2. Vincenzo Coronelli, Celestial Globe (1680–83), detail. Bibliothèque nationale de France.

outdoors into public squares and onto bridges to view the recurring bright spectacle in the night sky and inspired songs, sermons, poems, learned guidebooks, fashions, games, and much debate.[7] In December 1664, Marie de Rabutin-Chantal, marquise de Sévigné (1626–96), the formidable epistolary chronicler, writes to the marquis de Pomponne (1618–99) about a high-profile trial all of influential France is following. She concludes by recounting how a comet has likewise captured public attention:

> First of all, there is a comet that appeared four days ago. At first it was only discussed by women, people mocked them for it; but now everyone has seen it. Monsieur d'Artagnan stayed up last

> night and saw it very comfortably. Monsieur de Neuré the great astrologer said that it is of considerable size. I saw Monsieur de Foix who saw it with three or four *savants*. I, who am writing to you, am going to stay up tonight to see it too: it appears at three; I am informing you of it, you can have the pleasure or the displeasure of it.[8]

At first, talk of the comet is the domain of women and it is risible ("elle n'a été annoncée que par des femmes, on s'en est moqué"). Yet in just four days, curiosity has penetrated throughout society, as exemplified by Sévigné's enumeration of people, from the musketeer Charles de Batz de Castelmore d'Artagnan (1611–73) to the mathematician/philosopher Mathurin de Neuré (year of birth unknown; d. 1677) to the *savants* to the marquise herself, "Moi qui vous parle." It is the news that she sees fit to pass to her friend, exiled on his estate, to keep him in the loop and offer him the chance to participate in this act of collective wonder.

Visible to the naked eye, comets are part of the earliest astronomical records of diverse civilizations, from Greece to China.[9] As planets and stars were emplotted with increasing certainty, and as observational techniques improved, comets began to draw special attention. Published in 1627, Johannes Kepler's Rudolphine Tables charted with startling accuracy the past, present, and future position of planets and stars.[10] Kepler based his tables on the Copernican assertion that each planet and star followed a stable and predictable orbit around the sun.[11] The orbit of a comet is highly elliptical, however. These ellipses cannot be described using the tools of geometry; they require calculus, which would be proposed at the end of the seventeenth century by rival mathematicians Gottfried Wilhelm Leibniz and Newton. Without calculus, no matter how attentively learned people recorded and charted the successive locations of a comet, available mathematical models could not convincingly connect these dots into a discernable orbit. Unable to map a comet's observed path in a way that integrated all its successive positions into a recognizable shape, neither amateur nor professional astronomers or mathematicians could describe or predict its movement. Theories about the makeup, location, and motion of comets proliferated, as did dissent over models, measurements, and calculations.[12] It was not until Newton supplied the math, in his *Philosophiae Naturalis Principia Mathematica* (*Mathematical Principles of Natural Philosophy*), presented at the Royal Society in 1686 and published in July 1687 at Halley's urging and personal expense, that Halley could mathematically demonstrate that the comet he observed on his honeymoon in 1682 followed a regular and predictable elliptical orbit shared with comets observed in 1531 and 1607. Newton took notes on the comet in late 1664, and an illustration of his attentive mapping of the 1680–81 comet (which he hypothesized at the time was two different comets) figured in his *Principia*. Halley, who had been in France during the 1680–81 comet, published his *Astronomiae cometicae synopsis* in 1705, and it was not until decades later that his theory was universally embraced before being confirmed on Christmas 1758 with the reappearance of the comet he had predicted.[13]

Earlier in the seventeenth century, a comet in 1618 had drawn intense debate.[14] As other astronomical questions were settled over the following decades, the issues surrounding comets remained a recalcitrant problem.[15] Thus, while comets had attracted attention, excitement, and fear from ancient times, the comets of the later seventeenth century were some of the last holdouts in a sky increasingly tamed by observation and mathematics. As such, these heavenly bodies took on a very special significance as they drew the attention of individuals from seemingly all social conditions, regions, and walks of life, both for the showy display they put on nightly for weeks or months at a time and for their resistance to available models of knowledge.[16]

This resistance to knowledge or epistemological mastery was doubly fraught in the French context because in the early years of Louis XIV's reign, the cosmos was saturated with political and cultural significance. In these years, the sun was the master symbol, the most iconic image used to represent the young king and the luminous power for which he wished to be remembered.[17] Although previous monarchs, including the king's own father, had used the sun to represent their dominion, Louis XIV's claim on this iconography was made indelible in the pageantry of the 1660s and the architecture of the 1670s.[18] Louis was represented as Apollo in ballet performances, in statuary in Versailles's *Grotte de Thétis*, and in the painted decor of Le Grand Appartement du Roi, a series of six salons featuring mythological figures and episodes.[19] The image of the sun also became the recognizable imprint of the monarch on medallions, tapestries, and prints. Emerging in the late 1650s, and becoming ubiquitous after the carrousel of 1662, in which the king wore the emblem of the sun, the king's *devise* paired the rather cryptic inscription *Nec Pluribus Impar* with the image of a radiant sun, often positioned over the terrestrial globe.[20] As a dialogue by the Jesuit essayist and critic Dominique Bouhours (1628–1702) explains, the *devise* was the king's "true portrait," conveying his essence more truly than a mere likeness of his person.[21] The *devise* of the sun shining its rays over the earth, one of Bouhours's characters explains to the other, "puts that [idea] before the eyes, and produces an exact comparison."[22] It is a theory of power distilled and conveyed in the blink of an eye. Commentators such as André Félibien, who wrote a series of official descriptions of royal ceremonies, buildings, and works of art, depicted the king's dazzling éclat as a quasi-natural force, with the power to transfix (*éblouir*) those who encountered him.

In addition to providing mythological and symbolic figures for representing the king, the sun also offered a powerful spatial schematic: to understand social and power relations by mapping onto theories of heliocentrism, which, although still not officially sanctioned by Rome, were becoming increasingly pervasive.[23] The well-ordered cosmos presented a seemingly perfect symbolic structure through which to conceive a reign striving to be as expansive, centralized, and tightly orchestrated as the sky.[24] The solar system likewise gave shape to the regime's lavish and symbolic performances and constructions, from ballets, which Louis and his entourage refined and codified, to the circular bosquets, or garden rooms, that punctuated the Versailles gardens.[25] And figuratively, the sun and the solar system offered an analogical means of expressing (or hypostasizing) the new model of control to which the regime aspired. Louis XIV was celebrated as the source of all light and all resources, the center around which everything circulated, from the economy to social standing. In court ceremonies and entertainments, wherever the king sat determined the position of his courtiers, who arrayed themselves around the monarch according to rank and favor. Although, as Hélène Himelfarb and Gérard Sabatier have emphasized, after 1680 royal image makers moved away from solar imagery in representations of the king, many of the most enduring images of Louis XIV date from the period covered in this book, from the Great Comet of 1664–65 to the extraordinarily bright comet of the winter of 1680–81.[26]

In this book I explore what kinds of knowledge, texts, and social affiliations are created when people—Louis XIV's contemporaries, but also present-day critics and readers—turn their attentions to comets, a spectacular and seemingly erratic and elusive celestial body, rather than toward the sun and its organizing pull. Comets galvanized throngs of people captivated by a bright celestial apparition, yet the visual encounter with a comet organized an entirely different set of experiences and relationships from those choreographed in relation to the sun or the king. Viewers were transfixed by the bright heavenly body but did not know what it was, where it traveled, what it meant, or where, when, or if it might return. Comets did not assign individuals a defined place or subject role. Stories and depictions about their passage often note the heterogeneous jumble of people brought together on streets, bridges, and rooftops to observe and discuss.

Comets were as spectacular as they were unruly. And comets attracted attention, in print and in the public square, which compels the question: If we understand the era of Versailles to define and be defined by a tightly concerted regime of looking and knowing (knowing the spectacular center and one's place in relation to it), what happens when people ponder comets in Louis XIV's France from intractably diverse positionalities? If the king's authority manifest in the royal gaze constitutes a very particular subject, what sort of political subject and what sort of knowledge are constituted by looking at comets? What instances and experiences of seeing, knowing, and feeling does a comet assemble? What happens when people look at an object that they cannot understand and whose shape they cannot discern or predict? What kinds of affiliations, connections, and erratic shapes does such a spectacular object's chaos allow or occasion? *In the Sun King's Cosmos* addresses these questions by examining a wide variety of texts, from plays and ballets to a building, a theological treatise, and periodicals created in the context of the *cométomanies* of 1665 and 1681.

The comet widely visible from November 1664 through February 1665 inspired books, letters, images, and broadsides across Europe, Asia, and North America. It features in a famous sermon by Cotton Mather, a folio painting by Muhammad Zaman, and Samuel Pepys's diary.[27] The great Polish astronomer Johannes Hevelius (1611–87) published a description of the comet dedicated to the powerful French minister Jean-Baptiste Colbert (1619–83), along with a prospectus announcing his *Cometographia*, which would appear in 1668, dedicated to Louis XIV (see fig. 3).[28] The comet was first observed traveling rapidly from east to west but then reversed course and slowed. It appeared stationary before fading from view only to be replaced by an even brighter comet with a longer tail that was visible in March and April 1665. Observers around the world from Jesuits in China to ladies in the French provinces recorded sightings, but the opinions that they generated were cacophonous. There was not even agreement on the number of distinct comets that had been seen.

For a time, the Great Comet of 1680–81 shone so bright that it could even be seen in the daylight.[29] This comet likewise inspired an international boom in publishing and public interest. Writers debated whether comets were portents or simply stars; others disagreed about whether

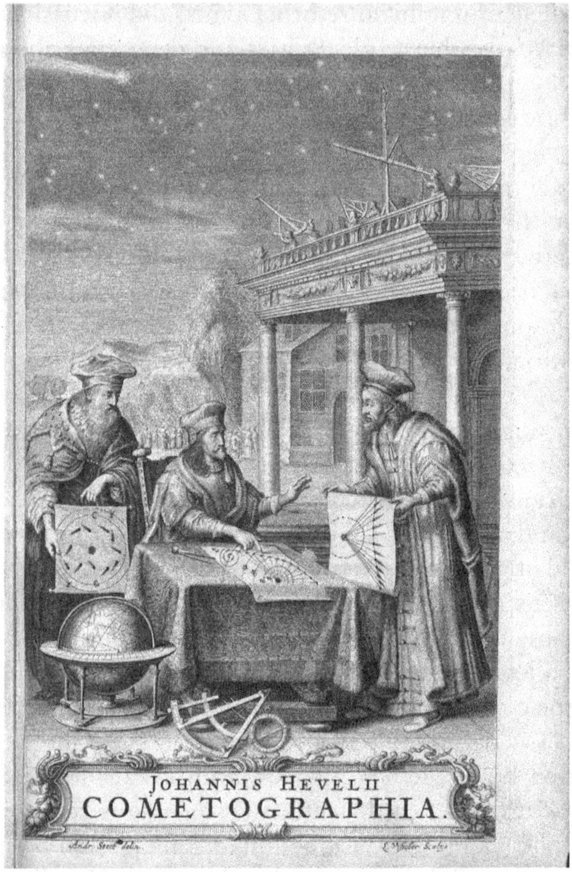

Fig. 3. Andr[eas] Stech delin., L[ambert] Visscher sculps. Frontispiece from Johannes Hevelius, *Cometographia* (Gdansk: Simon Reiniger, 1668). Library of Congress, Rare Book and Special Collections Division.

they were singular bodies (appearing once and then disappearing forever) or permanent and stable objects that travel in a regular if difficult to discern path. It wasn't clear whether the comet appearing in late 1680 and early 1681 was the same comet or two different ones. Still other writers used the comet as an opportunity to write sly love poetry or discuss etymologies. Lutinist and composer Jacques Gallot (ca. 1654–95), whose Parisian concerts and compositions inspired regular coverage in the *Mercure galant*, drew inspiration from the comet to compose a chaconne.[30] Natural philosophers offered various theories about comets: their composition and location in the cosmos, distance from the earth, source of light, and the reason for their fiery tails.[31]

Comets provided a focal point for debate and dissention as early modern cosmography became more institutionally grounded and more cohesive (if still contentious) regarding practices. For a long time, historians of science credited the development of telescopes in the early seventeenth century with putting an end to classical, especially Aristotelian, theories about comets (that comets were sublunar, circulating in the earth's atmosphere like the weather).[32] Comet books from the late seventeenth century, such as Claude-François Ménestrier's *Lettre d'un gentil-homme de province à une dame de qualité sur le sujet de LA COMÈTE*, published at the height of *cométomanie* in January 1681, however, still rehearse Aristotle's comet theory among others, while regretting that contemporary natural philosophers have not gotten much further in settling the question of comets. Ménestrier's gentleman narrator opens the epistle with the comet theories of "several ancient Philosophers who could not in fact determine what a Comet is, any more than the Philosophers of today, whose ideas are much different; and who in refuting ancient opinions, start to resemble some of them."[33] Aristotelianism persisted, assimilating new discoveries.[34] Scholars such as Andrea Gualandi and Sophie Roux highlight the great diversity of comet theories circulating in Europe in the mid-seventeenth century.[35] Throughout the century, that is, a multiplicity of new and classical theories about the nature, genesis, location, motion, and trajectory of comets coexisted and garnered new champions. Aristotelian, Pythagorean, Cartesian, and other theories were part of the same conversations. As Roux emphasizes, in France this battle only inflamed existing tensions between Jesuits and Cartesians.[36] While the comets of 1664–65 inspired a spate of publications, Roux concludes that they document a proliferation of theories and quarrels rather than any kind of synthesis or significant gain in understanding: "The maze of books that the 1664–1665 comets gave rise to disappeared almost as rapidly as the comets themselves. . . . Not only were these publications on comets ephemeral, but with no significant truth discovered, they seem to have also been forgotten as soon as they were published."[37] Comets generated conversations, wonder, and text, but they did not establish the foundation for any kind of stable cultural or epistemological edifice. They generated energy and chaos in equal measure.

While the astronomical status of comets drew great interest, this was far from the only thorny set of questions comets posed. Comets also drew attention to debates about the existence of portents and divine signs.[38] Sometimes the same source engaged both sets of questions—comets' movements and place in the universe (and, relatedly, the shape of that universe) and comets' meaning as divine signs or messages. Ménestrier's book offers a catalog of the former questions before concluding with reflections on the latter. Historians from Voltaire, writing in 1751, through Paul Hazard, publishing in 1935, have used the comets of late seventeenth-century France to index a supposed demystification of the modern world, pointing in particular to the comet visible in the winter of 1680–81, widely (if erroneously) thought to have inspired theologian Pierre Bayle's *Pensées diverses sur la comète*, as a definitive turning point away from superstition and toward rationalistic modernity.[39] Many of Bayle's worldly contemporaries viewed their own era in similar ways.[40] As demonstrated by Sara Schechner Genuth, however, Newtonian mathematics and the discovery of comets' orbits accommodated rather than refuted traditional modes of thinking. "Vulgar comet lore was absorbed into the elite astronomical work of Newton and Halley," and models that subsequent history has tended to hail as first iterations of modern scientific thinking were in fact capacious enough to assimilate classical and folk beliefs.[41] In other words, the eventual determination of comets' regular orbits did not neatly dispatch superstitious beliefs about comets' causes or their roles as portents. The triumphalist Latin ode with which Halley prefaced Newton's *Principia* famously claimed a definitive victory over comets in the newly totalizing "pattern of the skies." Halley proclaimed:

> Now we know
> The sharply veering ways of comets, once
> A source of dread, nor longer do we quail
> Beneath appearances of bearded stars.[42]

However, *Principia* did not effect a clean and rapid break in the way people understood the cosmos. Cartesian and other cosmologies persisted well into the next century.[43] Observation, the mathematization of the universe, and New Science were all a great deal unrulier and more contradictory—in Ofer Gal and Raz Chen-Morris's term, more baroque—than the idealized and aspirational image their practitioners publicized.[44]

French encounters with the comets of 1664–65 and 1680–81 are the focus of this study because of the significant cultural attention these two periods of comet mania generated. This book is not a work of history of science. Scholars such as Gal and Chen-Morris,

Roux, Schechner, and Tabitta van Nouhuys engage with more technical aspects of seventeenth-century debates around comets, revealing the complexities and contradictions inherent in what previous historiography often presented as the rational age of New Science.[45] Nor is this a thematic book tracing how these comets were represented in literature. Rather, the book uses the tools of close literary reading to delve into two brief and intense periods of *cométomanie* in order to gain insight into the cultural resonances the comets provoked and the relationships, networks, esthetics, and associations forged by observation, representation, and discussion of comets.

The two *cométomanies* I study bookended a period in which solar imagery was being cultivated as a way of theorizing—and naturalizing—the ways power, resources, and people circulated within a modern nation. In 1664 Louis XIV welcomed the court to the newly renovated Versailles gardens, while the conclusion in December 1664 of the treason trial of his former minister, Nicolas Fouquet, cemented the young king's dominion over law, the economy, and the arts. In 1681 workers were undertaking the expansion and renovations that would allow the court to occupy the Versailles palace the following year.[46] From Voltaire in *Le siècle de Louis XIV* to Louis Marin, who saw in Louis XIV's consolidation of "absolutist" power the origin of Althusserian ideology, the early years of Louis XIV's reign have long been a subject of critical interest for the ways in which distinctly modern forms of power and control—and concomitant modes of subjectivation—were installed through a regime of the spectacular.[47] Comets troubled the direct link that this narrative, and the narrative of modernity more broadly, posited between vision and knowledge, vision and order, and vision and control.[48] As some of the last intractable bodies in the sky, comets during the age of Louis XIV cannot be neatly assimilated to systems of order. They are brilliant like the sun, but they arrive in the dark, appearing nightly (and sometimes by day) for months at a time, in seemingly haphazard locations, with a dramatic tail in tow. Comets, and the attention people paid to them, allow us to see the seventeenth century in ways that complicate the narrative of a race toward rationalization, classicism, and modernity. They index, rather, a messy period during which the spectacular is inscrutable; comets, in other words, raise questions rather than offer answers. Exploring the late seventeenth century in France with an eye for comets significantly reshapes our understanding of the literary and cultural field.

I aim to discover the ways French people engaged with the bright and mesmerizing celestial apparitions in two politically and culturally key moments, when the spectacular was so political, and to explore the tensions that comets engendered in a context where visual opulence was being actively cultivated as an ordering cultural force. The comets seemingly defied the gravitational pull of an orderly cosmos, even as they offered a kind of nocturnal mirror image of the symbol that Bouhours called the king's "true portrait." I offer readings of literary, philosophical, and journalistic texts, images, and even architectural and administrative structures generated at these moments when comets drew eyes and sparked the popular imagination. In these readings, I explore the dynamics and concerns that crystalize around the instances of viewing that the comets inspired. In attempting to decenter critical attention, this book joins a constellation of critical voices looking differently at a formative period of political modernity and discovering a complex and often quirky world connected by loose associations, performance, esthetics, and compilation rather than by a coherent master plan.[49]

Chapter 1 examines the symbiosis between comets and the periodical press, the way comets at the close of the seventeenth century continue a "particular convergence of optics and news in the early modern period" that Eileen Reeves traces in the early decades of the century.[50] The book's next two chapters (chapter 2 on two comet ballets and chapter 3 on the Observatoire Royal and observational science) examine cultural artifacts related to the 1664–65 *cométomanie*, while the last two (chapter 4 on a play, *La comète*, and chapter 5 on Bayle's groundbreaking *Pensées diverses sur la comète*) respond to the comet of 1680–81. The first in each of these pairs of chapters offers a close reading of a dramatic text that depicts comets on a public stage, revealing some of the preoccupations and concerns comets indexed in the period and playing out some of the ways comets' problematic movements sparked social motion. The second chapter of each pair examines an important cultural monument inspired by a comet as a way of accessing the concerns comets raise about social organization and epistemology—how humans can know things and what they can know.

In chapter 1, I pursue what coverage of the 1665 and 1681 *cométomanies* in four early French periodicals—Jean Loret's newsy letters, republished as *La Muze historique*; Dennis de Sallo's *Journal des sçavans*; Jean Donneau de Visé's *Mercure galant*; and an *extraordinaire* of Nicolas de Blégny's *Journal des nouvelles découvertes conçernant les sciences et les arts qui font parties de la médecine*—can tell us about how the news was constituted as a textual and social practice in seventeenth-century France. Reading for comets in Loret's letters allows us to trace the networks of affiliation and exchange that developed through the burgeoning periodical press. Comets perhaps even offer a figure for this constellation of diffuse contact, which spread far and wide with no particular organizing principle or hierarchy. Following coverage of comets in seventeenth-century periodicals like the *Journal des sçavans* also highlights the formal structure of this press: a miscellany propelled by the constant irruption of the new. Just as comets in this era could not be made to fit into the orderly pattern of the sky, the ungainly method of composition used in periodicals seems anathema to the principles of classicism championed by literary institutions of the period. However, even as the institutions of centralization and order were gaining increasing power over the cultural field, just like the comet, periodical writing, however unruly, also held the attention of French readers during the era in which so many works that would become enshrined as symbols of the French canon would be written. Reading for comets in the *Mercure galant* prompts us to recognize and appreciate ways that classics—and the king himself—are entangled with comets, periodicals, and the sparkling, irregular, haphazard, and ephemeral esthetics that normative accounts of the era have rendered peripheral. The special issue of Blégny's medical journal has been remembered for including the first—and until the mid-twentieth century, the only—French translation of Galileo's *Sidereus Nuncius*. I will argue that the issue was inspired by the comet, which the volume attempts to leverage as a performance of observation in service of a larger project of establishing institutional, professional, and public authority. Unfortunately for Blégny and his journal, comets prove an ill-suited foundation for any of these aspirations.

The 1664–65 comet whose movement fascinated and perplexed people from all walks of life inspired and starred in two ballets, which are the subject of the book's second chapter. The *Ballet de la comète* was performed in February 1665 in northern France, while the comet was still visible. Six months later students of the Jesuit Collège de Clermont in Paris danced the *Ballet des comètes* at the school's annual prize ceremony. Dance offers a powerful way to pursue the cultural and embodied interrogation of form and movement that comets inspired. While dance, like comets, is ephemeral, printed programs and livrets conserve traces of some of the preoccupations that the ballets explore. My reading of these booklets discovers that the two ballets employ divergent strategies to resolve the problem of disorder that comets raise. The first ballet celebrates the disruptive potential of a comet to open up spaces beyond social, temporal, and spatial norms, while the second employs dance to resolve the kind of tensions comets bring to light, harmonizing the comets with a system privileging order and the sun.

The third chapter begins with the story of how the quest to understand comets helped spur the founding of the Académie royale des sciences and the construction of the Observatoire Royal outside of Paris. This building was far more than a home of scientific activity. It also functioned powerfully as a symbol for observational discovery.[51] I study descriptions and visual representations of the observatory in order to better understand the political stakes of the crown's performative investment in observation—or, rather, the crown's investment in the performance of observation. Ironically, Claude Perrault's magnificent building was much more effective as an architectural *representation* of the regime of observation, and the scientific quest to see and know, than as a site of actual astronomic observation, because, unfortunately, the neoclassical design severely impeded the observatory's functionality. Following the story of the Académie royale des sciences from the off-kilter perspective of a comet highlights some of the contradictions, complications, and failures of the project to link seeing and believing in Louis XIV's France.

Chapter 4 analyzes a one-act comedy, entitled *La comète*, published and first performed while the Great Comet of 1680–81 was still visible. This short occasional play delineates some of the fascinating questions the comet raised regarding the relationships between vision and knowledge, print and power. While at first glance the play seems to confirm the traditional narrative that the comet of 1680–81 marks a transition from the reign of superstition to an era of scientific inquiry, upon closer examination the text reveals as much suspicion of observational science and its methods as of traditional superstition. Rather than urging theatergoers to become disciplined and systematic observers of the comet's peripeties, the story concludes that the circulation that matters most is not that of the comet in the sky but that of bodies and objects on the ground: books, money, words, and daughters.

My final chapter engages with the most famous and most culturally significant work published in the wake of the 1680–81 comet, a work that stands in celebrated contrast to the unity, clarity, and order of the French classical ideal: Pierre Bayle's *Pensées diverses sur la comète*. My readings of the introductions and paratexts situate the text within the active and often contentious field of literary experimentation that

Bayle, a voracious reader, closely followed. The comet is the fictional conceit that offers Bayle an opportunity to pursue his bold text, and it also serves as a figure for Bayle's unorthodox and diffuse style. In this chapter I argue that *Pensées diverses* unseats authority not only in its multifarious critiques of astrology, superstition, and idolatry but also in the way it is written, in the form and stagings of its discourse. Bayle utilizes introductions and paratextual frames to reflect on the literary and theoretical questions of voice, narrative, and genre. Examining the narrative strategies employed in the introductory paratexts across the four editions of *Pensées diverses sur la comète* that Bayle published during his lifetime, I demonstrate that *Pensées diverses*, and the resistance to dogmatic logic for which Bayle became famous, proceed through literary means, including epistolary address and digression, and draw on forms and markers of truth and truthfulness deployed in literature and debated in the periodical press and salons. In other words, in his various introductions, Bayle experiments with—and very much performs—the literary tools and techniques that would come to be associated with novels. These stylistic features, which were central to contemporary debates about the status of truth and belief in prose fiction, contribute meaningfully to the most essential critiques of Bayle's *Pensées diverses*. For Bayle, ethics is like a novelistic fiction, grounded not in the bedrock of prior and immutable truth but in the consensual creation of communal norms. The world he imagines is diffuse and ad hoc. As a kind of modern solution to the problems of belief and knowledge in a demystified world, Bayle's comet proposes the diversity and verisimilitude of fiction.

Finally, in harmony with the subject of this book, I do not come to a single conclusion, geometrical and certain, but rather end with three *pensées diverses* of my own. I reflect on the networks and the *esquives*, or dodges, which structure all the stories told in this book. Then, for readers wondering what became of the globes I described at the book's opening, I finish the book with their cometic story.

Part One

Mediatic Comets

CHAPTER 1

"Si confondus à la tour de Babel"

Comets and the Commercial Press in Late Seventeenth-Century France

In December 1664, the French jurist Olivier Lefèvre d'Ormesson (1616–86) was deeply absorbed in political deliberations following the trial of disgraced financier and former minister Nicolas Fouquet, when he made note of a comet that was, like the trial, the talk of the town. As d'Ormesson documented the tense days leading up to the deliberations, and his own role in resisting the king's attempt to dictate the trial's outcome, he viewed the comet from his attic and from the street, heard discussions about it around Paris, and noted that it had already spawned several essays, which were circulating among the social elite.[1]

In his diary entry from Sunday December 21, d'Ormesson writes that "every night the Pont-Neuf and the large public squares are full of people to see it."[2] The following day, Fouquet's friend, Sévigné, includes the comet in the closing lines of another letter to Pomponne. Written a few days after the letter quoted in this book's introduction, this is the last in a series of missives detailing the court proceedings and offering a Fouquet family update, largely from information garnered from d'Ormesson, who figures in Sévigné's account. As in her December 17 letter, Sévigné concludes with information not about the trial but about the comet: "I saw the comet tonight: its tail is lovely and long; I'm placing part of my hopes in it."[3] Like the comet that Sévigné and the mixed masses of Paris had gone out to see, justice and the outcome of political negotiations are both spectacular and quixotic, transfixing and unpredictable. Just as no one can foresee exactly where the comet will appear tomorrow and what it will look like, no one can know in advance what will transpire next in court. All the marquise can do is hope on the fragile beauty of a passing comet that whatever happens tomorrow will spare Fouquet.

When d'Ormesson and Sévigné took note of the comet in the winter of 1664–65, France was embroiled in a decisive shakeup of political power. People were nervously awaiting the outcome of a trial that would define the adult reign of King Louis XIV. D'Ormesson's diary records a bustling and curious Paris. Acquaintances give meaningful salutations to convey their respect for his resolve, and friends encountered all over the city (for example, Sévigné's husband, the marquis) are eager to stop d'Ormesson to hear news of the trial developments. Sévigné's letters in effect publish the news d'Ormesson relayed. Traveling quickly via the newly reformed postal system, her epistolary dispatches keep other intimates apprised of the latest developments, before being passed, copied, and recounted along an expansive social web of people anxious to know the latest developments in the courtroom and in the capital.[4] As we will see in other more widely circulated and deliberately commercial print—the four journals at the center of this chapter—the comets of 1664–65 and 1680–81 appear as mobile embodiments of such transmissibility. Following these two comets in the late seventeenth century offers a glimpse into the development of the medium that we have come to think about as print journalism.

The comet d'Ormesson and Sévigné wrote about in late 1664 didn't just coincide historically with the deliberations over the Fouquet trial; the two events also shared some structural dynamics. Each was spectacular, attention-grabbing, fast-moving, evanescent, and totally inscrutable. Both the comet and the Fouquet affair created conversation, speculation, and unexpected community, drawing heterogeneous crowds of people together to witness, discuss, and speculate about what was happening and what might come next.[5] Both also spurred an avalanche of fast press, reports of the trial, and incidental pieces attesting to Fouquet's guilt or his innocence.

Much has been made of a supposed transition in Sévigné's tone between comet sightings in December 1664 and January 1681. At the appearance of the Great Comet in January 1681, Sévigné, who had "pinned her hopes" on the comet sixteen years earlier, now writes with arch skepticism about the significance of comets. In a letter to Roger de Bussy-Rabutin, she laughs at the expense of her more credulous contemporaries, recalling an anecdote about the dying cardinal Mazarin who scoffed at sycophants who claimed that a comet had appeared to herald his end. The cardinal, who as a student in the Jesuit College in Rome had given a lecture on theories about comets during the much-discussed appearance of a comet in 1618, quipped, according to Sévigné, "The comet does me too much honor." The marquise concludes, "In truth, we should come to the same conclusion as he did; human pride makes too much of itself, believing that big things happen in the stars when people die."[6] Is Sévigné's worldly skepticism in January 1681 so different from her sophisticatedly tempered expression of hope of December 1664? Or do both, rather, index—and, at the same time, deploy—the networks of conversation, speculation, and discussion that we now call the news?[7] While Sévigné did not publish her letters in print, her letters were *publiées*—made public—that is to say, read, shared, and copied within a network of people eager to hear about the events of the day in her inimitable voice.[8]

While Sévigné's letters were first published a quarter century after her death, it was during her lifetime that the news developed into a major print phenomenon. Comets offer a privileged, if idiosyncratic, opportunity to understand this development.[9] Mysterious and flashy, appearing every night in a different location and in a slightly different form, comets provided a natural subject for a commercial print sector that was cultivating new formats, genres, and readers in late seventeenth-century France.[10] In contradistinction with necrologies, the other abidingly popular subject of news coverage in this period, comets reappear bright and new each day—in an unprecedented location and with a new shape or color.[11] They stay long enough to generate and sustain interest but disappear quickly enough to be special. In the periods of their appearance in the seventeenth century, they offered a constantly renewing news resource, generating endless opportunities for observation, speculation, dissent, and conversation. Visible to the naked eye and seemingly resistant to attempts to chart and predict their motion, comets and the wide array of people who observed them night after night featured prominently in this far-reaching, temporally oriented, novelty-seeking press whenever they appeared.

Comets drew the eyes and the curiosity of viewers across the globe from the full spectrum of social positions, ages, and backgrounds. Objects of virtually universal fascination, comets offered the seemingly perfect subject to attract readers' attention in the competitive print marketplace, a familiar yet constantly self-renewing story, relevant to a community of anyone anywhere that could see and was curious. As objects of widespread fascination, comets tied people together in unprecedented constellations. They facilitated a variety of social formations, from the promiscuous mixture of people observing from rooftops, public squares, and bridges to discussions in academic colloquia, salons, and epistolary exchanges. As these scenes of social formation were represented in printed accounts about *cométomanie*, periodicals also effected novel networks, connecting a wide and diverse population of readers who were always also potential writers due to the interactive format of early periodicals.[12] While journalism shaped popular perception of comets, comets also helped define what journalism was in these early years of the periodical press. Both comets and the periodical are characterized by circulation that in the late seventeenth century was still haphazard. Just as the shape of a comet's path remained elusive, undiscernible in the absence of a mathematical paradigm that was not yet available, regular circulation, the periodicity after which periodicals are named, was still more an aspiration than a reality in these years. Regular publication and circulation was an ambition that guided publisher-editors in creating these new forms of print media and was often a feature that they announced in the prefaces or even title pages of all four periodicals I examine in this chapter, although this aspiration was difficult to realize.[13]

This chapter will explore how the early French periodical press created itself in no small part in the image of comets.

Comets and Fast Print

Comets are a figure for speed. In relation to print, they are a propulsive force, a motor for fast publication. As the narrator of Ménestrier's *Lettre d'un gentil-homme de province à une dame de qualité sur le sujet de LA COMÈTE* explains to his readers and his fictional letter's addressee, the way to ascertain comets' location and movements in the cosmos is to observe their precise position in the sky at a precise moment in time, from a precise place of observation.[14] This is also true of other celestial bodies, but since comets' location and movements were still a mystery in the late seventeenth century, their occurrences seem to have galvanized a very broad public to try their hand at observation as a way not only to record but also to perceive and experience the contemporary and evanescent event. The degree of attention to contemporary time and place that the fleeting event of a comet commanded meshed with the preoccupations and rhythms of the emerging institution of rapid print and ensured comets' prominent inscription within it. In the memoir he dictated after losing his sight, and published by his great grandson, astronomer Giovanni Domenico (Jean-Dominique) Cassini recounts at considerable length the passage of comets in 1652 and 1665. Cassini, who was summoned to France to become the first director of the Observatoire Royal Louis XIV had built (the subject of chap. 3), makes clear in his memoirs the degree to which publication was seen as an integral part of the observation process. In 1652, he was in Bologna. His patrons requested that he chart the comet's course and, he notes at two different moments in the anecdote, arrange for his observations to be printed in real time. "We brought printers from Modena to print my discourse as I was writing it," writes Cassini, emphasizing the simultaneity of the enterprise with the phrase "à mesure que."[15] The investment Cassini's patron made in bringing a team of printers thirty-four miles from Modena to Bologna indicates a desire to collapse the distance between observation, recording, and publication in order to capture and disseminate the event on paper in real time. Some thirty years later at the time of the 1680–81 comet, Cassini's patron was the French king, and his observations were printed and lavishly illustrated. Yet the rush to publish them as close to the event as possible was the same.[16]

Stanisław Lubieniecki's (1623–75) folio *Theatrum cometicum* manifests in book form the way that the discourses, conversations, and opinions that comets proliferated in turn spurred a rush to print.[17] Its first section reproduces a compendium of letters about the 1664–65 comets from illustrious thinkers all over Europe. The "theater" that follows includes a wide variety of graphic renderings of the comet's path and locations, using a variety of representational conventions.[18] These representations in the printed medium effect immediacy, allowing readers to see the comet alongside the famous observers whose opinions unfold, as if in real time, through their exchange of letters. Moreover, this immediacy is replicated by each reader and each printed copy. Even as printed representations reproduced the heavenly body as a simulacrum, the comet as a print event paralleled the immediate experience of observing the natural phenomenon. Several occasional texts from prior comets were reedited in subsequent comet years. At the advent of the 1681 comet, for example, Lubieniecki's *Theatrum cometicum* was republished in Leiden.[19]

Only a small fraction of publications recording the passage of comets were written by astronomers or for kings. The two ballets we will examine in chapter 2 and the play that is the subject of chapter 4 all feature cheeky comic characters who regard the comets as an opportunity to market almanacs, timely and ephemeral print products. The errant paths of roaming peddlers in the *Ballet de la comète* and of "Bohemians" in the *Ballet des comètes* evoke the movements of comets—unconstrained by any visible ordering principle. This aimless motion also highlights the mobility of the wares they are plying—almanacs printed to disseminate news of the comets' appearance and speculations about their causes and significance. The enterprising valet character in chapter 4's comedy *La comète* also aims to launch an almanac to capitalize on the current comet craze, a clever self-ironic touch for an occasional play likewise written to exploit public interest in the comet. Judging from the recurrence of these figures, almanacs and ephemeral printed products about comets must have been common and widely distributed during the *cométomanies* of 1665 and 1681. They were, however, extremely ephemeral. I have been unable to find any French almanacs from 1665 or 1681 mentioning or depicting a comet.[20] Doubtless, there were many booklets, prints, and *feuilles volantes*, ephemeral records from this fleeting celestial event, that never made it into any archival record. To some extent, the print record of early modern comet mania is a product of chance and susceptible to disappearing, like the comets themselves.

In his *Cométographie* of 1783 Pingré casts the rush of publications emanating from comets of the late seventeenth century in biblical terms: "The world was flooded with Writing on [the comet's] phenomena, its nature, its meaning."[21] Rather than the resulting profusion of simultaneous observations creating any kind of clarity, the publishing boom calls to mind a second Genesis story in Pingré's recounting: "I think that all the languages at the Tower of Babel were not as mixed up as opinions were about this infamous comet."[22] While comets observed in this period elicit, in Elizabeth Eisenstein's words, the "collaborative effort in simultaneous observation that was made possible by print," in the case of comets, unlike the astronomical phenomena Eisenstein studies, the accumulation of results did not yield any conceptual consensus, just a Babel-like tower of speculation that generated ever more type.[23]

This chapter examines coverage of the 1664–65 and 1680–81 comets in four early French periodicals, all instances of speedy texts that Pingré likens to a Tower of Babel. In each periodical's treatment of the comet, we will discover some of the epistemological and formal characteristics of this new print medium, which may not be equally present in other of the publications. Jean Loret's weekly rhymed letter, addressed to the Duchesse de Longueville but read far and wide, covered the 1664–65 comet only months before Loret's death and the precipitous decline of *La Muze historique*, the publication series created to capitalize on the popularity of his weekly dispatches. Loret's letter highlights the complex ways comets and the emerging print media that covered them created heterogeneous communities of readers and observers. Comet coverage also features prominently in the very first edition of Dennis de Sallo's *Journal des sçavans*, which débuted in January 1665, at the peak of a *cométomanie* (and two months before the *Philosophical Transactions of the Royal Society of London*, allowing the publication to claim the venerable title of Europe's first academic journal).[24] The inaugural issue of the *Journal des sçavans* opens with a preface announcing the periodical's ambitious project. The journal aspired not only to follow the progress of knowledge but also to participate in knowledge production by offering a place where the exchange of ideas could lead to the refinement of understanding. In his commentary on books about comets, however, the editor highlights the irreducible diversity of views and texts that comets engender. The following pieces about comets in the *Journal des sçavans* reveal this observation to be equally apt as a description of the genre of periodical literature itself. Despite the aspiration toward synthesis articulated in the journal's preface, reading for comets helps us recognize the formal structure of this press as a miscellany propelled, much like a comet, by the constant irruption of the new. Reading for comets in Jean Donneau de Visé's *Mercure galant* focuses our attention on the preoccupation with novelty and popular taste in periodical literature. Comets offer an example of how an edition of the *Mercure galant* is built from the delight of the shared experience of new and interesting things.[25] Finally, an October 1681 special edition (*extraordinaire*) of Nicolas de Blégny's *Journal de médecine*, a periodical that has been forgotten by history, includes a chapter on telescopic observations of comets as a marker of contemporary relevance and a signal of the journal's commitment to observation. The comet, which I speculate may have been the inspiration for this *extraordinaire*, reveals the commercial importance of timeliness in Blégny's ambitious and innovative entrepreneurial use of print media.

Comets gave people a powerful experience that they could then see written about in the emerging journals. Comets weren't newsworthy things that only other people had done or experienced—comets and their social and cultural epiphenomena were happening to them too! Readers were part of the news, could see themselves reflected in the news, and they shared this experience in unprecedented ways with newly emerging social constellations. Examining how comets are covered in French periodicals from the 1660s through the 1680s brings into focus a textual economy organized around novelty, diversity, speed, haphazard trajectories, and serendipitous, ephemeral, and nonhierarchical contacts. If the movement of comets was captivating because it was unpredictable—so unlike the careful ballet of the heavens and the court—the same was true of the (loosely) ordering principle of periodicals: miscellany. Comets both shaped and resembled the news; both were something different at every moment. Whether as a synecdoche or a model for the news, I would wager that the advent of a comet offered new ways of thinking about the shape of the news.

Following the periodical press through its engagement with comets reveals the chaotic and entrepreneurial side of seventeenth-century literary culture that Christophe Schuwey portrays in his study of the enterprising and inventive editor, writer, and assembler Jean Donneau de Visé. Donneau, as Schuwey tracks him, beginning with his commercially experimental publications of the 1660s and culminating in the *Mercure galant*, which he founded in 1672, is less an author than a *fripier*, a secondhand clothes dealer, a collector and purveyor of diverse amalgamations, whose wares aim to capture a growing literate public hungry for novelty.[26] For Schuwey, Donneau's commercially oriented publication strategy is "tentactular" in form, an assemblage of diverse pieces.[27] The comet as news item offers another metaphor that we might use to understand the propulsive energy and the structure of the *Mercure galant* as a print phenomenon. News like the comet showcases the shiny, brilliant, and ephemeral—no less fragile than the wares of the *fripier* and no less amusing. However, unlike the *fripier*'s rags, which were continually recycled, reused, or repurposed, each comet appeared ever novel, its purpose enigmatic, its origins and fate unknown.[28] Considering the news as cometic emphasizes its propensity toward the networked and connected, the multitudinous, the novel, the ephemeral—all the things the periodicals highlight.[29]

Dis-Organizing principles: Comets in Jean Loret's *La Muze historique*

Following Benedict Anderson's influential 1983 book on print media and the rise of national consciousness, literary scholars and cultural historians became attuned to the ways that media organizes people into imagined communities.[30] In earlier scholarship focusing on descriptions of Versailles and of royal fêtes in the 1660s, I, like many other scholars, have argued that these publications encouraged readers to imagine themselves in certain relationships to the monarch, to understand in new ways their identity as French subjects.[31] Printed descriptions provided a powerful image of a spectacular and highly centralized court that was the model for Louis XIV's state projects. Reading the emergent French press from the perspective of comets, however, offers a significantly different view of the period and, consequently, productive angles from which to appreciate questions of identity, subjective experience, and communal and national affiliation.

From 1650 until his death in 1665, Jean Loret wrote a weekly epistolary news dispatch, in rhymed verse, to the duchesse de Nemours, Marie d'Orléans-Longueville, who supported him with a regular stipend. Copied and shared, as correspondences were at the time, Loret's letters attracted enough interest that an enterprising printer decided to capitalize on their popularity by marketing a printed version.[32] Eventually Loret gained the *privilège* (copyright or license) for an edition of his own and contracted with the Parisian printer Charles Chenault to publish collections of his letters under the title *La Muze historique, ou Recueil de lettres en vers contenant les nouvelles du temps: Écrites à S.A. Mademoizelle de Longueville; Par le Sr Loret.*[33]

Loret ends his letter of January 3, 1665, with a postscript, excusing himself for not having covered "the Comet / that so many are discussing."[34] Although his readers are curious to hear about the comet, being "only an Author" and "Neither Philosopher, nor Doctor / Nor knowledgeable in Astrology," he has not dared to discuss it: "Sentant bien que ma veine / Ne coule, à present qu'avec peine, / Je n'ai pas voulu hazarder / D'en écrire.[35] The next week's letter, from January 10, 1665, capitulates to his readers' interest and begins with a long account of the comet presently visible. Rather than focusing on the comet, however, the account pays most attention to the debate and social scene that constellate around it. The variety of ways that people engage with the apparition manifests in the variety of forms their discourses take: "Lots of people have often spoken / Some in Verse, others in Prose" ("Bien des Gens ont parlé souvent / Aucuns en Vers d'autres en Proze").[36] We can almost hear the cacophony as the voices of "Bien des Gens," "Aucuns," and "d'autres" all pivot around the shared verbal phrase "ont parlé souvent." The comet occupies both "Les foibles et les forts Esprits" and generates a multiplicity of forms and genres of discourse. While "the low class speak [about it] like bumpkins. . . . The *savant* and the deep / Write admirable discourses about it."[37] Loret's verse wryly summons the hordes of serious thinkers ("plusieurs / Plusieurs de ces doctes Messieurs") who write about their "diverse opinions" and "Dialogues, / apparemment contentieux" in learned genres such as dialogues and debates, "Témoignant, dans leurs controverses / Être d'opinions diverses."[38] While these "doctes Messieurs," whose Latinized names he unfurls in a comically long list, hold "différents discours," and charlatans publish horoscopes, the account conveys an ironic detachment to the spectacle: "So, even if it killed me, I wouldn't know how to speak well about this fiery Comet / which is so renowned."[39] The vignette concludes with a verbal picture not of the comet but of the scene of the social theater that engrosses all members of the society: "The King, the Princes, the Princesses, / Marquises, Countesses, Duchesses, / Ignorant people, learned people, / Everyone wanted to see it."[40] Universally visible and an object of shared interest and curiosity, the comet organizes the whole society, creating out of the cityscape a kind of theater: "The roof tiles served as a Theater, [so did] dormers, high floors, / Terraces and clock towers."[41] The account depicts the entire social body united in observation of the comet. The verses end with a kind of syllogism, complete with an ironic (and comically unimpressive) flourish of rhetorical Latin:

> Et comme les nuits hivernales
> Aux Forains sont toujours fatales,
> (*Id est*, à ceux qui sont dehors)
> Plus de cent trente mille corps,
> Et possible encore plus grand nombre,
> Respirants un air froid et sombre,
> Sont enrhumez, en vérité,
> Tant dehors, que dans la Cité.
>
> And as winter nights
> Are always fatal to Revelers
> (That is, to those who are outside)
> More than one hundred thirty thousand bodies
> And possibly even more
> Breathing in the cold, dark air
> Have in truth caught colds,
> Both outside and within the City.[42]

Loret writes like a kind of natural historian of the social order he sees. The rhyming description catalogs the conditions and draws a conclusion. This conclusion, however, is that the entire populace, united in curiosity over the comet, is deserving of ridicule.

The verses disseminated in Loret's January 1665 letter and broadcast in *La Muze historique* depict the whole of French society—the monarch and his court, the lowborn, the learned, and even the scoundrels—crowded on rooftops, streets, and clock towers in rapt attention to a comet that they cannot make sense of and that in the end levels social differences and makes them all equally sick and ridiculous. This scene was addressed to the stepdaughter of the famous *frondeuse* as well as an extended worldly public of readers during the very moment when Louis XIV's cultural promotion machine was working to establish an entirely opposite image of French society, arrayed in ordered awe around a solar mythology that placed the king at the gravitational center of the realm. Just the previous May, the court had been convened to the newly renovated Versailles gardens in order to take in the multimedia, multiday spectacle of Les plaisirs de l'isle enchantée. Descriptions and illustrations of the fête published subsequently depict the king at the center of the action as a way of representing (and instantiating) the broad and all-encompassing influence he wished to exert over France. As scholars have explored, André Félibien's descriptions emphasize the guests' awe and wonder at the spectacle meant to represent the king as well as at the spectacle of the king himself. On the frontispiece by Israël Silvestre, Louis's laurel-wreathed solar escutcheon is flanked by an ellipse representing his courtiers' coats of arms of (see fig. 4). A tiny caption explains that the engraving reflects the order of the procession, not the order of rank. In other words, the immovable and indefatigable king, depicted as a sun with the motto *Nec cesso, nec erro* (I won't rest, I won't stray), shapes the orbit of power around himself according to the logic of his favor and will. In another image the king performs the hero's role on horseback, followed by Apollo's chariot and a ring of his courtiers while the rest of the assembled nobles look on from the bottom and foreground of the image. The majority of the engravings include this back-of-head view of the audience below the stage. When the king is not on the stage performing, he stands out at the center of the seated figures, with his courtiers arrayed around him according to their standing and rank.[43] The urban theater Loret's account depicts seems to contest the vision of the monarch and the French state that the Petite Académie and the structures of artistic production Colbert had established were actively cultivating. In this official narrative, the king, like the sun, is the center of attention and the organizing principle around which power, society, and resources were arrayed. He is the inspiration for panegyric, the source, patron, subject, and recipient of literary and artistic works. In the scene depicted in Loret's dispatch, people of all ranks and social conditions mix on the undignified rooftops and towers to watch the comet. They produce catalogs and lists

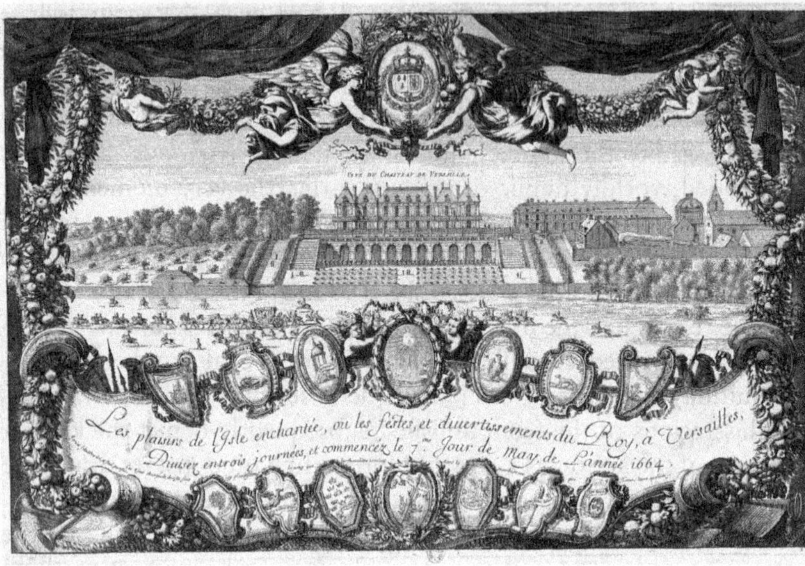

Fig. 4. Israël Silvestre, Frontispiece engraving for *Les plaisirs de l'isle enchantée* (Paris: [Imprimerie Royale], 1673). Bibliothèque nationale de France.

of discourse, but no knowledge, prestige, or order seems to accrue. Likewise, this sometimes irreverent accounting of the news defied any imperative to order, traveling as it did among readers and listeners in a virtual version of the scene these comet verses depict.

This ephemeral account of a passing event traces the interweaving of news, letters, and print in seventeenth-century France. The rapid and far-reaching connectivity of the late seventeenth-century's epistolary networks provided not only the form and vocabulary of the epistolary novel that developed during this period but also the format for Loret's rhymed accounts and the print phenomenon they spurred in *La Muze historique*. By the time the next *cométomanie* galvanized France, another popular and irreverent periodical was there to follow the comet's movements and its observers' foibles. While Loret's dispatch, like Sévigné's, originated as actual letters, Donneau de Visé's *Mercure galant* was conceived from the start as a mass-print project, even as it was styled in an intimate epistolary voice. Before we explore the ways comets streak through the witty miscellany of the *Mercure galant* in the early months of 1681, however, we will follow the comet of 1664–65 to witness the creation of an ambitious publishing project that would become the first, and most enduring, scholarly periodical in Europe.

"Tout ce qu'on dit de ce phénomène est douteux et incertain": Comets and the creation of the *Journal des sçavans*

Denis de Sallo's innovative and ambitious periodical project, the *Journal des sçavans*, began publication in January 1665, weeks after a comet grabbed public attention in France and abroad.[44] According to a preface by the publisher addressed to the reader ("L'imprimeur au lecteur"), the *Journal des sçavans* aimed to share with its readers news from the international community of scholars and seekers known as the Republic of Letters.[45] The projected content included book reviews; death notices; reports of experiments, discoveries, inventions, and observations; important legal rulings; and anything else of general interest. "In sum," the prospectus concludes, "we will try to make it such that nothing happens in Europe worthy of the curiosity of People of Letters that one could not discover in this Journal."[46] The editor also added that the *Journal des sçavans* would contribute to the advancement of knowledge by offering a place for people to publish new ideas, which they might hope to perfect with the collaboration and collective insight of the journal's far-flung readership.[47]

Comets fall under the third kind of inquiry that the new journal proposed to further, a very broad category roughly encompassing human understanding of the natural world, including physical and chemical experiments, machines and mathematics, "celestial observations, the observation of Meteors, and what new things Anatomy can discover about animals."[48] Like meteors, which the editor specifically mentions as a phenomenon adjacent to the solar system if not quite part of it, comets were seen as an anomaly, something that fell outside of understood systems of knowledge, and were for that reason the object of special curiosity. Prepared and published while the comet captured the attention of the kind of inquisitive individuals who were the journal's target audience, the first several issues of the *Journal des sçavans* include five pieces inspired by the comet and the books and conversations it occasioned. As they accrue, reviews of books the comet has inspired and reports on learned society's engagement with it show up comets as a kind of engine that multiplies points of view and opinions and creates questions rather than consolidating knowledge to "contribuer à la perfection des choses," as the preface optimistically proposes would result from the exchange between readers.

A review of a Samuel Sorbière's *Discours de M. de Sorbière sur la comète* in the journal's inaugural issue draws attention to the way the comet serves as a motor for publication and discourse.[49] It is a *nouveauté*, both novelty and news: "There are so many Discourses on the Comet that everyone will find one that suits him. Some are written for Astronomers, others for people who never bothered to observe the Sky, but who are nevertheless curious about the novelties [*nouveautés*] that happen by."[50] The evaluation of Sorbière's book, an elucidation of Pierre Gassendi's understanding of comets, is positive in so far as it judges Gassendi a "grand Philosophe" whose conclusions are "très curieuses," which is to say they spark curiosity, one of the era's highest compliments.[51] The intrinsic interest in Gassendi's philosophizing notwithstanding, however, the article notes that the philosopher's understanding of comets remains but speculative, "ses raisonnements vont à faire voir."[52] In fact, the book's conclusions are judged to be dubious. Rather than consolidating knowledge, the review concludes, comets raise questions and uncertainty. Where they do help contribute to the human state of knowledge is through their signature disruption—their ability to dismantle bad ideas, rather than accrete positive understanding: "Everything that people say about this phenomenon is doubtful and uncertain; this discourse will serve to combat received ideas rather than to establish any new ones."[53]

The proliferative energy comets engender is reflected in "Diverses opinions de la comète." Occupying the final eight pages of the January 26, 1665, issue, this article reports on a "grand assemblée" that took place January 10 at the Jesuit Collège de Clermont in Paris before members of the court, the Grand Condé (Louis de Bourbon, 1621–86), his son Henri-Jules de Bourbon, duc d'Enghien (1643–1709), and his younger brother the prince de Conti, Armand de Bourbon (1629–66).[54] The stated aim of the conference was to ascertain "the causes and effects of Comets"; however, Sallo's article, in offering a précis of each presentation, highlights with the phrase, "on y rechercha," the lack of scholarly consensus despite the conference's synthetic aims.[55] The experts included Jesuits Arroüis and Jacques Grandami (1588–1672), the mathematician Gilles de Roberval (1602–75), and a Flemish doctor named Phelippeaux. The report expands as each presenter offers a different and contrary explanation of comets, their origins, makeup, and movements. Roberval was a Copernican, Phelippeaux was a Cartesian, and Arroüis presented the theories of Democritus. Grandami expounded the theory that he supported at the time of the last great comet in 1618 and that he would recapitulate in two illustrated works published in the coming months.[56] The article ends by introducing Adrien Auzout's attempt to demonstrate the periodicity of the comet (the assertion that comets followed a regular orbit that could be mapped and thereby predicted).[57] Although Auzout was not present at the assembly, the journal article summons Auzout's virtual voice into the proliferation of opinions that the conference assembled. This framing highlights the uncertainty of Auzout's claim and contextualizes it within a crowded field of equally uncertain statements. Even as the report conveys Auzout's observations and hypotheses, it pointedly undermines Auzout's claim to authority, noting that his projection is based on only five or six observations.[58] The article's conclusion communicates an utter lack of confidence in the ability of natural philosophy, mathematics, and observation to definitively solve the riddle of comets, yet it celebrates the creativity inherent in the various attempts: "In any case, if these conjectures do not succeed, this attempt will not cease to be glorious to its Author; as for issues where there are so many difficulties, and so little help in surmounting them, it is always worthy to have tried."[59] Comets proliferate theories and attempts, books, and observations, with no end or certainty in sight.

The article, which concludes the issue, ends with a description of various maps, charts, and globes illustrating comet sightings, including a map presented to the king by the duc de Saint-Aignan (1607–87), a high-ranking member of court, literary patron, and member of the Académie française, with observations by astronomer-mathematician and future member of the Académie royale des sciences, Jacques Buot (1623–78). The journal includes a textual summary along with coordinates, and Sallo, the editor, concludes the issue with a promise to share further observations and theories as he receives them: "When we are able to gather the different opinions and observations that were made abroad, we will not fail to include them in the pages of this Journal."[60] Anticipated further opinions and observations promise above all to generate more pages and future issues.

A short piece in the February 2, 1665, *Journal des sçavans* highlights the way that every facet of the comet, even its linguistic representation, generates pages.[61] The article, entitled "Si l'on doit dire le Comète ou la Comète" (Whether we should say *le Comète* or *la Comète*), focuses on the word "comet" and how people cannot decide the noun's grammatical gender in French. Sallo begins by revisiting his assertion that the comet generates endless dialogue and debate. While astronomers trace its movements and attempt to identify its place in the cosmos, and the ignorant masses fear whatever punishments it foretells, "those people who are neither enlightened enough to discover its nature nor superstitious enough to fear it debate its gender," in other words debate whether to say *le* or *la comète*.[62] This comment lands like a quip, but in fact offers a definition of the new reading public—urbane, stylish, informed—that watches comets and reads periodicals. He notes, "These types are not insignificant in number and have made a great deal of noise [ont bien fait du bruit] in Paris."[63] These readers constitute a significant social block, "[ils] ne composent pas la moindre partie du monde." Representatives of the worldly, *le beau monde* (*le monde* for short), their opinions dominate conversation in the capital. They care neither about cosmology nor about superstition, but they care deeply about words, etymologies, grammar, and the refinements of language, and they are also deeply interested in categories of gender and sex. With his flippant tone, Sallo demonstrates his urbane wit, and thus his place in this increasingly important cultural milieu, whose members constitute a major part of his readership. Despite the ironizing framing of the question, the editor concludes by conceding, in a very passive and indirect formulation, that the very fact of public interest is what makes something worthy of inclusion in the journal: "Since people discerning in literature did not find this question unworthy of their interest, we were persuaded that it would not be found out of place if we reported in this Journal what we could collect about this debate."[64] With this self-aware if torturously articulated remark, the article seems to offer a definition of the news that abides, sometimes uncomfortably, to this day. It does not need to be something that a specialist, authority, or editor would deem important or noteworthy. Rather, the news is what engages people.[65]

The emphasis on the energy that comets leverage to generate pages and diverse discourses returns in the *Journal des sçavans*'s coverage of the Great Comet of 1680–81. By this time, the journal had achieved considerable success under the editorial direction of Jean-Paul de La Roque (ca. 1630–91). The journal was available for sale at new locations, in new formats and reeditions, in response to greater demand and competition from unauthorized editions.[66] Coverage of the comet appears under the title "Nouveautè du commencement de l'année. COMETE," noting, "Everyone is talking about the Comet, which is without a doubt the most considerable news [*nouveauté*] of the start of this year."[67] The editor shares those observations and projections that have reached him, from Cassini at the Observatoire Royal and an observer in Strasbourg, and invites the "Curious both within and outside of the Kingdom to send us [their discourses and dissertations] as soon as possible" for publication in a planned special issue, or *extraordinaire*, on the comet.[68] Journal issues over the next seven months include reviews of published observations, reports, and theories of comets: Anthelme Voituret's *Explication de la comète*, Ménestrier's *Lettre d'un gentil-homme de province à une dame de qualité*, Claude Mallement de Messange's *Dissertation sur les comètes*, and Jean-Dominique Cassini's *Abrégé des observations et des réflexions sur la comète.*[69] In May, a letter from M. Gallet, provost of the Église Saint-Symphorien in Avignon, shares his observations of another comet seen at the end of April.[70] The June issue reports on a book published in Turin and available for sale in Jean Cusson's Paris shop.[71] A July 1681 review of Father Jean de Fontaney's telescopic observations from the Jesuit Collège de Clermont notes that all of these various reports, observations, and hypotheses manifest different points of view: "It is quite right that in such a large and *savant* City as Paris, the Public should receive more than one type of observation of the Comet," echoing the journal's assertion from a decade and a half earlier that comets spark theories but not consensus.[72]

In one article, "Observations by Almost All the Astronomers on Earth of the Comet of 1681," what is newsworthy is the fact of the universal interest in the comet: "Never has a Comet been observed by a greater number of Astronomers. We received news from Spain that it was seen in Tangiers and elsewhere on the Coast of Africa. We have Reports from observations in Constantinople: and even America was not lacking in Observers for this occasion, nor all Places in Europe, as you will see in the continuation of these Observations."[73] The description travels around the globe, to Africa, Asia, and America, as well as all over Europe. However, what the comet coverage documents is not the kinds of discoveries or experiments promised in the first issue's programmatic essay but rather a record of a web of common questions. The comet actualizes a global network of interest and communication based on curiosity and shared uncertainty.

Sparkling and always new: Comets, news, and the ephemeral in the *Mercure galant*

Founded in 1672 as a quarterly (which ran for two years before a two-year hiatus), and reinvented as a monthly in 1677, Jean Donneau de Visé's *Mercure galant* aimed to share with a far-flung community of readers everything new and stylish, whether in the domain of politics, literature, art, culture, style, technology, court goings on, or philosophy.[74] During the winter of 1680–81, when people around the world were transfixed by the Great Comet's nightly spectacle, the comet also appeared in multifarious forms sprinkled throughout the pages of the *Mercure galant*. The poetry and prose, discourses, and news dispatches about the comet in the *Mercure galant* issues of January and February 1681 were rapidly collected and published at the height of the comet mania.[75] In several ways, the comet represents the epitome of a *Mercure*-ready news item: It was bright, shiny, and unexpected. Utterly of the moment, it interested, and puzzled, everyone. The *Mercure galant* is dizzyingly multiple. It is a universe composed of issues, both monthly and *Extraordinaires* (printed differently by different printers in different cities), each composed of many pieces in many genres and media by many authors depicting a dazzling array of subjects.[76] While it is easy to ascertain the shape of one of its many engaging and scintillating parts, like a comet's orbit from the perspective of the late seventeenth century, a real sense of the whole remains elusive. Following the comet through the January and February 1681 issues offers a sampling of some of the interests and curiosities that propel the heterogeneous project. As we do so, we also ascertain the periodical's affinities with the comet it depicts: its irregular shape, relationship to the present, and esthetic and commercial orientation toward the captivating, the new, and the fugitive.

The comet instantiated a physically embodied version of the public sphere as it drew heterogeneous crowds of people out onto streets, bridges, and squares to get an unimpeded view of the sky and to discuss, debate, and fret, while the periodical's public sphere was constituted by a community of readers who solved the published enigmas, debated the proposed conversation topics, sang the printed songs, imitated the outfits described in detail and sometimes even illustrated, and sent in pieces of their own.[77] In the January 1681 *Mercure galant*, an *avis* near the beginning of the issue (following a table of contents and a four-page catalog of articles advertising the *Extraordinaire*) assures readers that any letters they submit will be published, no payment needed: "We will put them all in, as long as they do not incommode any Individuals with satirical remarks [*traits satyriques*], and if the Stories sent in do not affront in any way the modesty of Ladies."[78] The *Mercure galant* was a novel media experience for its readers, so features like sending letters to be published, or obtaining subscriptions, needed to be explained clearly. Readers are instructed to stamp and properly address letters, and above all to write neatly. Sometimes contributors' names end up misspelled because they were too difficult to decipher, "There are also Pieces that we do not put in, because they are too difficult to read."[79] Authors are asked to be patient: "The first received are always printed first, unless the new material is so timely that it cannot be delayed."[80]

The *Mercure galant* is regular in its circulation (after the enterprise achieves stability after 1677). The promotional text at the beginning highlights the speed with which new issues will be dispatched straight from the printer to readers all over France and beyond "as soon as it is printed . . . he will take care to bundle them right away [*sur l'heure*] and to have them taken to the Post, or to the Messengers subscribers have designated."[81] Although there are some regular features, such as the Enigma, the articles track what people are interested in right now.

The comet makes its first appearance about a third of the way through the January 1681 issue, following a series of obituaries of notable figures, which followed a handful of opening articles related to the king or to official institutions. Although deaths are a current event of considerable interest to *Mercure galant* readers, necrologies are fundamentally a backward-looking genre. They stabilize the memory of the recently dead, situating their stories vis-à-vis other notable families, individuals, institutions, or historical events, and tie the present to the past. A collection of pieces on the comet abruptly redirects the issue's retrospective orientation, introducing precisely the sort of new and unknown phenomena the *avis* described as "so timely we cannot delay it" ("tellement du temps, qu'on ne puisse différer").[82] To switch gears from high society dead to the comet, the journal's epistolary narrator addresses the *Mercure*'s inscribed female reader: "I do not doubt, Madame, that you have seen from your Province the Comet that appears here."[83] From disparate locales in France and beyond, a community is cemented by two shared experiences: looking at the sky to see the same bright streak and reading the timely *nouvelles* published about it while they are still current.

Whereas, the editor notes, the comet's nightly apparition "makes the Weak tremble" ("fait trembler les Faibles") and drives the "*Peuple*" to prognosticate ("le Peuple, que les choses peu communes ne manquent jamais d'épouvanter, en a tiré de fâcheux présages"), it spurs the journal's readers to send in sonnets and impromptus such as the piece sent in by a reader identified as Mr. Philibert from Antibes, which inverts the scene of observation, imagining the comet observing the crowds on the ground. In this upside-down perspective, the French reader is given her own society to gaze at as the object of the *comet's* attention.[84] It seems like the ultimate Parisian bias, to imagine that even the Great Comet would be transfixed by the city that was in the process of becoming the City of Lights. In fact, an article about Gabriel-Nicolas de La Reynie, the powerful lieutenant of police who innovated the use of streetlights for public safety, follows directly after comet-related material.[85]

Philibert's comet scoffs at the addle-brained Parisians, "comme des Cerveaux troublés," who go to great lengths to find lookout spots ("Ils cherchent les lieux écartés / Ils montent sur les éminences").[86] The comet ridicules natural philosophizing of the type depicted in the *Journal des sçavans* ("Je conclus de leurs conférences, / Qu'on doit nommer absurdités / Ce qu'ils appellent connaissances") and at the same time critiques folk predictions by "de faux Savants" about comets as portents of war or the death of an important leader. The poetic comet thus clearly shares the same tastes as the public, who care more about language than about cosmology or superstition.[87] The comet goes so far as to repeatedly demur as to its power to affect anything at all. "My body is bright and luminous," it insists. "If people only opened their eyes, / They would see that my face / Does not express any fury / That foretells the least storm. . . . I do not wish anyone's death."[88]

While the humble comet keeps his attention trained on the foibles of the French capital and its denizens, at the climax of the poem he narrates some dramatic and transformative earth-spanning movement. From the perspective of witness rather than actor, the comet testifies to Louis XIV's martial exploits. "LOUIS LE GRAND" irrupts into the narrative in all capital letters as the comet describes the vigorous monarch traveling the globe. "Even the farthest Climates / conquered by his formidable Arm," Louis, "indefatigable," establishes his domination over both "the tepid regions and the freezing ones."[89] In the poem's conclusion, the comet declares himself an effect of human projection, a simple scapegoat for the results of human actions. In this poem, Mr. Philibert from Antibes likewise deflects the light of attention away from the comet that is transfixing his contemporaries. It seems that he is training that light squarely on the monarch, but even that aspect of the poem consists of rather formulaic encomiastic flourish. Ultimately, he deflects attention sparked by the comet to his own literary efforts, as a named contributor to the *Mercure galant*.

While the comet clearly dominates contemporary attention enough to inspire these submissions from readers, one of the ways attention is stylishly manifested in the *Mercure galant* is with studied disinterest. Like readers of the *Journal des sçavans*, the *Mercure galant* reader is curious, and it is this curiosity for new and interesting things (ideas, songs, fashions, tidbits of information) that propels her through the issues.[90] Yet, there is also an aspect of studied sophistication, a desire to find out the latest while already seeming to know everything, that figures in the *Mercure galant* reader's engagement with the news and that distinguishes her curiosity from that manifested in the *Journal des sçavans*. Readers of the *Mercure galant* do not necessarily read in pursuit of knowledge. Rather, they read for something to discuss. Both the monthly editions and *Extraordinaires* are full of questions to discuss and opinions and propositions to debate. In this context, the comet confers a timestamp of relevance, a kind of news value, radical contemporaneity, or relevance to another subject to which attention is diverted. So to look at the comet in the *Mercure galant* is sometimes also an invitation to look at something else. Even in the attention the comet garners it is less an organizing principle than a kinetic agent.

The reader-submitted madrigal that follows Philibert's poem also name-checks the comet and then proceeds to change the subject: "When I want to plumb the harshness of my fate, / Two more charming Stars are my two Oracles; / And to threaten me with my Death Warrant, / Phyllis's lovely eyes serve as my Comets" ("Lorsque je veux sonder la rigueur de mon sort, / Deux Astres plus charmants sont mes deux Interprètes; / Et pour me menacer de l'Arrêt de ma mort, / Les beaux yeux de Phyllis me servent de Comètes").[91] The comet that the poem's narrator treats with such suave disregard serves as the yardstick of his hyperbole: the inconsistent and irregular lights he has his eyes trained on are his mistress's eyes. Others may wonder about what the comet portends, but this poet has thoughts only for the portent of her eyes, whether she is looking at him, or not, which would mean his imminent death. In the February edition, a reader pens a similar sentiment. The unpredictable movements that he tracks are those of his lady love, not those of the comet in the sky: "Her bleak departure troubles and alarms me. / This is what should be feared, not the comet" ("Ce funeste départ me trouble et m'inquiète. / C'est là ce qu'il faut craindre et non pas la comète").[92]

Yet another poem manages to reference two current events with the same dry wit, as the editor shares in an impromptu about the comet portending a death of "un Grand"—who turns out to be an elephant.

Be certain that the comet
Heaven's dire actor,
Always announces the death
    of a Great one.
Is there not to behold, at
    Versailles,
Stretched out on the straw,
Just deceased, an Elephant?

Assurez-vous que la comète,
Du ciel la funeste interprète,
Prédit toujours la mort d'un
    Grand.
Ne voilà-t-il pas qu'à
    Versailles,
Étendu, couché sur la paille,
Vient de mourir un
    Éléphant?[93]

A spontaneous poem, created in an unplanned instant, is a perfectly apt performative form to mark a comet's passage. Creating one is an act of *sprezzatura*, studied casual brilliance, and the poem's irony and droll manipulation of hyperbole and understatement contribute to its witty shrug at the Great Comet. The other tour de force of understated stylishness is the tension the poem creates by manipulating expectations. Following the first couplet (comète/interprète), the first full sentence ends with "la mort d'un Grand." It seems as if the poet is describing the death of a Prince of the Blood foretold by the comet. The dead man is rhetorically summoned in an appropriate context: "Ne voilà-t-il pas qu'à Versailles," and then, in the couplet's closing, in an utterly scandalous and inappropriate one—"Étendu, couché sur la paille"—before the tension is released with the deferred rhyme. The "Grand," whose death has occurred under the sign of the comet, is an elephant, not a prince. The impromptu flash of inspiration and the comet inspire these sly interrogations of hierarchies, while also ironically deflating superstition.

The Mercure galant's epistolary narrator uses the apostrophized reader's interest in the comet to propel her through various other texts, authors, genres, and registers. Different pieces are proffered for the apostrophized reader's assumed approbation, appealing to categories of taste such as gallantry—"Here I add a Madrigal, which you will find gallantly turned"—or *esprit* or analytical intelligence ("After having seen this material treated in Verses with such *esprit*, I think, Madame, that you will not be angry to see an treatise [*Dissertation*]") to make the connection between very disparate kinds of engagements with the comet.[94] The January and February *Mercure galant* issues feature reports of comet sightings and related *faits divers* as well as a motley assortment of poems. On the serious side, the January issue includes a *dissertation* by Claude Comiers, who concludes that comets are portents of neither good nor bad events. Comiers was also a contributor from 1676 to 1678 to the *Journal des sçavans*, yet the *Mercure galant*'s heterogeneous and mostly literary interventions have no vocation to form readers into knowledgeable natural philosophers or geometers. The editor presents the piece as "aussi curieuse que savant." He uses similar language to inform his readers about an address that the abbé Fleche made at the conclusion of a meeting of the Académie royale d'Arles on "la nature des comètes": "He presented the opinions of the antique and the modern philosophers, in terms that equally reflected the force of his intellect, and the curiosity that this subject must have sparked."[95] This lecture was no doubt more serious than Philibert's poem reprinted earlier in the same edition, yet both items share a similar dynamic. In his review of Fleche's learned presentation, the *Mercure galant* article diverts attention from the academic quest to understand the comet to the brilliance of the speaker and the interest of the subject matter.

It is this sparkling subject matter that the editor harnesses to propel the reader's eyes promiscuously from one piece to the next, from one issue to the next. After the elephant poem in January 1681, the editor uses interest in the comet—and the kinds of stories it has spurred—as a way to pique interest in the next month's edition. He promises more witty and stylish pieces inspired by the comet, "If my letter were not so long, I would share with you here an Adventure of the Comet, that I will put off for another time."[96]

"La comète"—a nouvelle *in the* Mercure galant

The comet wends its way erratically through the January and February editions, cycling through each with several reprises. Reading the *Mercure galant* with an eye out for comets draws our attention to the diversity of the kinds of pieces included in the periodical. It also highlights the organic organization of each issue, a kind of stream of consciousness attuned to each post delivery, new event, or whiff of fresh gossip. Along with contributions that we would now recognize as news of some sort (political, cultural, sartorial, etc.), the periodical regularly features short works of fiction, also called *nouvelles.*[97] The "Aventure de la comète" announced in January's letter and included in February's is a novella about fear and the desire for the new. The tale is actually less about the comet than a rather typical *Mercure galant* short story, involving a well-born yet impoverished beauty in service as a companion to a generous and benevolent lady. The beauty aspires to marriage, while her admirers have other ideas. One day, her newest rich and independent admirer suggests that they go up to the garret to see the comet. The moment the comet first enters the narrated world of the story deftly runs on two tracks: "It had been visible only for the last two days; and since all extraordinary things make a strong impression, this new apparition made all of Paris talk. The Beauty who had already heard that people had seen a Comet expressed a great desire to see for herself what this long Tail that scared so many People looked like."[98] The narration uses the comet in a hilariously ribald, if not unexpected, way to make phallic innuendos, while avoiding anything overt enough to transgress the editor's policy of avoiding any subject "qui blesse la modestie des Dames."[99] At the same time, the exposition highlights some of the signal characteristics of the comet, such as its novelty, sparkle, speed, and long tail. Indelibly impressive, it generates talk and inspires desire, fascination, and perhaps fear.

Each of these two readings of the comet points to a potential reading of the tale. On the one hand, it is a standard farce. The young couple go up to the attic to observe the comet. Although the exposition of the story has insisted on the young woman's virtue, the double entendre is plentiful, from the phallic innuendo to the depiction of the beauty's voracious curiosity ("La Belle voulut contenter sur l'heure l'humeur curieuse qu'elle avait marquée"). They go up (*monter*) to the very top of the house where she can look comfortably (*commodément*) at the comet.[100] When the candle goes out, the young woman realizes it looks bad. Hearing voices on the stairs, she panics and locks the poor man in the attic. Discovered the next morning sleeping in the hay, he is taken for a thief and beaten. The mistress of the house settles the whole misunderstanding by arranging their marriage. The young woman gets the husband she was hoping for; the man, having planned for a conquest, gets a wife. The *Aventure* ends with a moral: "You see there that it's not right to say that Comets bring nothing but misery, because this one Contributed to the happiness of a Person, to whom Fortune had not been favorable until this point," concludes the narrator.[101]

If the description of the comet is taken in a straight, unironic way to be the description of a comet, not the description of a phallus, the exposition still provides a key, but in this case to a story not about sex, but one about a comet, and the news: The comet is brand new. It is a shared preoccupation. It is here now. It motivates and propels desire, and even a little fear. It brings people together in unexpected ways in unexpected places, and in doing so it shakes up the social order.

Comet Eggs: The novelty of a hoax

A final example will allow us to observe how comets trace a path through the *Mercure galant*, appearing at unpredictable intervals to mark, each time, a present moment when a curious thing is happening, when an event occurs that engages everyone—inspiring them to talk, write, read, and especially wonder—wonder, for example, what other people might write in the pages of a periodical. This example will also highlight the comet as a figure of replication and multiplication and the almost fun house quality any attempt at assaying its truth thus assumes.

Three-quarters through the January 1681 issue of the *Mercure galant*, comet news arrives unannounced. It is not mentioned in the table of contents or demarcated by any narrative buffering. A mere paragraph break serves as the transition from a lurid piece on a tragic fire that has killed a newly married couple and the priest, also a relative, who had married them the night before. Following the description of the "extremely lugubrious spectacle" of the three charred bodies being carried out of the house on a single litter, separated, again, by only a passage break, the text delves into another "spectacle," albeit a less gruesome one. The page ends with the portentous announcement, "The Comet having appeared" (fig. 5).[102] Doubtless an accident of typography, the isolation of this phrase at the bottom of the page almost feels like a performance of a comet's surprising and dramatic irruption into the sky following another fiery event. This performative aspect of the mise-en-page is enhanced by what follows. At the turn of the next page, a comet appears! Or the reader sees the engraved images of the comet inside two egg forms, provided the bookbinder followed the instructions at the start of the issue specifying the engraving should be bound exactly here (fig. 6). After this cometic irruption, it takes a few moments to figure out what this story is about. In fact, in the Bibliothèque nationale de France (BnF) copy of the *Mercure galant* that has been digitized for Gallica, a reader has glossed the story in the margin, "Oeuf où l'on voit la figure d'une cométe" (fig. 7).

If this story starts off in a cometic fashion, it is only indirectly about the comet. The qualification that introduces the story, "everyone figured [*raisonnait*] according to the power of their intellect [*esprit*], or their natural timidity," hints that the episode may also be comedic.[103] Sent from Rome, the engraving represents not the comet, as the incomplete phrase "La Comète ayant paru . . ." seems to indicate, but rather two different renderings of the same chicken egg, laid at midnight by one of the Marquis of Maximi's hens. The paper recounts how the bird "wanted to make people talk as much about her as about the Comet."[104] The hen, who had never before laid an egg, made so much noise that she woke the household. The episode's dramatic recounting borrows from the language of wonder and prodigy to comic effect. "At midnight, she began making herself heard in an extraordinary fashion," and "the great noise she had made aroused curiosity." Holding the egg up to the light, the *Mercure galant* recounts, the observers comment on what they see: "Some say the figure of the Comet and others something similar" ("la figure de la Comète ou quelque chose d'approchant").[105] Even the original eyewitnesses are not sure what it is that they see.

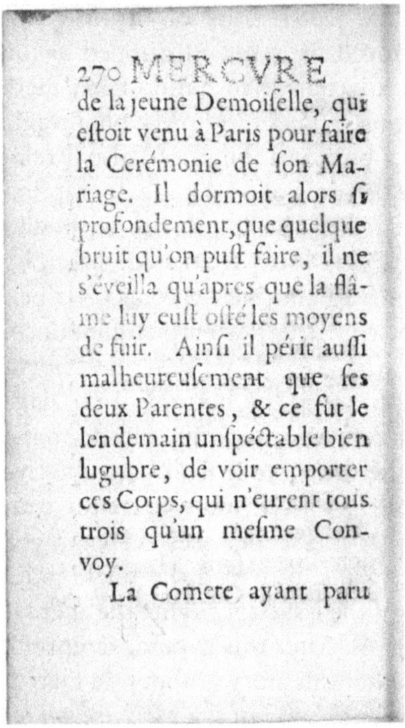

270 MERCVRE
de la jeune Demoiſelle, qui
eſtoit venu à Paris pour faire
la Cerémonie de ſon Ma-
riage. Il dormoit alors ſi
profondement, que quelque
bruit qu'on puſt faire, il ne
s'éveilla qu'apres que la flâ-
me luy euſt oſté les moyens
de fuir. Ainſi il périt auſſi
malheureuſement que ſes
deux Parentes, & ce fut le
lendemain un ſpéctable bien
lugubre, de voir emporter
ces Corps, qui n'eurent tous
trois qu'un meſme Con-
voy.
La Comete ayant paru

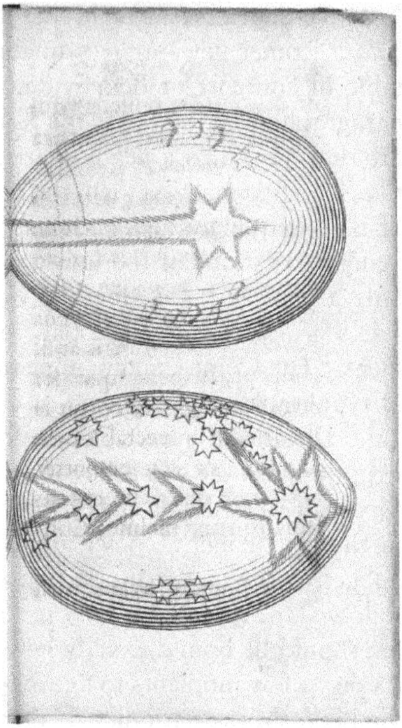

Figs. 5 and 6. Left, irruption of the comet into the narration. Right, illustration of two eggs sent from Rome. *Mercure galant* (January 1681), 270 and n.p. Bibliothèque nationale de France.

The story's placement following more serious *faits divers* and the use of a costly full-page illustration seem to promise that some type of demonstration is about to follow. We may assume that the image is going to double and verify the text it precedes. What ensues, however, is an anecdote that repeatedly emphasizes the ambiguity and likely sham nature of the marvel that is depicted and self-reflexively draws attention to the dynamic of reproduction and transmission spurred by comets' irresolvable mystery and effectuated by print and periodicals. The image that opens the story is not a comet, and not even the comet on an egg, but a print depicting two different renderings of the same comet on the same egg. The doubling of the image makes the reader witness to the artifact's unreliability, and to the infinite reproducibility of print, and comets: "Several different ones arrived from Rome. I am sending you two that I had engraved. This will show you how difficult it is to be certain of anything in this world, the sentiments of Men

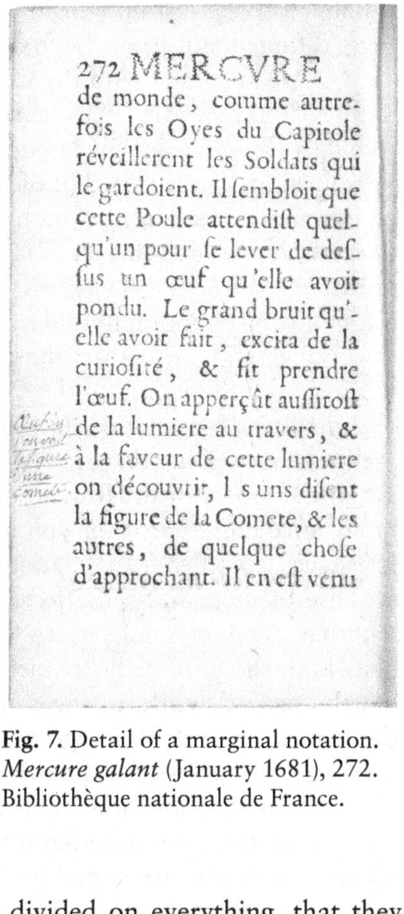
272 MERCVRE
de monde, comme autrefois les Oyes du Capitole réveillerent les Soldats qui le gardoient. Il sembloit que cette Poule attendist quelqu'un pour se lever de dessus un œuf qu'elle avoit pondu. Le grand bruit qu'elle avoit fait, excita de la curiosité, & fit prendre l'œuf. On apperçût aussitost de la lumiere au travers, & à la faveur de cette lumiere on découvrir, l s uns disent la figure de la Comete, & les autres, de quelque chose d'approchant. Il en est venu

Fig. 7. Detail of a marginal notation. *Mercure galant* (January 1681), 272. Bibliothèque nationale de France.

being so strongly divided on everything, that they are even divided on facts, which they should never be, especially regarding eyewitness reports."[106] The question the comet raises here is not that of superstition but of the irresolvable incertitude of the comet's path and the multiplicity of possible representations and interpretations.

As if to demonstrate how these interpretations can multiply infinitely, some fifty pages later in the same (January 1681 Paris) issue, the editor notes a late-breaking dispatch. In the time elapsed between dispatches, the volume has accreted novellas and news: an abjuration and festivities, family trees, deaths, notices about plays (notably one by the magazine's editors), and the popular Enigma feature, including the solution of the previous month's puzzle and the names of those who had written in to share their solution. Now he has received yet another image of the same egg. He recounts this event with the present tense verb and the word *présentement*, emphasizing the immediacy of the breaking news: "I am receiving just now from Rome itself another Figure of the same egg, different still from the first two."[107] "It is surprising how freaks of Nature appeared with the Comet, if they are in fact real freaks of Nature."[108] If there are tricks (*jeux*), they are human, not of nature. But even if the "oeuf en comète" is an international hoax and not a natural phenomenon, by the very dint of its repetition, replication, and spread, it becomes news. The next month's *Mercure galant* includes a much more elaborate print—this one a large foldout—of the third Roman egg (print).

The article describes how the temporal unfolding of the news—new things seen (in the sky or in letters)—structures the object that becomes the *Mercure galant*. "I received the Figure of the Comet, and of three extraordinary Eggs . . . as I was finishing my last Letter. For that reason, I contented myself with describing them for you. I send you today this Figure, engraved on the model of the Roman one, so you can see for yourself what I may have represented for you imperfectly."[109] The *Mercure galant* article stages a moment of fast-print transmission, conjuring for readers a sense of immediacy: the letter just received, the previous one so recent that its ink was not yet dry. What the editor has done, however, is shift both the gaze of his readers and the location of primary fact from the natural world to the printed page. Instead of contemplating the comet, or even its image on the egg, the journal's readers scrutinize imported prints. The French copy of the Italian engraving keeps all of the inscription and publication information in Italian, putting on display the composite nature of the periodical. Things collected and sent from afar are added to the ever-expanding body of the gallant mercurial text.

Following the comet, we see how the periodical accrues. We also observe its fragility and vulnerability to fracture. The illustration from Rome that the text refers to with such declarative certainty is missing from many library copies of the January 1681 *Mercure galant*. Foldouts are easily ripped and torn, or users take them out for other uses—pin them up, wrap food in them. . . . They get taken and sold separately. They are like comets, difficult to pin down and keep in any one place.

And then they are gone . . .

Comets dominate the news until as suddenly as they appeared they fade from view. There are no more comet stories or references in the March 1681 *Mercure galant*. The only remaining spark is publicity

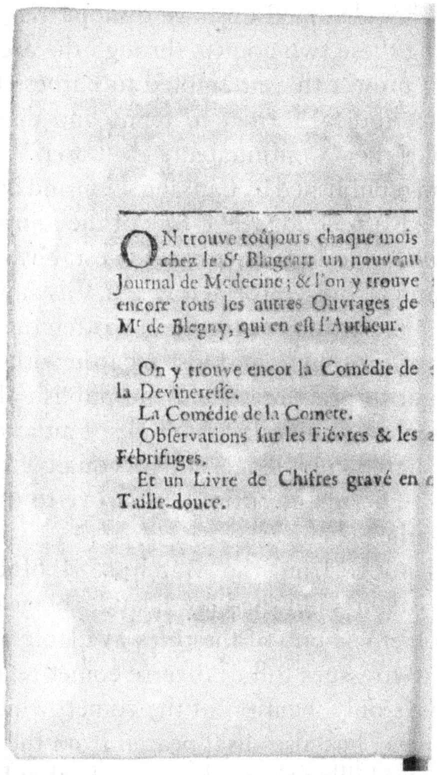

ON trouve toûjours chaque mois chez le Sr Blageart un nouveau Journal de Medecine; & l'on y trouve encore tous les autres Ouvrages de Mr de Blegny, qui en eſt l'Autheur.

On y trouve encor la Comédie de la Devinereſſe.

La Comédie de la Comete.

Obſervations ſur les Fiévres & les Fébrifuges.

Et un Livre de Chifres gravé en Taille-douce.

Fig. 8. In the March 1681 issue of the *Mercure galant* printed by Blageart, the only remaining sign of a comet is in an advertisement for *La comédie de la comète. Mercure galant* (March 1681), n.p. Bibliothèque nationale de France.

(see fig. 8) for the printed edition of a play, whose star would likewise soon fade and which is the subject of this book's fourth chapter.

Comets and the Authority of Observation: Galileo and Nicolas de Blégny

In the period between the comets of 1664–65 and 1680–81, alongside Sallo and Donneau de Visé, an untold number of other ambitious writers and printers attempted to exploit innovations in printing, post, and publicity to grab and hold the public's attention. Most of the publications they dreamed up have disappeared, barely leaving a mark.[110] Following these two comets through the archive, I found the trace of one such project that attempted to harness the collaborative potential of the periodical format to refine and disseminate medical knowledge. Unlike the "Continuateurs de Loret," whose dispatches were collected and republished by banking scion and bibliophile Nathan James Édouard de Rothschild (1844–81), or the *Journal des sçavans* or the *Mercure galant*, which each managed to establish a readership and create a distinct and lasting editorial formula, this ambitious periodical project has been almost totally forgotten, written out of the history of science after its author fought and lost a battle with the University of Paris over the right to practice, teach, and publish about medicine.[111] The journal presents a fascinating example of an attempt to establish a community of shared knowledge, and an equally fascinating illustration of the fragility of an enterprise that strove to establish authority using the tools of the ephemeral.

A commercial notice from the March 1681 official Paris edition of the *Mercure galant* published by Claude Blageart advertises *La comédie de la comète* as one of the titles available at Blageart's shop (see fig. 8). After two issues full of diverse comet-related features, this is the March issue's only mention of the comet, which, no longer visible to the naked eye, has also disappeared from the news. The notice also gives prominent billing to another periodical of Blageart's production, leading with the news that "every month at Blageart's shop you can find a new Journal of medicine; and you can also find the other Works by Mr. de Blégny who is its author."[112] It sounds very much as if Blageart assumes that readers of the *Mercure galant* will also be familiar with Blégny, his monthly *Journal de médecine*, and the author's (numerous) other books.[113] In this last part of this chapter, I try to discern how in this forgotten journal the 1681 comet appears as the pivot that connects a strange constellation of disparate discourses including a translation of Galileo, a catalog of the past century's astronomical discoveries, and a treatise against the use of astronomy in medicine. With its erratic nature, a comet is an appropriate body to tie these varied subjects together. However, due to this same erratic nature, the comet is unable to provide the kind of stable foundation Blégny was seeking as a means to establish institutional, professional, and public authority.

The title page of the October 1681 *Extraordinaire* of the *Journal de médecine* that Blageart published in collaboration with Laurent d'Houry (or d'Hourry) advertises a combination of subjects that seems

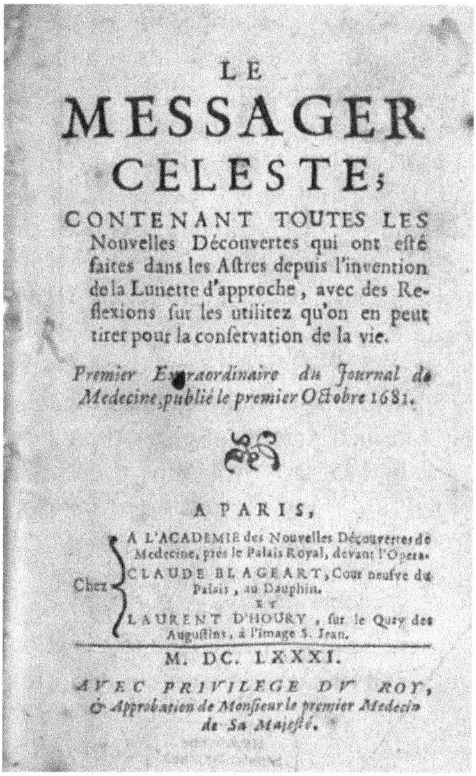
LE
MESSAGER
CELESTE;
CONTENANT TOUTES LES Nouvelles Découvertes qui ont esté faites dans les Astres depuis l'invention de la Lunette d'approche, avec des Reflexions sur les utilitez qu'on en peut tirer pour la conservation de la vie.

Premier Extraordinaire du Journal de Medecine, publié le premier Octobre 1681.

A PARIS,
Chez A L'ACADEMIE des Nouvelles Découvertes de Medecine, prés le Palais Royal, devant l'Opera.
CLAUDE BLAGEART, Cour neufve du Palais, au Dauphin.
ET
LAURENT D'HOURY, sur le Quay des Augustins, à l'image S. Jean.

M. DC. LXXXI.

AVEC PRIVILEGE DU ROY, & Approbation de Monsieur le premier Medecin de Sa Majesté.

Fig. 9. Title page of *Messager céleste* . . . (October 1681). Bayerische Staatsbibliothek Münchin.

unexpected to a modern reader.[114] The masthead proclaims "Le Messager Céleste" in the largest font (see fig. 9). This is the first published French translation of *Sidereus Nuncius* (1610) in which Galileo Galilei described his pathbreaking telescopic observations of the moon and stars. As we can see in the second order of text on the title page, this "messager céleste" brings French readers more than solely Galileo's (by 1681 somewhat classic, less than timely) astronomical "news." Announced directly under the work's famous title, the volume also "contains" contemporary additions. The first, "All of the New Discoveries Which Have Been Made in the Stars since the Invention of the Telescope," seems like a logical appendix to the *Sidereus*, where Galileo recounts his telescopic discovery of the satellites of Jupiter, which he named, after his patron, the Medicean stars. The second addition, "Reflections on the Utility That One Can Take from This for the

Conservation of Life," makes the connection to the *Journal de médecine*, of which, the title page announces, this volume is a special issue.

This volume has entered library collections and scholarly consciousness as the first and only (and incomplete) French translation of *Sidereus Nuncius* until the mid-twentieth century. More than simply a first and incomplete translation, however, the *Messager céleste* of 1681 is a quirky, hybrid enterprise very much of its own time and context. Galileo and his ideas circulated widely in France during the author's own lifetime, in the published work of Marin Mersenne and Pierre Gassendi and in the disputes and exchanges of the learned community.[115] In the seventy-year interim between Galileo's text and the 1681 *Messager céleste*, telescopic astronomy was no longer a novel experiment but a widely ramified practice. Rather than a consecration of a classic after the tides had shifted toward the widespread acceptance of telescopic astronomy and the Copernican worldview, the 1681 *Messager* engages actively in struggles over more temporal authority—the professions of medicine and surgery; the relative place of academies, public conferences, and the university; and the role of the emergent periodical press in collecting, diffusing, and sanctioning knowledge and values. In a rather diffuse constellation of parts, the comet of 1681 appears at both the physical center of the compilation and also at the center of its polemical stance, as a locus for performing the dedication of the author, and thus the associated journal and Académie des nouvelles découvertes, to a regime of observation.

Creating a corpus of medical knowledge: Nicolas de Blégny's journal, 1679–81

The October 1681 volume was published as a special issue of the *Journal des nouvelles découvertes sur toutes les parties de la médecine*, a periodical edited and largely written by Nicolas de Blégny (ca. 1642–1722) that first appeared at the end of January 1679 and ran for three years, with some changes in name and experimentation in format.[116] In the *avertissement* of the inaugural issue, the editor NDB (Nicolas de Blégny) describes his project: "To collect all of the Discoveries, Experiments and Observations that he could concerning medicine that he finds curious, useful, and new."[117] The editor proposes an interactive model similar to the exchange of ideas imagined by Sallo in his preface to the inaugural issue of the *Journal des sçavans*. Blégny invites reader-specialists across the medical spectrum, "Doctors, Surgeons, Galenic or Chemical Apothecaries from France or elsewhere"—to send in accounts of their discoveries, either observational or theoretical. These contributions would then be published, earning the authors the accolades and respect of their colleagues.[118] The *avertissement* imagines the project as a well-regulated interface of interlocking systems. The post will bring in *mémoires* from readers and send them back out in periodical form. At the end of each month, without interruption even for holidays, readers will receive two gatherings of twelve folios each, "in which will be included everything that will have been discovered during the course of this same month."[119] This publication project imagines medical knowledge to be evolving so rapidly that, like current events or even more like fashion, it risks losing its luster if it is out of date: "To make these Discoveries as useful and agreeable as can be, we will take care to publish them before they have lost the grace of novelty."[120] Blégny depicts the time of medical knowledge as regular and orderly, yet the organic time he evokes here is not the rhythm of the seasons, of the heart and breath, of health and sickness, or of youth and aging. It is, rather, the time of the periodical press; the final organism resulting from his collections is not the body (*corps*) restored to health but the bound volume of a duodecimo format book. At the end of the year, when the speed, mobility, and novelty of the time-sensitive discoveries have faded, the *feuilles volantes* (loose, literally flying, leaves) can be bound into a more consecrated, fixed, and authoritative *corpus* as twelve months of twelve-leaf packets are designed to be bound together. The journal's annual issues, containing a year's discoveries, will accrue into a continuously paginated small book, with the prefatory material published in each January's edition and the table of contents in December: "So that the gatherings can be better kept, and made portable, they will be given the form of a duodecimo Book and everyone who buys them will thus have the ease of making them into a complete Book, by having them bound together at the end of every year.[121]

While Blégny's journal has been almost entirely forgotten by medical and publishing history, in the late seventeenth century it seems to have been read far and wide. In May 1679, Blégny name-drops Monsieur le Prince, La Rochefoucauld, and Madame de Lafayette; He references correspondence with Turkey and "les deux Indes" in July.[122] His 1679 volume was republished in Hamburg in German translation and in Geneva in Latin a year later.[123] During the first year of publication, Blégny signed copies to distinguish them from counterfeits and offered a reward of twenty Louis d'or for anyone reporting unauthorized printing.[124] In 1680, the entrepreneurial and highly successful printer Claude Blageart joined d'Houry in printing and selling the journal, renamed

Le Temple d'Esculape, ou Le Dépositaire des nouvelles découvertes qui se font journellement dans toutes les parties de la médecine.[125] The new partners experimented with different printing formats.[126] They also debuted a distinctive elegant double-line typeface, created at Blégny's personal expense, meant to distinguish the journal from imitators and pirates.[127] Blageart had used this strategy starting in 1678 with the popular *Mercure galant.*[128] Alain Riffaud emphasizes, speaking of the *Mercure galant*, the incredible investment required to commission the design and casting of a distinctive typographical brand, a set of font to be reserved for only a select few texts printed in Blageart's shop.[129] The 1681 volume, labeled "Vol III" on the January title page continues to use the signature typeface but reverts to the less confusing title *Le journal des nouvelles découvertes, concernant les sciences et les arts, qui font parties de la médecine.*[130] Bound volumes of the journal can be found in medical and university collections across Europe and the United States from the Royal Society to the Philadelphia College of Physicians.[131]

Messager céleste, *a translation of Galileo?*

The *Messager céleste* of 1681 retains a place, although miniscule, in intellectual history as a small footnote in the story of the dissemination of Galileo. However, Galileo's name does not appear on the title page, which instead emphasizes novelty ("contenant toutes les Nouvelles Découvertes qui ont été faites dans les Astres depuis l'invention de la Lunette d'approche") and utility to medical practitioners. To understand why these "pieces" (pièces) are published together by Blégny in 1681, it is helpful to look at the volume's first paratext, an italic introduction, styled as a letter from Blégny to an unnamed gentleman (see fig. 10).[132] Much as the *Mercure galant* employed as a framing device a letter addressing an unnamed provincial lady, who wished to know about all the goings on in the capital, or as Ménestrier (or, we will see in chap. 5, Bayle) styled his comet book as a letter written to satisfy the curiosity of a fictional epistolary interlocutor, the letter to an unnamed gentleman was a conceit Blégny used in the journal, often as a way of publicizing his own publications and projects and sometimes also as a way of referencing other publications, including the *Journal des sçavans*. In this *Extraordinaire* he responds as if to a request to an order for prompt information on "Astronomical News."[133] This reference to *nouveautés* seems at first just a promotional flourish. Nothing in the volume seems to be particularly new. The Galileo text was a classic, as was the critique of medical astrology. Upon examination, however,

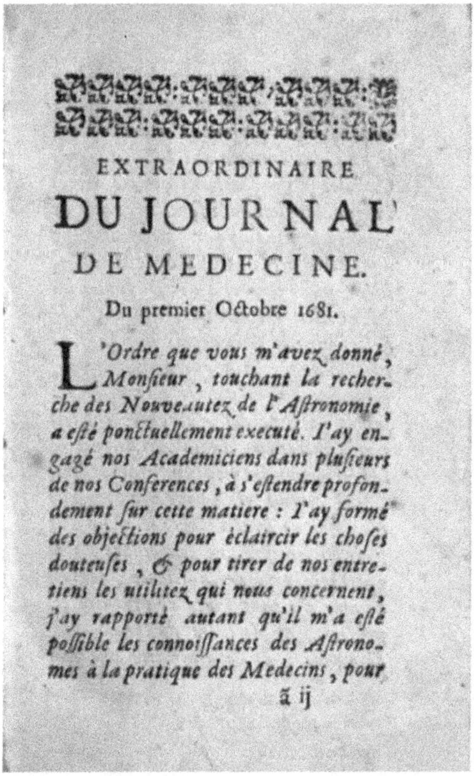

EXTRAORDINAIRE

DU JOURNAL

DE MEDECINE.

Du premier Octobre 1681.

L'Ordre que vous m'avez donné, Monsieur, touchant la recherche des Nouveautez de l'Astronomie, a esté ponctuellement executé. J'ay engagé nos Academiciens dans plusieurs de nos Conferences, à s'estendre profondement sur cette matiere : J'ay formé des objections pour éclaircir les choses douteuses, & pour tirer de nos entretiens les utilitez qui nous concernent, j'ay rapporté autant qu'il m'a esté possible les connoissances des Astronomes à la pratique des Medecins, pour

ã ij

Fig. 10. Editor's note, signed "De Blegny," at the start of the October 1681 *Extraordinaire du Journal de médecine*. Bayerische Staatsbibliothek Münchin.

the volume does engage with one novel celestial event, something that was timely but that the publication date had missed. The middle *pièce* culminates in a chapter on comets, the ultimate celestial and telescopic novelty and, just a few months prior to the publication of this issue, a huge engine of public interest and the subject of best-selling books. With his books on venereal disease, wound care, and coffee, tea, and chocolate, as well as his innovative interactive medical journals (all of which he actively cross-publicized), Blégny was clearly a writer with his finger on the pulse of public interest. The fact that this seems to be the only of Blégny's works that engages astronomy in any way, and that it was published in October 1681 with references to astrological novelties that it has just missed, all point to the Great Comet of 1680–81 as the event that sparked this issue.

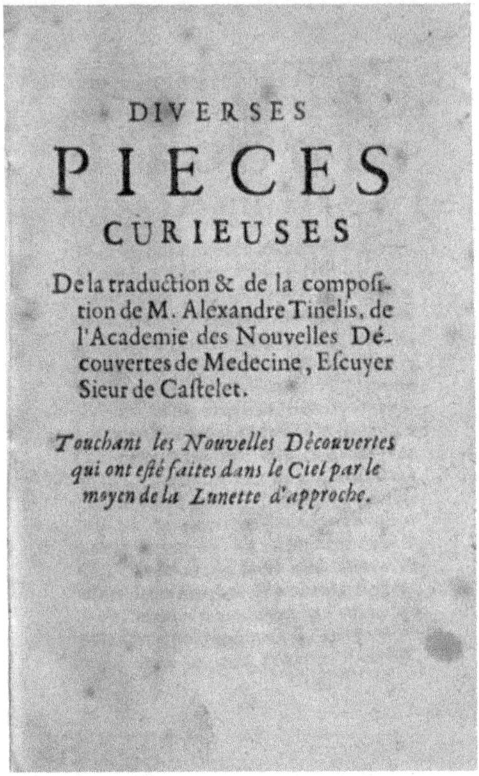

DIVERSES

PIECES

CURIEUSES

De la traduction & de la compoſi-
tion de M. Alexandre Tinelis, de
l'Academie des Nouvelles Dé-
couvertes de Medecine, Eſcuyer
Sieur de Caſtelet.

*Touchant les Nouvelles Découvertes
qui ont eſté faites dans le Ciel par le
moyen de la Lunette d'approche.*

Fig. 11. The volume is composed of "Diverse Curious Pieces." *Messager céleste* (October 1681). Bayerische Staatsbibliothek Münchin.

A second internal sort of title page announces "Various Curious Pieces. Translated and composed by M. Alexandre Tinelis, of the Academy of New Discoveries in Medicine, Esquire, Lord de Castelet" (see fig. 11). An italic subtitle advertises "New Discoveries that have been made in the Sky by the Telescope." Whereas the Galileo translation seems to be the only reason Blégny's journal has found a space, albeit small, in the consciousness of modern scholarship, at the time of its publication Galileo's text is but one of many "diverse curious pieces." When we stop looking at this 1681 French *Messager céleste* merely as a translation of Galileo, we begin to discern how the three "parts" of the volume authorize the project of the journal and of the associated Académie des nouvelles découvertes de médicine as one based on the bedrock of observation by drawing on the credit of both the historical figure of Galileo and the contemporary interest in telescopic observation, most vividly exemplified by the widespread popularity of observing the comet only a few months prior to this edition's publication.[134]

Whereas this title page does not mention Galileo, whom we now recognize as a hero-author of the scientific revolution, it trumpets the status of author/translator Tinelis as a member of the Académie des nouvelles découvertes de médecine, which evokes a certain kind of institutional authority, although the institution referenced is no longer familiar to contemporary readers. *Les adresses de la ville de Paris avec le trésor des almanachs, livre commode en tous lieux, tous temps et toutes conditions, par Abraham Du Pradel, astronome Lyonnais* (1691) introduces the public conferences in medicine held by the Académie des nouvelles découvertes de médecine as a royally sanctioned institution "established by order of the court and organized on the model of royal Academies," where "new discoveries are verified." The entry also lauds Blégny and advertises his publications.[135] The next year's Paris guide, slightly renamed, takes further liberties with the "society's" name, praising Blégny's connections and calling it "La société royale de médecine," "a sort of Academy where one moves from rules to practice . . . composed of Philosophers, Doctors, Surgeons, Artist apothecaries . . . holding public Conferences . . . publishing volumes of Observations and Experiments . . . working without pause to make new discoveries . . . disseminating and sharing with the provinces, with the shared goal of perfecting the most important of all the Arts."[136] These appreciations draw on a mixed bag of markers of authority. On the one hand, the academy and the public conference articulate the social instantiation of knowledge via site- and time-specific performance; on the other hand, the learned network and the periodical articulate a theory of knowledge that accrues through distribution, speed, publication, and sharing.

We do not know very much about Tinelis de Castelet the author/translator of the 1681 *Messager céleste*.[137] He is listed in astronomical bibliographies of the eighteenth and nineteenth centuries for the *Messager céleste* (as a translator of Galileo "avec des additions") and Moréri's *Grand dictionaire historique* invokes his opinions on the disputed invention of the telescope.[138] In his preface, Blégny highlights the author's modesty and wish to avoid attention, polemic, and strife. Above all what we discover in Tinelis's own preface is his paranoia about critics and cabals and his strong anti-Cartesian orientation, which he casts as a kind of crusade against "authority" and "dogma" and for "observation." As deployed by Tinelis, Galileo becomes a representative of the cutting-edge observer who risks censure in order to fulfill the manifest destiny of publication, while the comet is evoked as a celestial event that had recently provoked a widespread popular interest in observation and that invites potentially limitless future telescopic observation.[139]

In "An Empire Divided: French Natural Philosophy (1670–1690)," Sophie Roux offers an overview of the fractious field of natural philosophy during this period when Blégny set up shop with his academy and journal. Against traditional narratives of the scientific revolution, but also in response to the fragmentation resulting from exceedingly local and anecdotal histories, Roux shows that, notwithstanding a rich spectrum of ideas, in the field of natural philosophy, actors tended to range themselves into opposing old and new camps. She demonstrates how the "old" philosophers (Aristotelian, of the schools) start to critique new (mostly "mechanical") thinkers, whom they unsubtly and often inaccurately lump together under the rubric of "Cartesians" for being too theoretical, not rooted in experience.[140] Despite popular conceptions to the contrary, the ancient/modern (Aristotelian/Cartesian) debate took place not only, not even principally, in the university and in scholarly Latin tomes but also in salons, and in duodecimos in French like the ones produced by Blégny.[141]

As one of the "ancients" Roux describes, Tinelis de Castelet rails against Cartesians, whom he sees as holding to "authority" against observation. He does not elaborate the most nuanced critique of Descartes and Cartesians, although it is sometimes hilarious.[142] Over against the mendacious, childish, and risible "authors" such as Malebranche and the bad, dogmatic Cartesians, Tinelis depicts Galileo as the paradigmatic and pioneering astronomer-observer in terms that echo Galileo's own famous rhetoric: "It is better to read in the Book of Nature than in chimerical speculations of Authors: Everyone agrees that Galileo is the person who has best opened this beautiful Book, and who was the first to break the almost infinite obstacles that distance us from its design: It is time that I cede my place to this illustrious Astronomer. It is he who is going to address you in the first part of this Book; next it will be my turn to speak to you again."[143]

Even as Tinelis opposes Galileo's observation-based astronomy to what he depicts as Cartesianism's slavish devotion to textual authority and celebrates Galileo for having "opened up" the "Book of Nature," Tinelis de Castelet does not really cede the floor to the translation as he promises. He repeatedly interrupts with his own italic editorial comments to signal passages he deems no longer relevant.[144] For example, the contemporary author/translator cuts out some of Galileo’s observations of the satellites of Jupiter: “I thought it useless to place them in this Work, as in fact this Astronomer himself declared that he only wrote these sorts of observations to help exactly determine all the things that depend on the movements of these little Planets. The second Part of this Treatise will show you that this hope of Galileo’s is today entirely accomplished, and that on this subject there have even been discoveries that have surpassed his hopes, and those of several other *curieux*.”[145] For Tinelis, the great astronomer has been superseded; it is in the *seconde partie* of the volume, about the last seventy years of telescopic observations, that Galileo’s promise finds fulfillment, casting Tinelis’s essays as a kind of New Testament to Galileo’s Hebrew Bible.

Graphically, this *seconde partie* of Tinelis de Castelet’s composition seamlessly follows from the translation of Galileo, which he truncates and interrupts. The title “LE MESSAGER CELESTE” continues to unfurl at the top of the pages in the journal’s distinct, elegant double typeface (see fig. 12). At the moment of transition, he echoes Galileo’s famous theater metaphor, yet in promising “a second Scene, with a much more magnificent set than the first” he cites the maestro in a gesture that closes the curtain on him, for telescopy, in its infancy in Galileo’s time, has made so much progress.[146] According to Tinelis’s editorial presentation, reading Galileo does not so much instruct au courant astronomers of 1681 but rather allows them to take greater pleasure in new discoveries and astronomical observation by imaginatively projecting themselves into Galileo’s perspective as a kind of virgin viewer:

> There is not a *curieux* today who should have missed this Book of Galileo, where he lists everything the telescope revealed [*a fait voir*] when it was only in its infancy . . . and when one knows about these first discoveries, one learns with much more pleasure about more recent ones, and in this way one feels in a certain sense the satisfaction that we easily imagine a person having, who having only had the telescope that Galileo used, perfected it more and more and saw it reach the state where we presently admire it.[147]

Playing on the word *justesse*, which means both apt and precise, Tinelis signals that the more appropriate or correct focus of our admiration should be the triumphant present tense.[148] Today’s myopes, crows Tinelis, are “incomparably more clairvoyant than the Arguses of past ages.”[149]

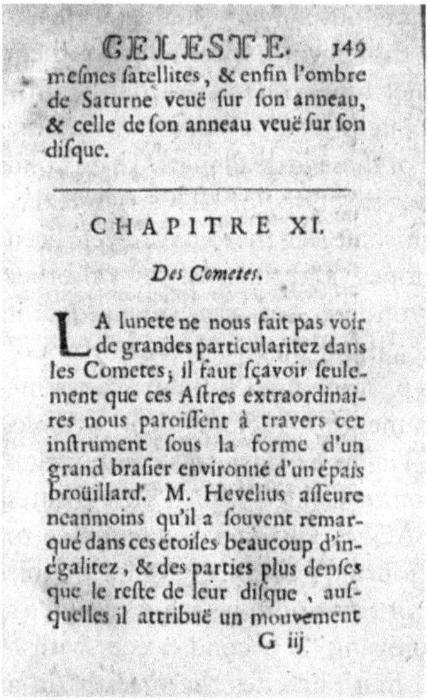

CELESTE. 149

meſmes ſatellites, & enfin l'ombre de Saturne veuë ſur ſon anneau, & celle de ſon anneau veuë ſur ſon diſque.

CHAPITRE XI.

Des Cometes.

LA lunete ne nous fait pas voir de grandes particularitez dans les Cometes; il faut ſçavoir ſeulement que ces Aſtres extraordinaires nous paroiſſent à travers cet inſtrument ſous la forme d'un grand braſier environné d'un épais broüillard. M. Hevelius aſſeure neanmoins qu'il a ſouvent remarqué dans ces étoiles beaucoup d'inégalitez, & des parties plus denſes que le reſte de leur diſque, auſquelles il attribuë un mouvement

G iij

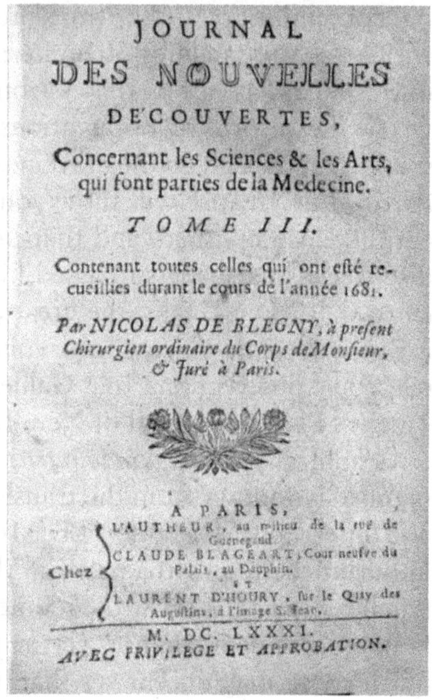

JOURNAL

DES NOUVELLES

DECOUVERTES,

Concernant les Sciences & les Arts, qui ſont parties de la Medecine.

TOME III.

Contenant toutes celles qui ont eſté recueillies durant le cours de l'année 1681.

Par NICOLAS DE BLEGNY, à preſent Chirurgien ordinaire du Corps de Monſieur, & Juré à Paris.

A PARIS,

Chez L'AUTHEUR, au milieu de la rue de Guenegaud.
CLAUDE BLAGEART, Cour neuſve du Palais, au Dauphin.
ET
LAURENT D'HOURY, ſur le Quay des Auguſtins, à l'image S. Jean.

M. DC. LXXXI.

AVEC PRIVILEGE ET APPROBATION.

Figs. 12 and 13. The text at the top of the pages connects Tinelis's essays (left) to Galileo's as if a single project. The signature typeface also links the volume to Blégny's journal (right). Bayerische Staatsbibliothek Münchin and Österreichische Nationalbibliothek.

The *seconde partie* offers a history of telescopic innovation, followed by a narrated voyage through the solar system: from the sun through the planets, stars, and moon, each its own given chapter. Tinelis compiles observational highlights of the preceding seven decades, citing directly and indirectly from works by major European astronomers. Along the way, however, Tinelis's text does not become more certain or as eagle-eyed as the Argus he triumphantly summons in the preceding pages. Instead, the textual planetarium becomes murkier and murkier. Close observation seems to lead to uncertaintly. Tinelis casts into doubt Copernican cosmography, promising to soon disclose, but never in fact disclosing, his own empirically based system. He ends the narrative journey with a final chapter on comets.[150]

Comets correspond perfectly to some of the exigencies of Tinelis's project and of periodical publication in general: a comet sighting is just the kind of "nouvelle découverte" liable to be shared speedily in the journal or a meeting of Blégny's academy. Also, wide swaths of the public were curious about comets, so during the early months of 1681, at the moment this issue seems to have been conceived, there was great interest in observing comets with telescopes, an activity that combined two novel and fashionable things and at the same time signaled a commitment to observation that the *Extraordinaire* seems to have been assembled to showcase on behalf of the Journal *des nouvelles découvertes* and the associated academy. The chapter "Des comètes" emphasizes observation as an idealized conduit to discovery; however, it also reveals some of the challenges observation comes up against in a real-world setting, when the kind of full-throttle accretion of new observations conceived in Blégny's program does not add up in the way he planned.

The chapter entitled "Des comètes" begins with the very unpromising assertion that "the telescope does not reveal [lit: does not make us see; *ne fait pas voir*] great particularities in Comets."[151] Nevertheless, due to the abiding question marks that comets pose—what are they, where are they, what is the shape of their motion—comets invite the kind of interest that spurs people to pick up telescopes: "When people are talking about Comets in ordinary conversations, someone almost always says that Mathematics instructs us to determine their position above or below the Moon by means of parallax and since few people know what parallax is, I thought it would be apt/timely to explain it."[152] In this comment we can catch a glimpse of the *Journal des nouvelles découvertes*'s readership, knowledgeable enough to touch on parallax in "ordinary conversation" but not necessarily certain about what parallax is.

Tinelis de Castelet does a laudable job explaining the concept in terms that would allow readers of the journal to take measurements and make calculations of their own, using their here and now as the way to potentially unlock some piece of the comet's mysteries—and thus participate in one of the most highly discussed scientific questions of the moment. As he says later, "I gave them a clear and satisfying idea of something that had long been harped on around them, but that they might still not have known."[153] Only, by the time the *Messager céleste* was published, the Great Comet was no longer visible, so the possibility of sighting a comet and taking its positions had been relegated to the realm of old news. Tinelis ends the chapter with a roundup of publications that have weighed in on "this recent Comet that appeared at the start of this year."[154] Rather than helping his readers make observations of their own, he is inviting them to read. Furthermore, he reports a contradictory set of assertions from various observers. Cassini, who last saw a vanishingly faint trace of the comet on March 22, asserts, contra his own predictions, that the comet did not retrograde, while Father Antheleme saw it retrograde on April 21 and 22 and even predicts that it will be back next January and February.[155]

Tinelis de Castelet's translation and continuation of Galileo appears as part of Nicolas de Blégny's project to harness the rapidly developing periodical press as a kind of epistemological model—a model shaped by "nouveautez journalières," the constant and regular arrival of novel pages, meant to be bound together, to create an ever-expanding and progressive authoritative corpus of knowledge that could explain the universe and the human body alike.[156] But Blégny's foundational premise—that the accretion of novel observations coalesces into stable bodies of knowledge (the book, the academy)—ultimately proved elusive. As exemplified by the case of comets, observation can drive the periodical press; The comet of 1680–81 was a journalistic phenomenon. However, comets are transitory and, in 1681, could not yet be assimilated into any graspable coherence. Ultimately, the forms of authority marshaled by and around this volume collapse one after the other. In 1683 a rival *Journal de médecine* appeared, edited by Jean-Paul de La Roque, who had served as editor from 1675 to 1678 of the *Journal des sçavans*. In 1682, while he was planning his own medical journal, La Roque successfully lobbied to have Blégny stopped from publishing his *Journal des nouvelles découvertes* with an order of counsel (*arrêt du conseil*). With the protection of Monsieur the king's brother, superintendent of police La Reynie, and other powerful patrons including the king's first physician Antoine d'Aquin, Blégny continued his publications for a year before ceding the project to a Huguenot physician named Gautier, who was living in Amsterdam.[157]

Blégny's Académie des nouvelles découvertes, of which Tinelis de Castelet was a member, and under whose aegis this text was produced, was also less solid an institution than Abraham Pradel's Paris directories lead us to believe. In fact, Pradel, "*philosophe et mathématicien*," was a pseudonym for Blégny himself. This early city guide was another of Blégny's innovations in the realm of commercial print.[158] The archives show traces of Blégny's repeated clashes with the University of Paris over medical authority. The year 1693 finds the innovative author/publisher embroiled in trouble again, but this time, the king's physician cannot save him; D'Aquin has been exiled and Blégny is sent to For-l'Évêque Prison. Blégny had his side of the story printed.[159] The document lays out a baroque plot of enemies and jealous competitors. It is not clear what forces helped Blégny emerge from prison in 1699 with permission to retire to the Papal City of Avignon, where he lived until his death in 1722.

Even as the practice of periodical publication became increasingly common, Blégny's innovations were disowned by his successors.[160] The eighteenth-century authors and editors who established a history of the previous century's literature have nothing but contempt for the use that Blégny made of the periodical format. Jean Berryat's 1766 *Collection académique*—which copiously reprints both the *Journal des sçavans* and the *Mercure galant* as part of the "Extraits des meilleurs ouvrages périodiques"—and eighteenth-century and nineteenth-century editions of Moréri's *Grand dictionnaire historique* both pile scorn on his contributions to the development of periodical literature. The editors of the *Collection académique* call Blégny and Gauthier's *Mercure sçavant* a "work full of bile and ignorance" ("Ouvrage plein de fiel et d'ignorance").[161] The only way Blégny figures in their history is for the impetus his journal provides Pierre Bayle, as a negative example against which Bayle conceives of his *Nouvelles de la République des Lettres*: "Our illustrious author was indignant that an invention as useful as periodicals became, in vile hands, an instrument of calumny and pettiness, and thus turned into a source of shame for literature."[162] We will meet Bayle in the fifth and final chapter of this book. When Blégny's signature typeface reappears in 1710 in the headings of the *Mercure galant* issues edited by Charles Du Fresny following Donneau de Visé's death, and published by Daniel Jollet, Pierre Ribou, and Gilles Lamesle, it is without mention of Blégny. There is no containing comets.

Part Two

Cométomanie, 1664–65

CHAPTER 2

Comet Ballets

In late seventeenth-century France, dance was a social practice thoroughly imbricated in people's daily lives and educations as well as in festive and solemn occasions. The art that we now know as ballet in English (and in French, "ballet classique") was codified under Louis XIV in one of the first official acts of his adult reign with the publication of *lettres patentes* in March 1661 establishing the Académie royale de danse.[1] However, Julia Prest and Mark Franko both emphasize that although seventeenth-century ballet is considered the origin of classical ballet, performances from that era were spectacular and often strange multimedia affairs. Prest characterizes seventeenth-century ballet as "a world of illusion and allusion, of spectacle and symbol, of marvel and mystery. More than anything else, it is a world of diversity."[2] Franko evokes "a potentially chaotic mélange of music, decor, costumes, props, declamation, distributed librettos, and the audience's active participation, as well as dancing, acting, improvisation, miming, and mumming. Its illusionary universe was much closer—in material density if not in philosophical outlook—to twentieth-century performance art than to classical ballet."[3] Dance was used by diplomats, sovereigns, and pedagogues.[4] Ballets and festival entries and later operas staged the king's glory and enacted theories of kingship in the lived, embodied space of multimedia events.[5] Marie-Françoise Christout calls dance a "privileged media interpretation of power" designed "under the veil of allegory. . . . Its secret aim was to educate and to inform (while amusing), to strike the imagination and to thus impress essential ideas."[6] Dance also served as a powerful rhetorical trope: often referencing classical texts from Plato to Pythagoras, the metaphor of dance provided a conceptual framework that helped people conceive the order in the cosmos or in human institutions. As a somatic practice inscribed in space and time, dance places bodies in relation to each other and thus experientially models and investigates the dynamic relationships that evolve between them. Likewise, the experience of dancing allows humans to imitate, on earth, the harmony of celestial order.[7]

When descriptions of comets first caught my attention in letters and diaries chronicling Louis XIV's early reign, I assumed that since there was such wide public interest in comets in 1664–65 and 1680–81, and such lavish use of the cosmos in royal ballets, from the young king's star turn as the sun in 1653's *Ballet de la nuit* to his final planned role as Apollo in *Les amants magnifiques* (1670), I would discover comets in the kinetic representations that presented the king to his courtiers and nation.[8] I was thus surprised to find comets nowhere in these extravaganzas, although there are records of discussions of comets in learned contexts proximate to the court.[9]

This surprising absence from France's spectacular monarchy, at a cultural moment when comets galvanized so much public attention and inspired so many and such varied publications, was one of the questions that spurred me to search for traces of the comets in the archives of the era. What might comets have represented to a French public in these years when Louis XIV, and France, were being reimagined using cosmological terms and visual symbols? In this chapter I examine the records that remain of two ephemeral dance performances from 1665 that put comets on a dance stage. In the very limited archival records of these two ballets, I look for traces of some of the ways French people of the late seventeenth century reacted to comets' spectacular appearance and seemingly irrational motion. The portrayal of comets through the medium of dance is especially interesting for the way it inevitably must confront the issue of the chaotic movement that set comets apart from other celestial apparitions.

Looking for comets in the archive draws our attention far afield, decentering scholarly attention much in the way comets' seemingly random motions drew observers' eyes in directions that seemed to defy the tight logic of the solar system. In studying the often-fugitive texts and performances inspired by *cométomanies*, we can discern some of the ways people in seventeenth-century France engaged with disorder. We can also find traces of some of the connections, experiences, and relationships this disorder produced. The two ballets I have found with the generous assistance of colleagues manage comets' disruptions in divergent ways.[10] *Ballet de la comète* danced in Soissons in February 1665 stages the comet as an invitation to pay attention to bodies, forces, and activities outside the orbit of the usual. In contrast, the *Ballet des comètes*, a Jesuit school ballet, stages cometic transgressions in order to subsume them into forces of order.

Comets Dancing Toward Disorder in the *Ballet de la comète*

Performed in the ancient town of Soissons, several days' ride from Paris and even longer from Versailles, the *Ballet de la comète* uses dance to explore social disorder enabled and made visible by the light of a passing comet.[11] Taking the comet's celestial disruption as an occasion to consider social disruptions, the performance might have resembled the baroque ballets, which, according to Franko, incarnated disruptive and contestatory social forces. The ballet seems to operate entirely outside of the realm of control established with the promulgation in March 1661 of the *Lettres patentes du Roy pour l'établissement de l'Academie royale de danse en la ville de Paris*.[12] The royal academy founded by this decree established norms for technique and restricted choreography and instruction to a royally sanctioned group of experts. Franko emphasizes the political implications of this royal oversight established over dance: "Having danced in ballets since childhood, Louis knew their ideological potential and intended to exploit it to enhance his monarchical prestige. That prestige necessitated the confiscation of ballet from the hands of would-be competitors: the noble class. Subsequently, nobles would continue to dance in ballets, but they would cease to organize and perform them without royal surveillance, especially in the public sphere."[13] The *Lettres patentes* charge the academicians to "deliberate on the means for perfecting [dance], and to correct the abuses that can have been introduced or that could be introduced."[14] In other words, the academy was established to protect dance from outside influences and to make dances and dancers conform to centrally dictated forms.[15] Since it was danced so far from Paris, the *Ballet de la comète* is perhaps a transgression of the king's recent monopoly over dance, or perhaps it is simply occurring outside the center of gravity of the court and Paris, where the regulations were most strenuously imposed.[16] In either case, the ballet program offers a rare glimpse of a contemporaneous enactment of a comet in motion on stage. The themes and concepts the ballet engages fall outside the domain being carefully demarcated for ballets post-1661, and it is possible, although we will likely never know, that the choreography and dancing did as well.

The *Ballet de la comète*, staged while the winter 1664–65 comet was still visible, has been preserved in the archives thanks to a single copy of a sixteen-page program: *Ballet de la comète, Divisé en deux Parties: et dansé à Soissons le 9. Février 1665*, published and ostensibly sold in

Soissons by Nicolas Asseline, Imprimeur et Marchand libraire.[17] The sole copy listed in the international WorldCat union catalog is held in the collection of the Bibliothèque Municipale de Lyon in a *recueil factice* of seventeenth-century theater ephemera from the library of the Jesuit Collège de la Trinité in Lyon. While a minimalist (there are no references) entry in the CÉSAR database of ancien régime theater performances lists a "Ballet de la Comète" as having been danced at the Collège Saint-Nicolas in Soissons, the ballet published by Asseline makes no mention of the location of the performance, and there is nothing in the program to suggest that this *Ballet de la comète* is a college ballet.[18] The dancers, identified only by honorific and initial, seem to have included eleven male performers and five or six married and unmarried women, some with nobiliary particle, some without. With its thematization of social disruption and timing in late winter, the ballet may have been part of the local carnival celebration.

There are no mentions of the performance of the *Ballet de la comète* in the contemporary periodical press, and I have been unable to find any descriptions or references in letters or diaries. The program preserved in the Jesuit *recueil*, however, records many significant narrative details of the dance performance, although it leaves esthetic details (music, tempi, dance forms, choreography, costumes, setting) and all sense of experience inaccessible.[19] Asseline's printed program includes an *argument* summarizing the ballet's storyline and then, for each of the ballet's twelve *entrées*, a short description of the scene and action and the initials of the dancers, all in roman type. These are followed by rhyming verses for each dancer, in a format similar to Benserade's court ballet programs. The *argument* begins with a meditation on ballet's particular ability to create order out of chaos: "In order to present in an orderly fashion all of the people shown in this BALLET, and who by accident or design saw the COMET, the ballet was divided into two Parts."[20] The typography of this sentence seeks to balance "BALLET" and "COMET." "BALLET" corresponds to the imperative to "faire paraître avec ordre," while "la COMETE" represents an element of *hasard*, or chance. The order established by the ballet's tidy division into two symmetrical parts, each composed of six *entrées*, harmonizes with the theatrics of ballet itself, which organizes the movements of bodies and among bodies. The first part takes place at night, in the town, as urbanites of various social strata in the streets after dark pursue romance, encounter thieves, wander aimlessly, peddle almanacs, and observe the sky—all by the light of the comet. Part 2 is set in the countryside before daybreak. Peddlers, so-called "Bohémiennes," hunters, milkmaids, a musician, shepherds, and shepherdesses also variously pursue mischief, crime, and romance in the comet's exceptional early light. The ballet's bipartite construction balances night and morning, town and country, men and women, high and low status, law-abiding and criminal elements. All of this well-balanced composition draws attention, in contradistinction, to elements of *hasard*, the errancies of people and the unpredictable appearance of the comet, the (dis)ordering force, which sets the ballet's drama in motion.[21]

As the description of the opening scene makes clear, a comet painted on the scenery provides a visual focal point for the disparate danced tableaux and galvanizes the attention of the dancers. The pamphlet emphasizes the mise en abyme of an audience transfixed by the spectacle of the dancers transfixed by the spectacle of the comet: "The Comet is visible in the Sky, and draws the eyes of everyone who appears on the stage."[22] The ballet opens with a prologue spoken by the comet. While other dancers are identified by title and initials, there is no attribution for this role, although the text makes clear that the character is female. She declares that she has not come to foreshadow the regular disasters, like famine and war, as her "sister comets" usually do.[23] Rather her appearance "marks" another kind of *malheur*: "those that I come to mark are of another sort, / And pertain to certain Lovers / A thousand times more worthy of cruel torments / Than of a pleasant fate."[24] Just as the comet herself might be seen to stand in contrast to nature's order, her propositions are also "d'une autre nature," a counternature marked out (*marquer*) on stage by the dancers who perform the entrée. The comet foretells woe on a catalog of unworthy lovers: lovers who complain constantly, indiscrete ones, inconstant ones, inconsiderate ones, insolent ones, and, especially, faithless ones.[25] The comet's anaphoric pronouncement of *malheur* on this litany of undeserving *Amants* sets a somber tone that is diffused by the prologue's final arch quatrain. If any lovers are spared, it is only because if all of the unworthy ones were struck down, no one would be left on earth ("Si le Dieu que lance le Tonnerre / Les sauve tous d'un infâme trépas, / C'est que pour cette fois sa Bonté ne veut pas/ Dépeupler tout à fait la Terre").[26]

The comet represents the inspiration from which the ballet ensues and the conceit that ties it together, but the comet is a slippery *fil conducteur*. Unpredictable movement is a difficult organizing principle. For all the symmetry the libretto imposes, and the despite the opening promise to "faire paraître avec ordre," the comet repeatedly introduces randomness or chance. It is both the double of the sun, and its opposite. As an unexpected light in a sky that should naturally be dark, the comet draws people outside when they normally would not be, to do things they probably should not be doing. An interruption in the natural balance of day and night, the comet provides both the condition of possibility and the alibi for a variety of socially transgressive behaviors that constitute the ballet's twelve entrées.

Part 1 opens with Mercury, god of travelers, trickery, commerce, and crossroads, urging lovers and thieves out at night to take advantage of the stolen time he offers them: "Fly Scamps, fly Lovers, / Help yourself to all of the moments / That my attention spares you."[27] "Volez": He urges lovers to rush, and, with the same formulation, urges crooks to steal. Or is he ordering lovers to steal (hearts) and thieves to rush (to steal)? In the self-referential doublespeak of drama, he is commanding both pairs of characters to fly, in other words to dance. This double entendre continues: to defeat the forces of order, "A Thousand Archers, and a Thousand Jealous Men," the trickster god coaches them to "Be dexterous like me."[28] In part 1, the ballet, described as a dexterous danced flight that Mercury inaugurates, happens in stolen time and as a challenge to forces of order. The prologue makes clear that the dancing that happens in the comet's untimely light will be quite unlike the stately and ordered procession over which the sun would preside, with each planet and star moving according to its proper role in a harmonious ensemble.

While descriptions of royal ballets tend to claim the king's omnipresent energy to a paranoid degree, the *Ballet de la comète* does the opposite, undermining the ostensible subject of the spectacle from the very start.[29] At the beginning of the first entrée, the two gallants admit that "it is certainly less for the Comet / that we come to this place / Than to see some Coquettes," who appear and dance in the next entrée.[30] The scamps (*filoux*) who feature in the third vignette admit that they "are not without a plan" ("ne sont pas sans dessein"), but their planfulness likewise has nothing to do with astronomical observation. They "will rob the Coquette . . . maybe even worse" ("détrousseront la Coquette . . . peut-être encore pis").[31] A *trousse* is both a purse and an undergarment, so the threat implies the possibility of rape. The intrigues (*desseins*) that draw the flirts and hoodlums out into the street in the dark of night might evoke for listeners, even as they defy it, the homophone *dessin*, the harmonious line that should unite a ballet, circumscribe social order, or describe the motion of planets and stars. Comets, perceived as shapeless and unpredictable, failed to conform to any such regularity. Likewise the characters whom the comet's light draws out in the night. The program admonishes the spectators to "See their actions. . . . Woe to anyone that falls into their hands / these Gentlemen think that rules do not apply to them."[32]

M. De la P. in the role of the *oublieux*, or forgetful (perhaps senile or demented) person, continues part 1's decent into disorder. He performs and personifies confusion, wandering the streets only at night. He explores the city when he should be in bed dreaming and sees by the light of his lantern sights that do not conform to the reality that daylight would normally reveal. "The Sun will not let me see / The things my Lantern shows me," he declares.[33] We might picture the dancer's lantern moving erratically around the stage, following neither the logic of a sane mind nor the chronometry of a proper day, and in this way acting as a kinetic double of the comet painted on the scene's backdrop.

In part 1's final pair of entrées, first two peddlers plan to capitalize on public interest in the comet to sell almanacs to the credulous (fifth entrée), and then the astrologer they summoned for a reading takes the stage with "pedantic gravity" (sixth entrée). These two scenes present a comic commentary on the comet. It may be the subject of this ballet, but it is a silly one that inspires funny business and scams, laughs, confusion, and misdirection of various sorts. The peddlers unsettle the temporal chronology of the ballet. Moving in the direction of increased entropy rather than toward resolution, they promise almanacs that predict the past with great accuracy, and the future with none at all ("Dans l'avenir, pour ne vous tromper pas, / Nous confessons que nous ne voyons goutte / Si c'est pour le passé, ne faites aucun doute / De prendre de nos Almanachs.").[34] This comic verse turns spectators' attention to the way that comets represent a temporal disruption, irrupting unpredictably and with no visible schedule or pattern. Even the most serious astronomers and mathematicians could not make sense of their time or place, and could only look back to the past to describe their inscrutable appearances.[35]

The astrologer has nothing to say about the comet, on which he comments in only a negative construction ("Par la remarque que j'ai faite, / J'ai reconnu que la Comète / Ne montre rien dans son aspect / Qui doive nous être suspect").[36] However, in the ultimate comet-shaped move, he quickly diverts his gaze, and the ballet's narrative thread. His *dessein* lies elsewhere. "But is it only about this Meteor / That I plant to talk to you?" he asks, pivoting to another subject entirely. He looks away from the comet, toward other stars and planets, and turns to praising the local dignitaries: the d'Estrées family, in particular the ninety-two-year-old duke François-Annibal I d'Estrées (1573–1670), former marshal of France, retired diplomat, appointed governor of Soissons—and father of Cardinal César d'Estrées, who commissioned the Coronelli globes.[37] After quickly dispatching the comet, the astrologer spends a long stanza foretelling the family's good fortune. The balance of dancing, politicking, and the planets leaves the comet to the side as the soothsayer brags about his prophesies in language that also references his dancing. Despite little preparation or foresight, "I don't look clumsy; / And some might say in seeing my discourse, / That next to me the most learned Sybil / can only do half-turns."[38] We have no way of knowing if the dances of the gallants, coquettes, crooks, or comet were balanced and orderly, or exaggerated and comic. It is possible that the two tendencies alternated, as was sometimes the case in court ballets. No matter what the meter and dance style, the entrées are inscribed under the sign of a comet that illuminates, (dis)orients, disturbs chronology, and provides an alibi for the characters' nocturnal perambulations.

The second part of the ballet is set in the countryside, where rural people catch a look at the comet in the early hours before dawn. In a prologue, Mlle. L. in the role of Aurore, the dawn, draws our attention to the strange atmosphere. Even though the sun has not yet appeared, "everything shines brightly." She wonders, "Who, then, can be here, doing the Sun's job?"[39] The comet takes the place of the sun, supplying light when normally there would not be any. Instead of the daylight rays that symbolize knowledge and truth, the incarnation of chaos and confusion lights up the sky, setting the stage for six more danced entrées.

At the conclusion of her prologue, Aurore addresses the women in the audience. In the place of Apollo, who embodies harmony, reason, and moderation as he presides over daylight, dawn presents a feminized and unpredictable presiding spirit and source of light: "It is not he [the sun] who, / bringing back the light to this place, banishes the dark night; Lovable Beauties, today Nature / Owes this rare Miracle to the sparkle of your eyes."[40] The narrator offers up the *Ballet de la comète* as a literalization of a familiar poetic trope, aligning the light of lovely female eyes with that of the comet, both volatile entities that invite the adventure of unruly lovers. Consequently, in the six vignettes that follow, everyone seems to go about their early morning occupations, but they are out hours earlier than they normally would be and are all focused on love instead of work. Peddlers show off their wares yet dream of seducing young beauties (entrée 1). "Bohémiennes" appear as racialized figures of hypersexualized femininity like the black-up characters Noémie Ndiaye traces in baroque ballet.[41] The first is "dangerous and deft" ("dangereuse et subtile") at stealing hearts and purses alike. "Whatever precautions someone takes / to save himself from misery, / The cleverest cannot defend himself from it. / When it pleases me, I have his purse and his Heart."[42] The hunters wonder what Destiny foretells for their hunt but actually care more about amorous pursuits: "But to speak frankly, / The heart of a loveable Shepherdess / Is for us the most charming prey."[43] A pair of milkmaids trading innuendo about their creamy wares in the fourth entrée are followed by a seductive accordion player (entrée 5). Finally, a ballet of shepherds and shepherdesses "after having considered the Comet for a while, perform the last and final Scene."[44] The concluding verses urge the dancers to enjoy youth and springtime: "And if you believe me, think only of dance, / of games, of laughter, at every hour, in every place."[45] Just as the final hemistich extends the space of exception created by the comet beyond the night depicted in the ballet, "à toute heure, en tous lieux," according to the *argument* at the beginning of the program, the performance opened onto the Grand Ballet.[46] The final tableau harmonizes in dance the bodies of performers and spectators, even as it extends the contradictions inherent in the ballet as a whole. This dance takes place within the liminal space of a comet-lit night and has been occasioned by an extraordinary and irregular apparition. This final ballet creates a community of dancers outside the bounds of normal balance, in a spectacle that celebrates ephemeral and unpredictable spectacularity.

Postscript to the *Ballet de la comète*

The copy of *Ballet de la comète, Divisé en deux parties: et dansé à Soissons le 9. Février 1665* collected in the *recueil factice* in the Lyon public library offers a fragile record of a ballet danced outside the newly developed norms that governed dance at court and in the capitol. The ballet's author has also left barely a trace, although on the basis of a short obituary in the *Mercure galant* I am able to identify him as Charles Bertherand (1631–1700), whom the *Mercure galant* celebrates as a leading citizen, founding member of the local academy, and an excellent poet, who nonetheless published only this ballet and one other work.[47] We have no information on the ballet's music or the choreography, the place of performance, or the full names of the dancers. However, the printer's name appears on the pamphlet's title page. According to his profile in the catalog of the BNF, printer and bookseller Nicolas Asseline had worked for Nicolas Fouquet and relocated to Soissons during the highly public and political trial of his former patron.[48] While in 1660 he published an ode on the royal entry to Paris, in 1665, after Fouquet's sentence and the dispersal of the literary and artistic world the minister had cultivated at Vaux le Vicomte, Asseline's *Ballet de la comète* preserved an otherwise ephemeral ballet celebrating a comet's disruption far from the king's vision or center of power.

Only two years later, Louis XIV's reign would create an unprecedented police administration designed to curtail, at least in the capital city, the kind of unruly night the *Ballet de la comète* enacts and celebrates. As part of his campaign to impose law and order, and control the city of Paris, the king's chief of police La Reynie issued an ambitious order for the illumination of the streets.[49] By the time the comet became visible in November 1680, the capital's night skies were illuminated by close to three thousand identical lanterns. The king's light shone, regular and steady, in the night as well as during the day.

Pedagogical Comets at the Collège de Clermont: the *Ballet des comètes*

If the program Asseline printed in Soissons for the *Ballet de la comète* documents a performance celebrating a comet's disruption, the livret from the *Ballet des comètes*, danced six months later at the Jesuit Collège de Clermont, celebrates a return to order. The *Ballet des comètes* is a pedagogical work. An important skill at court and in other prestige settings, ballet was a fundamental part of the education of the children of noble and well-to-do families and a particular focus in Jesuit institutions.[50] From the return of the Jesuits in 1618 until their suppression in France in 1762, dance formed an important part of a curriculum that emphasized performance and theatrics. In his 1685 pedagogy manual *De ratione discendi et docendi*, Father Joseph de Jouvancy (also spelled Jouvency, 1643–1719) includes ballet among the subjects to be taught at Jesuit institutions, judging it "a pleasure worthy of a well-educated man and a useful exercise for the young. In addition, dramatic ballet is a sort of silent poetry, expressing in skillful movements the feelings that the poets express in verse."[51] Every August during the annual prize distribution ceremony, students from Jesuit collèges performed a ballet between the acts of a drama. According to Ernest Boysse, "ballet was the real originality of Jesuit (pedagogical) theater. Ballets were the principal attraction of their spectacles, and it was there that they deployed all the resources of their imaginations."[52] What Boysse refers to as the resources of the Jesuit imagination included lavish costumes and theater machines as well as elaborate sets (these last shared with the drama to which the ballet's parts served as *intermèdes*. Jesuit collèges also employed professional dancers as dancing masters and choreographers and to accompany the students in their performances.[53]

Jesuit collège theatrics participated in the order's cultivation of social and political influence.[54] The annual prize-day performance at Clermont, housed on an outdoor stage in the collège courtyard, attracted a huge audience, including members of the court and of the royal family.[55] In 1665, these resources were leveraged to mount a ballet that put comets on stage in front of a large audience of powerful people at the most prestigious educational institution in Paris.[56] On August 6, 1665, students at Clermont danced a ballet about two comets at the school's annual prize-giving ceremony. The performance, intercalated between the acts of a tragedy called *Irlande* written by the Jesuit father François Diez (1629–87), depicted what observers interpreted as a pair of comets: the first observed from December 1664 until February 1665 and the second in March 1665.[57] The ballet is described in a four-page livret, *Balet* [*sic*] *des comètes: Pour la tragédie d'Irlande*, and the tragedy is described in an eight-page program *Irlande tragédie sera representée au collège de Clermont de la Compagnie de Jésus: Pour la distribution des prix fondez par Sa Majesté*. Both livret and program would have been given to spectators to help them follow the action on stage and possibly in advance as invitations to the celebratory performances. La Gravette de Mayolas, who continued the *Lettres en vers* after Loret's death three months prior, reported the event and praised the erudite Diez, whom he credits for composing the tragedy and training the student actors and dancers.[58] Unfortunately La Gravette's description, while laudatory on all fronts, contains few of the atmospheric details Loret had been able to fit into his rhyming descriptions of current events. All we are able to glean is that Diez has applied great erudition to the subject and that the dancers incarnate their roles with discretion (unless the writer chose the word *discrètes* principally for its rhyme with *comètes*).

The storyline of the *Ballet des comètes* holds only the loosest connection to the tragedy, a lost-child romance about an innocent and wronged Crusades-era queen named "Irlande," which the ballet tracks by building suspensefully toward a crisis in part 3 (between acts 3 and 4) but does not reference in any concrete thematic, historical, or symbolic way.[59] Jesuit Claude François Ménestrier, (a few years later author of the letter on the comet), in *Des ballets anciens et modernes selon les règles du théâtre* (1682), which theorizes ballet as a dramatic art conforming to the classical rules of theater, specifies that the ballets at college prize-day performances usually relate to the theme of the tragedy: "At the Collège de Clermont, where there is always a big Tragedy performed during the distribution of Prizes given by his Majesty, the subjects of the Ballets are usually connected to the Tragedy's."[60] In his discussion of the *Ballet des comètes*, however, Ménestrier stipulates that "people are not always bound by these connections. Sometimes a subject was inspired by current events, as in 1665; when two Comets appeared that year, they did the *Ballet des Comètes*."[61] As Ménestrier signals, the *cométomanie* of recent months provided a current event dramatic and attention-grabbing enough to furnish the ballet subject during the collège's most important public performance of the year. The other "ballets de circonstance" of the era represented major events, from the king's marriage to military victories.[62] The decision in 1665 to opt for one of "les sujets que le temps fournissait," rather than build on one of the themes of the tragedy, indicates that comets were a subject that the Jesuit fathers deemed worthy of engaging on the public stage. Once performed at Clermont or one of the other collèges, a drama and ballet were often produced in successive years at Jesuit institutions in other localities, but neither *Irlande* nor the *Ballet des comètes* appear to have been reprised.[63] By choosing to create a lavish performance based on a current event, the Jesuits were investing in a conversation of the moment. A comet was important enough to engage with that they were willing to forgo the possibility of future reuse. In this way, the *Ballet des comètes*, like a comet, was composed outside the regular circulation of spectacles passing from one Jesuit institution to the next, connecting the various student bodies via a shared repertoire.

For Laura Naudeix, Jesuit collège ballets functioned as a kind of enigma, a pedagogical exercise that required spectators to actively puzzle through the interpretation of what was happening on stage and what it might mean.[64] Considered from this vantage point, the livret, *Balet* [*sic*] *des comètes*, highlights the performance as a double enigma: a presentation of a spectacular and enigmatic natural phenomenon via a genre that is intrinsically spectacular and enigmatic. The ballet poses the question of what it means for humans to look at something as dazzling, unpredictable, and unknowable as comets, and it does so through a highly abstract means of expression that both thematizes and employs visual spectacle. The ballet does not simply recapitulate comets' enigmatic irregularity, however, but rather resolves it in a way that supports forms of monarchical hierarchy being consolidated in these years. The performance displaces the unsettling phenomenon of comets from the sky to the Parisian stage. Even while it draws attention to the disruptive chaos of a comet's motion, the ballet imposes a choreography that brings comets into harmony with the social and epistemological regime that Jesuit pedagogy supported and with the kind of highly symmetrical stage decor the Jesuits preferred.[65]

The school ballet offered to an elite Parisian public as a way to engage with and think through the the *cométomanie* of 1665, a spectacle of the unknowable, the unknown, or the irregular, principally by offering a pedagogical allegory that trains the gaze back toward the powerful ordering force of the sun. In order to create a narrative out of the comets, the ballet gestures to the traditional understanding of comets as portents. However, the depiction and critique of superstition in the ballet is a red herring, and an incoherent one at that. Rather than present the comets—or the divine—as the engine of this conflict, the ballet locates the sun's emotions and agency at the origin of the drama, transforming the danced performance of the comet into a kind of royal apology. Through its attention to the tension between comets and the sun, and its depiction of the kinds of human actions spurred by the appearance of the comets, the ballet sets up a contest about what people are meant to be looking at—the ordered and diurnal, or the chaotic and nocturnal.

The four-page printed livret begins with an *argument* that explains that the ballet was inspired by a simile that likens the sky to a dance: "Since the Poets speak of the movement of the Heavens like a Ball, where the Stars are made to dance every night, it seemed justified to make a Ballet of the two Comets that appeared this year."[66] In early modern France, dance both instantiated and rehearsed the complex social relations of a courtly society.[67] This metaphor that serves as the conceit for the ballet prompts the audience to consider the sky as a kind of reflection or extension of the social and power relations specific to a court setting, which played out dramatically in dance performances and in instances of social dancing. At the same time, the simile naturalizes the social choreography of court society by suggesting that power negotiations played out in courtly dance followed the kind of immutable natural order observed in the cosmos.

Court dynamics center around who is in favor and who is out. This ballet's narrative arc casts the two comets seen over the past half year in this familiar drama of affirmation and rejection. However, some fancy footwork positions the sun, rather than either of the comets, at the center of the story. The summary of the ballet that opens the four-page pamphlet explains that since the first of the two comets traveled from east to west, "far from the Sun" it was presumed to be a bad omen, while the second "only promised good, for the opposite reasons." While the winter comet brings war, famine, and pestilence, the spring comet is depicted as favorable for the deference it shows to the sun. Traveling from west to east, it "preceded the Sun's rise every morning, as if to serve as a forerunner [or portent] for this Prince of the Stars, and by its light to draw unto itself the eyes of all of the nations that its novelty attracted to itself."[68] Even as the comet attracts interest for its novelty, it modestly subsumes itself as a kind of satellite whose role is to draw attention to the sun. In doing so it reestablishes order and harmony in human society.

The printed livret details the ballet's four parts, each composed of four or five entrées, which each might include dances, music, and songs.[69] The ballet opens with a depiction of human interest in celestial observation: mathematicians from the four corners of the world observe the night sky (part 1, entrée 1). These figures highlight the shared character of natural phenomena and the universal nature of human curiosity and intelligence: Comets can be seen everywhere on earth and everywhere people aspire to calculate their movements. At the same time, the figures incarnating the four corners of the earth recall the iconography of royal architecture and performances and the global ambitions of the Jesuit order and the French king. This is not a celebration of universalism, but of French royal power.[70] The attention the mathematicians pay to the night stars (entrée 2) inspires the sun's jealous ire in the third entrée: "The Sun becomes angry at getting less attention than [the stars]."[71] Consequently, in the last entrée of part 1 (entrée 4) the sun creates a comet "rather recognizable due to the irregularity of its movement."[72] This dancer's irregular movement corresponds to the perception that comets traveled in an irregular fashion, without pattern, and leaves a modern reader, with access only to the pamphlet's summary, wondering how a dancer depicting a comet would move, especially compared with the sun, who features in the preceding scene, or the stars of entrée 2. In contradistinction to what Judith Rock identifies as the "celebration of limits, precision, and small intricate complexity" performed by the serious and noble characters, the dancer performing the role of the comet might have adopted the "grotesque" or comic style used by comic and lowborn characters according to Rock, and, as Ndiaye explores, also ascribed to racialized figures. This style of movement marked characters or allegorical figures whom the pedagogical drama displayed in a critical light: "The Grotesque style pointed to a social norm and carried a social judgement."[73] In other words, the erratic dancing of the comet signaled that it transgressed the order that the pedagogical drama wished to instill in the students dancing the ballet and the audience watching the performance.[74] Writing of baroque dance popular in the decades prior, Franko argues that, in contrast to geometrical dance, "antitextual, or burlesque" dance asserted not only an esthetic challenge but also a political one.[75] While the comet's erratic movements on stage do not necessarily evade linguistic description like the baroque ballets Franko examines, the resistance to order this figure performs nonetheless shares in the "subversive" defiance of the choreographic order, and by extension summons on stage the "political resistance" performed by baroque dance.[76]

The fourth and final entrée of part 1 returns from sky to earth, documenting how the comet inspires the mathematicians to make almanacs to measure, track, and describe the comet and also to make predictions based on the comet's appearances. No mention is made in the livret of how these dancers move. The ballet might juxtapose burlesque with serious (orderly) entrées, or, given the fact that human figures are depicted acting in the comet's thrall, it would also be reasonable to imagine that the dancing would be keyed to a critical view on these human actions inspired by the comet.

In the second part, the action returns solidly to the terrestrial realm, depicting human dynamics set in motion by the comet. After the comet has lurched across the stage, people scrutinize its erratic dance in an attempt to transform observed movements into insight or knowledge. The resulting part 2 is a danced cautionary tale on the dangers of visual education badly founded and executed, as various characters advance hermeneutic insights, inspired by observing the comet, in the form of various (yet all suspect) economies of knowledge. So-called "Bohemians" sell almanacs and make predictions (entrée 1). Peasants buy their almanacs (entrée 2), succeeded by sailors (entrée 3), astrologers, and doctors (entrée 4). At the origin of this daisy chain of circulating knowledge is the comet, and also the "Bohemians," theatrical and ballet racialized stock characters who incarnate suspect and rootless physical movement and traffic in untrustworthy predictions (in marked contrast to the Jesuits and Louis XIV, who both represent themselves as expanding networks of order and civilization over France and increasingly distant lands).[77] Over the course of part 2, the *bon aventure* that the "Bohemians" peddle makes its way up the ladder of prestige and economics. It gains a slight veneer of credibility when published in the pages of almanacs, whence it travels overseas (with sailors) and across social boundaries (with astrologers and doctors). By presenting in succession all of these stock characters, who each embody a stereotypical form of knowledge, part 2 of the ballet stages the social penetration of the flawed understanding of the world that the comet sets in motion.

At this point in the drama, the ballet's critical position is complex. Since the ballet was written for and performed by students at a prestigious Jesuit institution, it seems obvious that the ballet offers a critical stance on prognostication. As seen in paintings such as *La diseuse de bonne aventure* (ca. 1630) by Georges de La Tour, visibly foreign and female soothsayers personify the way superstition was manipulated to dupe and fleece the ignorant. The itinerant "Bohemians" echo the comet as the very symbol of untrustworthy and unpredictable mobility. The almanac's astrologers and doctors seem to represent more sophisticated incarnations of the market for superstition, perhaps a critique of the market-driven production and consumption of information, a commodified simulacrum of true faith and its transmission, whether through theatrical exercises like this one or the broader Jesuit mission of which it was a part. Beyond the clarity of this critique, however, comets' status as messengers of divine anger or mercy propels the ballet even as the ballet denounces human attempts to prognosticate. Some of this ambiguity is never resolved. The first comet might be bringing disaster, or it may simply communicate the sun's (or divine) displeasure at humanity's attention to the stars, for which it will be punished even as the comet further galvanizes human fascination. The second comet portends, mediates, or simply marks divine misericordia.

At the end of part 2, the people try to assuage the antique gods by praying to their statues. Inspired by extremely current events (i.e., the *cométomanie* of the past year), the ballet's anachronism stems in part from the era's—and the genre's—pervasive tendency to classicize.[78] The displacement of a contemporary event into a classical past may also be read as an allegory of the theological idolatry that comets inspire: the erroneous human desire to see comets and stars as causes of human misfortunes rather than the effect of divine creation.[79] The transposition into a classical setting furthermore allows the ballet to sidestep the theological appropriateness of interpreting comets as portents, which the ballet seems to denounce in showing the populace's supplications to be in vain but also seemingly embraces when it casts the second comet as a portent and agent of peace reestablished.

After the ballet's third part depicts "Les Effets de la Comète," war (entrée 2), famine, pestilence, and death (entrée 4), the final part opens with a brutal winter resulting from the comet's triumph over earth and ocean. "The Comet, carried by the winds of Winter, triumphs over the Earth and over Neptune, whom the Winter holds captive."[80] In entrée 2, however, the sun intervenes to free land and sea of the comet's ill effects. "The Sun, in pity, melts their chains."[81] The sun, in the next entrée, creates a second comet. "Carried by Springtime and borne on Zephyrs," this new comet "comes to combat, through its order, the first."[82] The program to the tragedy refers to this fourth part of the ballet as "Combat of Two Comets," while the livret for the ballet calls it "The return of public Joy through the meeting, and the Victory of a second Comet over the first."[83] While it is impossible to know how the combat between comets figured on stage, the ballet livret provides a clue that illuminates the most important esthetic value in the ballet's pedagogy: order. The second comet "vient s'opposer par son ordre à la première." In other words, the orderliness of the second comet, communicated ostensibly by orderly movements, vanquishes the influence of the irregularly moving one that had fascinated the ballet's human characters, inspired their ill-begotten books and ill-considered movement and trade, and brought them suffering and misfortune. In conclusion, "the victorious [comet] brings back Peace, Health, and Abundance, and returns to the Sun in order to guide the eyes of all peoples there" ("va se rejoinder au Soleil pour y conduire les yeux de tous les peoples").[84] The conclusion celebrates the reestablishment of an esthetic of order firmly centered around the sun, who is represented as the cause of all celestial and terrestrial phenomena and the jealous but clement rightful focus of human attention.

This moral is typically Jesuit in the way that it accommodates moral and religious pedagogy to political conditions, here Louis XIV's consolidation of political, social, and economic power in these early years of his personal reign. December 1664, the month of the first comet's appearance, coincided with Fouquet's sentencing and exile. In the months that these comets appeared in the sky, the king required his subjects' full attention. He continued renovating Versailles, and his minister Colbert continued centralizing economic, political, and cultural power. As served by the second comet, the spectacle of the king's magnificence exerted an organizing force on his nation. To combat the irregular rhythms of a comet that would distract and disrupt, the sun creates a comet of his own, to reestablish order and draw all eyes to him.

Although no mention is made of this in the four-page livret, it would have been typical for the prize-day performance to end with a ballet general. The tragedy of *Irlande* concludes with the recognition of the rightful prince and the banishment of the play's villain. The dancers and actors would have gathered on the stage in a final celebratory dance, which would have culminated in the presentations of prizes—highly symbolic marks of the king's generosity to and approbation of this school, which he would soon claim as his own.[85] The prize-day choreography merges the harmony reestablished in the *Ballet des comètes* with the ballet of an orderly society, a social ritual shot through with symbolism. Comets have danced, but order and the primacy of the sun have been reestablished.

Postscript to the *Ballet des comètes*

The record of this 1665 performance is scant—a four-page livret, a mention in a rhyming account of current events in the hands of a replacement writer less insightful than Loret, and a citation in Ménestrier's history of ballet. These sources nevertheless document attitudes and social dynamics that crystalized around comets. Even if the chaos it sows is eventually resolved, a comet, and the irregular movements it inspires, occupied the stage for the major part of an important public ritual, temporarily interrupting the order that the regime worked to project and that Jesuit education sought to celebrate.

As they did for contemporary readers of the seventeenth century, comets can alter our perspective, pointing us to a vision of the 1660s as the era not only of Louis XIV and Versailles but also of information systems that ran counter to the heliocentric and the hierarchical, the kind of publication network exemplified by both Jean Donneau de Visé and Nicolas de Blégny.

While the livret for the *Ballet des Comètes* does not specify which dancer performed which role, the list of performers—eight Parisians, one English student, a pupil from La Flêche, and a pair of brothers the pamphlet refers to as "from America" ("de l'Amérique")—inscribes the performance within the expanding geography of Louis XIV's France.[86] Gaspard de Ricous would go on to become a diplomatic envoy to the court of the elector of Bavaria.

Antoine-Joseph Poncet de la Rivière came from a family that had invested early in the Compagnie de la Nouvelle-France. He was possibly a younger cousin of Joseph-Antoine Poncet de la Rivière, a

Jesuit priest and former student at Clermont, whose adventures in New France from 1639-57 included stints in a Huron mission, Quebec, Montréal and Trois-Rivières; and being captured, tortured (and adopted) by Iroquois before being returned to the French in a prisoner exchange. Following his brief return to France, which coincided with the years Antoine-Joseph Poncet de la Rivière appears on the cast list for this other Clermont performances, Poncet was posted to Italy and finally Martinique, where he died in 1675.

It seems likely that the two "Americans," Jacques-Dennis and Louis de Nambuc, are indirect decedents of Pierre Belain d'Esnambuc (1585–1636), sometimes rendered "Nambuc" in books of the period.[87] The younger child of a debt-encumbered Norman nobleman, Belain d'Esnambuc became a privateer, claimed half of Saint Christophe (present-day St. Kitts) from the English, and established the French colony of Saint-Pierre on the island of Martinique, leading to the massacre of the indigenous Karib population and the importation of the island's first enslaved Africans. He died in 1636, leaving no direct descendants or spouse listed on official records but having risen to the position of governor general of the French Antilles and installed members of his extended family in important positions. The French control of Martinique, which he established, continues to this day. A statue of Belain d'Esnambuc, erected in Fort-de-France in 1935 by the French government in the sunset years of French colonialism, was toppled in June 2020 after years of protest.[88]

The nascent history of French North American and Caribbean exploitation, a history that would eventually account for an astonishing share of French wealth, evoked in the elusive details of a ballet about comets, shares a certain affinity with the comets we are tracing in this book in that it draws attention away from the magnetic center of gravity that is the metropolitan French perspective on world events. This cometic decentering critiques a well-ordered and dazzling perspective by making visible (allusively, in the person of Poncet de la Rivière and the two "American" dancers) the multifarious forms of violence employed to establish the seemingly natural harmonies of Louis XIV's domain. Reading for comets—to find comets—also prompts us to read like comets, from decentered positionalities freed from the organizing tractor beam of the sun's rays.

CHAPTER 3

Beyond the Eye of Absolutism

Claude Perrault's Observatoire Royal and the Intractable Challenge of Comets

Located on a hill to the south of the city in the era of its construction (today central Paris of the Fourteenth Arrondissement), the neoclassical Observatoire Royal remains an abiding symbol of Louis XIV's rule. There were many cultural and historic factors that contributed to the regime's decision in 1666 to establish a scientific academy and to build an observatory and library to house its activities. However, the story of how Adrien Auzout (1622–91) and his *Éphéméride du comète de 1664* inspired the project has taken on the status of a foundational myth.[1] This story is repeated in books, articles, and exhibits about the observatory and its history, and even on the institution's home page, which celebrates how Auzout, "French astronomer and physicist, instigated the creation of the Paris Royal Observatory."[2]

Three and a half centuries after Auzout published his pamphlet, the observatory (now known as the Observatoire de Paris, or simply the Observatoire) is still standing, and so is this story about its creation. Both—the observatory and the stories that have been spun around it in visual depictions and in descriptions—are the subjects of this chapter. The story of Auzout's quest to master comets and the architecture of the Observatoire Royal each articulate, albeit in different ways, a cultural investment in observation and, more generally, a privileging of visual cognition, which was an essential element of the spectacular political modernity that Louis XIV's reign established.[3] The foundation of the academy and the construction of the observatory both articulate the promise that scientific institutions and rigorous, well-organized and well-equipped observation offer the power to domesticate the kinds of uncertainty and irregularity associated, for example, with comets.

Close readings of seventeenth-century encounters with the Observatoire Royal will reveal ways in which the building and institution could not always make good on these promises.[4] This story of the comet and the observatory brings into focus the misalignment of the observatory building as a potent political symbol of knowledge, on the one hand, and the needs of actual scientific discovery, on the other.

Organizing Observation: The Comet and the Foundation of the Académie royale des sciences

In 1665, while the comet was still visible nightly in the sky, Auzout published his *Éphéméride du comète de 1664*, a table charting the comet's positions. Auzout prefaces his table with an epistle to the king proposing the establishment of a royally sanctioned and supported *Compagnie des sciences et des arts*.[5] Such a *compagnie*, argues Auzout, would provide the structure, funding, and tools needed for France to make strides in science. The formal organization Auzout outlines would in some ways represent the continuation of the informal kind of associations already active in France and abroad. Working outside the traditionalist realms of university and court, private learned societies and academies brought together the curious and the learned to pursue experimentation, correspondences, conferences, discussions, and debates on a wide range of topics, including philosophy, culture, and politics.[6] Such associations included the circles around monk, theologian, and mathematician Marin Mersenne (1588–1648) and physician (and pawnbroker and innovative journalist) Théophraste Renaudot (1586–1653) in the first part of the century, followed by the more formally organized Académie de Montmor (1657–64), which had dissolved amid constant infighting the year prior to Auzout's publication. There was likewise a strong tradition of pursuing and sharing knowledge in the vibrant salons of the period, often hosted by women and drawing a public of male and female participants from the aristocracy and bourgeoisie.[7] While some of the French academies offered lectures to fee-paying audiences, and others benefited from the largess of noble and well-to-do members, private associations generally struggled to finance equipment, space, and personnel.[8]

The hope that Auzout articulates for official support from the crown represents both a continuation of an old tradition of aristocratic patronage and a significant change in status for science and scientists. In these first few years of Louis XIV's personal reign, Colbert was orchestrating a massive project of centralization that included activities and domains as diverse as forest and water, justice, and artistic production.[9] The crown had established the Académie des inscriptions et belles lettres and had renewed investment in existing academies such as the Académie française (est. 1635) and the Académie royale de peinture et de sculpture (est. 1648).[10] An official company for scientific research seemed like a logical extension of this push toward the centralization of resources and cultural and political oversight. The possibility of acquiring for the first time the stability and research conditions that *savants* craved seemed tantalizingly within reach.

Following the failure of the Académie de Montmor, Auzout was part of a cadre who aspired to found a well-equipped, productive, and focused association of experimenters.[11] In his 1665 pamphlet, he exploited interest in (and possible concern over) the comet as a way to encourage the king to establish a royal scientific company with the necessary resources and facilities. In a textual performance that models the relationship of royal patronage it endeavors to achieve, Auzout's letter depicts his own scientific curiosity and labor as inspired by the king's desire to see the comet and know its orbit: "The curiosity, SIRE, that everyone knows Your Majesty has demonstrated to see it and to know its movement inspired me to work to discover it."[12] This scene of viewing entails a triple recognition: the public recognizing the sovereign ("tout le monde sait"), whose manifest curiosity ("a témoignée") to see and know ("pour le voir et pour savoir") in turn motivates Auzout's scientific work ("m'a fait travailler à le découvrir"). The word "discovery"—etymologically, uncovering—signals the primacy of visual knowledge in observational science.[13] The king's curiosity is a kind of spectacle that propels his subjects toward both admiration and scientific inquiry. Promising "I will continue to consecrate my work to Public Utility and to the Glory of Your Majesty" and signing off as a "very humble, very obedient, and very faithful Servant and Subject," Auzout depicts his research as a kind of civil service to king and state.[14]

While ephemerides are usually a table of observations charting the movements of celestial bodies, Auzout's table was expanded to include a speculative projection of where he predicted the comet would appear.[15] Auzout describes his *Éphéméride*, in which he has endeavored not only to track but also predict the path of the present comet as an offering to the king, "a project never before undertaken by any Astronomer and that most will judge rash [*téméraire*] if I am not fortunate enough to succeed in my conjectures."[16]

While popular opinion would characterize comets as "irréguliers," Auzout is "téméraire"—audacious, courageous, risk-taking, foolhardy, someone who conquers new territory—and "dares hazard a prediction": "Up to now, everyone has believed the Movements of Comets to be so irregular that they did not think they could subject them to Laws; and they were content to observe exactly the locations they passed through, but no one that I know has yet dared to hazard a prediction of where they will travel, and where they should cease appearing."[17] In the body of the text, Auzout repeats his claim that "everyone thinks that there is nothing so irregular and so extravagant" ("il n'y a rien de si irrégulier ni de si extravagant") as a comet.[18] Comets create social consensus in the superlative ways that they do not fit in.

Auzout proposes that perfecting the conditions and practices of observation will enable him to develop a schema that will not only follow where comets appear and describe, in the past tense, where they have been, but will also accurately predict where they will travel in the future.[19] He signals futurity by using the modal verb "devoir," which means "should" but also, in its noun form, "duty," thus revealing the transformation that such a breakthrough would represent for the status of comets. In order to accomplish this expansion of the rule of law over the heretofore unruly comets, however, Auzout maintains that better conditions are needed for observation:

> Even though my Observations were only made with strings, rulers, set squares, and sticks, I was obliged to make do with them. Had I had a more suitable place and the large instruments necessary to make very exact Observations, I would have done so, and I do not doubt that they would have helped me to succeed better than I did. But SIRE, it is a misfortune that there are not any of these things in Paris, nor that I know of anywhere in your Kingdom, that I would trust to measure precisely the altitude of the Pole, which is perhaps the reason why there is not a Realm in Europe with such faulty Geographical Maps, and where the location of places is so uncertain.

Auzout emphasizes the material deficiencies of the current scientific climate, in which he, or any other independently acting individual working alone or in the informal networks of private academies or salons or under private patronage, is "constraint de me contenter" with mediocre conditions and mediocre results. The conditional perfect of speculative futurity ("Si j'avais eu un lieu plus propre et les grands instruments nécessaires pour faire des Observations très-exactes, j'en aurais fait et je ne doute pas qu'elles ne m'eussent aidé à rencontrer mieux que je ne ferai") summons a potential future where progress and success tantalizingly lie ahead and also evokes the future perfect tense found in descriptions of the king's monuments.[20]

Auzout's project of building a "suitable" and well-outfitted facility aligns the regime's symbolic and political needs with its practical concerns. The same conditions of possibility that will allow astronomers to tame the symbolically resonant errancy of comets, explains Auzout, will also provide the good measurements critically needed to make accurate maps of French territory. Such maps will in turn facilitate the centralization of taxation and resource management as well as military and territorial expansion.[21] A royally sponsored scientific company, a company of arts and sciences, argues Auzout, would be "so glorious to the State, and so useful to the Public."[22] Doubtless gesturing across the channel to England's Royal Society, established in 1660 and chartered in 1662 and again in 1663, and to the Medici-sponsored Accademia del Cimento founded in 1657, Auzout depicts celestial observation as a pillar of national pride and a site of international rivalry.[23] Establishing an observatory and an academy assures

> that when there are other new things to observe in the Sky, the French do not cede in this to Foreigners, as Your Majesty sees that they do not cede to them in all other things, and so that they can contribute like other Nations through the most exact Observations that one could wish for, to determine those things that the curiosity of the learned has inspired them to pursue for such a long time. It redounds, SIRE, to the Glory of Your Majesty, and for the Reputation of France, and that is what makes us hope that [your majesty] will commission some place to make all sorts of Celestial Observations in the future, that [you] will have it supplied with all the Instruments necessary to this effect.[24]

The right facility and the necessary instruments, argues Auzout, would reflect "la Gloire de Votre Majesté, et de la réputation de la France" and would place France firmly among those nations poised to lead the way toward defining the as yet unknown scientific curiosities of the future, aptly indicated in Auzout's text by the indefinite relative pronoun phrase "ce que." ("qu'ils puissent contribuer comme les autres Nations par des Observations les plus exactes qu'on puisse souhaiter, à déterminer ce que la curiosité des savants leur fait rechercher depuis si longtemps"). Far from an obscure scientific or court document, Auzout's observations and campaign to link the comet to his plans for a royal academy entered the public conversation. The January 1665 *Journal des sçavans* reported at length on Auzout's findings, closely echoing Auzout's rhetoric in the dedication letter of the *Éphéméride*.[25]

Auzout's epistle may have gained an outsized role in the histories of the Observatoire Royal, the Académie royale des sciences, and the reign of Louis XIV not because it was unique or pioneering in asking for royal support of scientific research (indeed it was not), but because it so deftly aligns the needs of science with the needs of the crown at a crucial early moment in Louis XIV's personal rule. The image of science promoted in Auzout's letter harmonizes with the political aspirations of the regime because the observational method that Auzout compellingly argues for seems to complement the crown's orientation toward the symbolic potential of vision and visibility. In a regime where political symbolism rested squarely on the faculty of vision (the king as both master seer and brilliant spectacle), the promise of an institution and a gleaming royally sponsored facility consecrating the acquisition of knowledge through observational science seems a perfect manifestation of Louis XIV's emerging mode of royal power.[26] The comet presented an ideal opportunity to demonstrate this power. It had drawn the attention of all of France. Bringing the spectacular and refractory outlier—a phenomenon resonant with the political upsets of the Fronde, supporters of Fouquet, and other unruly populations—into the orderly space of scientific law would be simultaneously an expansion of scientific rationality and a demonstration of the efficacy of centralized government.[27]

In the months following the much-discussed comet and Auzout's publication, plans to create a royal academy of sciences began in earnest and moved fast and secretly.[28] Colbert began recruiting members for the academy, including Auzout and some foreign luminaries, notably the eminent Dutch mathematician Christiaan Huygens (1629–95) and the Italian astronomer Giovanni Domenico Cassini (Frenchified to Jean-Dominique; 1625–1712).[29] The king officially established the Académie royale des sciences in 1666, and the appointed members held their first meeting in late December on rue Vivienne.[30]

The Observatory Commission and Charles Perrault

In close proximity to the residence of Colbert, the academy's patron, the rue Vivienne address seemed a perfect locale for the new academy. The block was becoming a nerve center of the regime's intellectual investments—with the king's library and cabinet of collections moving to the same building and, the following year, the installation of the Imprimerie Royale and engravers a few doors down.[31] However, the academy did not stay in Paris. Although the tendency of Louis XIV's France strongly pulled toward centralization, the need for what Auzout had termed "a more suitable place" for observations was such an important priority for the new academy that a site was found south of the capitol, with views unobscured by the city's smoke.[32] Preparation began, only ten weeks after the academy's first meeting, to build a permanent home for their collections, experiments, and observations.[33]

The royal administration gave little attention to the structure or aims of the Académie royale des sciences other than recruiting a roster of members; however, great energy and expense were marshaled quickly to build an observatory. David Aubin, Charlotte Bigg, and H. Otto Sibum emphasize that an observatory was not only a site of astronomical observation, it was also "a theater in which the state, science, and empire were displayed. Cosmological narratives were crafted and delivered there, representing the place of man in the universe, the conquest of new territories and populations, and scientific and technological progress."[34] The observatory building, one of the first completed royal buildings dating from the first decade of the king's personal reign, was conceived to be a monument to this symbolic, material, and administrative conquest.[35] René Taton calls the Observatoire Royal "a special royal building, destined to both satisfy the prestige of the monarchy and the needs of science," thus emphasizing the double function of the architectural project—to house scientific activity and to represent monarchical power in ways consonant with a particular presentation of scientific inquiry.[36] According to Roger Hahn, "The building became the effective symbol of the Crown's persistent efforts on behalf of science."[37]

A very peculiar knot of issues regarding the status of visual observation becomes concretized in the building of the observatory. In one sense, the building demonstrates the deep investment of Louis XIV's reign in an ideology of vision, an ideology whereby vision is consonant with knowledge and discovery. Careful study of representations of the observatory's imposing edifice, however, ultimately also reveals some of the complications and internal contradictions in this ideology. When Auzout writes to the king of an imagined project, the needs of science and politics are in complete harmony. This synergy is harder to maintain, however, once the institution and the building are realized, and we see the requirements of politics and symbolism diverge from and even impede the functioning of science. Aubin et al.'s observation about nineteenth-century observatories, that "tensions could arise from the observatory's dual role: emblematic and pragmatic," also holds true for this early edifice.[38]

While the name "observatory" evokes for us a specialized building dedicated solely to the pursuit of observational, especially telescopic, astronomy, in the 1660s, the practice and tools of astronomical observation were rapidly evolving and there were still very few facilities in Europe purpose-built to house this kind of activity. Most observation took place on rooftops using the kinds of simple tools Auzout mentions: sticks and strings, and if possible a timepiece.[39] The new Observatoire Royal was conceived to house all of the equipment and activities of the Académie royale des sciences. The plan to gather all scientific pursuits under this particular roof, far from other centers of royal activities, shows the preeminence of astronomical observation—figured specifically by the symbolically resonant case of the comet—in the identity of the Académie royale des sciences. Ultimately, however, Colbert's idea to unify all scientific inquiry under the roof of the observatory, while long on symbolic value, was shorter on pragmatic or operational grounds. The academy ultimately never adopted the observatory as its base of operations. As symbolically apt as the plan was, the observatory building was too removed from the center of administration, scientific inquiry, and society to be an attractive location for most of the academy's members.[40]

If Auzout's bold plan to better see and measure, and thereby know and master, the heavens in general—and, with the king's backing, to observe and subject even comets to law—figures in the enterprise of the Académie royale des sciences, then the observatory building is the edifice that concretizes this enterprise, its values, and aspirations. In a short text titled "Pourquoi et comment l'observatoire a esté basty" (Why and how the observatory was built) dated August 1667, Charles and Claude Perrault note "that for an observatory, which astronomy could not do without, they needed only choose a site they judged proper from which to observe well, and immediately a building would be constructed there that not only would surpass in grandeur, in beauty, and in usefulness [*commodité*] the observatories of England, Denmark and China, but which, most tellingly, would in some way match the magnificence of the Prince who was having it built."[41] The reference to Christopher Wren's Greenwich observatory, which was commissioned in 1675, reveals that this piece was written later than its 1667 attribution date; but whatever the date, the text clearly signals the observatory's primary purpose ("ce qui était tout dire"): to serve as a reflection of the king's magnificence and to establish his superiority vis-à-vis the great monarchical powers of the day, from England and Scandinavia to the Qing dynasty. This building, which irrupted immediately at the king's wishes—"ils n'avaient qu'à choisir un lieu qu'ils jugeassent propre pour y bien observer, et qu'aussitôt il y serait construit un édifice"—would be grand as well as well-planned and functional as denoted by the term *commodité*. This word appears over and over in descriptions of the observatory and describes attention to new conceptions of comfort emerging in this era, for example, in purpose-built spaces and furnishings.[42] Most importantly, the building would serve as a reflection of the king's magnificence, "répondrait en quelque sorte à la magnificence du Prince qui le faisait bâtir."

The commission for the observatory building was given to Claude Perrault (1613–88), a physician and lecturer in anatomy and physiology at the University of Paris, an inventor of mechanical machines, and the kind of polymath *savant* or *curieux* exemplary of scientific inquiry of this period.[43] He was the fourth of five surviving brothers in a close-knit, erudite, well-connected, and highly aspiring bourgeois family.[44] Claude probably owed the commission to his youngest, and best connected, brother, Charles.[45] Remembered today as the author of some of our best-known fairy tales from Cinderella to Little Red Riding Hood, Charles Perrault (1628–1703) was a close early associate of Colbert, who appointed him assistant comptroller of royal buildings and secretary to the Petite Académie, which Colbert established in 1663 to oversee the symbolic representation of the king.[46] In 1664, the year after his younger brother joined the Petite Académie, Claude Perrault became involved in royal architecture via the contentious Louvre colonnade design project.[47] He would have an important role in shaping the Académie royale des sciences, the Observatoire Royal, and the French architecture of Louis XIV's reign.

In his own time, and posthumously, the polymath Claude Perrault was strongly associated not only with his contributions to Louis XIV's persona and reputation but, above all, with doing so in ways that posit vision and observation as a means to knowledge, progress, and power. An engraved portrait memorializing Perrault's death in 1688 at age seventy-five (he contracted an infection while dissecting a camel for the *Histoire naturelle des animaux*) labels him as a member of the Académie royale des sciences, a doctor at the Faculté de Paris, a famous architect, and in the quatrain on the architectural pediment beneath the oval portrait, a masterful but modest observer: "There is

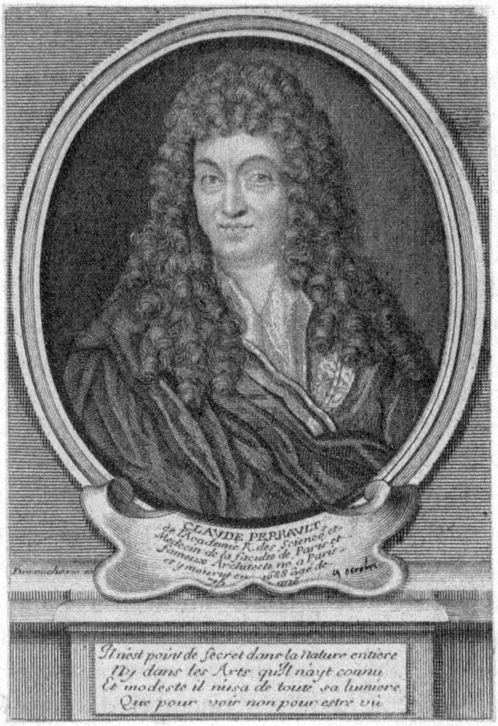

Fig. 14. Desrochers ex., *Portrait of Claude Perrault* (late seventeenth century). Bibliothèque nationale de France.

no secret in all of Nature / Nor in the Arts that he did not know / And modest, he only used all his intellect [literally: lights] / To see, not to be seen" (see fig. 14). This equation linking Perrault's active participation in the academy, his medical background, and his architectural practice celebrates vision as the faculty through which the secrets of Nature may be uncovered. The quatrain's initial rhyme, "Nature entière"/"lumière," employs the image of sunlight to celebrate Perrault's probing vision as a tool to master the natural world, while the final enclosed rhyme, "vu"/"connu," emphasizes the degree to which sight offers the key to knowledge. The last verse also uses the stress culminating from the long enjambed phrase—"Et modeste il n'usa de toute sa lumière / Que pour voir"—to highlight the infinitive "voir," to see. This term both resonates and contrasts with the poem's last word, the participle "vu" at the end of the final short hemistich. This pair of alliterative and related terms celebrates Perrault's judicious use of his faculties to examine and discover, rather than for the vain theater of court performance.[48]

In 1665 or 1666, at the same time plans were being made for the Académie royale des sciences, Colbert charged Claude Perrault with the monumental task of translating *De architectura*, Marcus Vitruvius Pollio's foundational text of architectural theory and the sole extant architectural treatise of classical antiquity, into French. This translation was a priority for the regime because it was meant to provide the builders of Louis XIV's monuments with the grounding in classical techniques, theories, and idioms that would signal the regime's classical antecedents and assure the king a monumental presence in the present and future. Picon speculates that Claude Perrault was given this symbolically freighted translation commission for several reasons. Known for his single-minded focus, Claude Perrault was a trusted quantity from a family closely tied to Colbert. He furthermore possessed an excellent mastery of Greek and Latin from his long medical studies. Finally, as a physician and someone interested in mechanics, he might have seemed to incarnate a modern scientific sensibility, in which knowledge of a mechanical universe was presumed to be gained through the faculty of observation.[49] Dating from the first century BCE, Vitruvius's *De architectura* is dedicated to Augustus, whose territorial and administrative expansion and cultural and artistic imprint served as a model for the ambitious French monarch eager to found a golden age as memorable as Augustus's Pax Romana. Perrault makes this connection clear in a dedicatory epistle addressed to the king: "The glory that this beautiful Science receives today equals that which was lavished on her in olden days. . . . Indeed, SIRE, to return this mistress of the Arts [Architecture] to the shine she had in the age of Augustus, it was necessary that she meet a Prince, who through extraordinary conquests and Virtues deserved her most beautiful and superb monuments."[50] Linguistic translation from Latin to French neatly figures the ambition for cultural translation from the classical world to Louis XIV's France, just as the crown's building projects aspire to establish France as the new Rome. A significant difference emerges between Augustus and his seventeenth-century scion, however: the book the modern "interpreter" renders benefits from the technology of the printing press.[51]

In 1666, while still at work on the Vitruvius translation, Claude Perrault was named an inaugural member of the Académie royale des sciences and commissioned with the ambitious and important task of building a permanent home for this new academy.[52] In its first decades, the Académie royale des sciences did not have a charter. The observatory building quickly being erected—and a series of images, publications, and descriptions—represented the academy's work and aspirations. Some half a century later, the academy's secretary, Bernard Le Bovier de Fontenelle (1657–1757), articulated the academy's ambition to establish an observational science:[53] "We have left behind a sterile physics that has remained at the same place for the last several centuries; the reign of words and terms is past; what we want is things. . . . Because we took it into our heads to consult, regarding natural things, nature herself instead of the ancients, she reveals herself more easily; and rather frequently, pressed by new experiments that have been created to probe her, she accords the knowledge of one of her secrets."[54] Fontenelle's much-commented passage posits, in strikingly gendered terms, the dream of a new science that observes, probes, and experiments on the physically embodied female gendered nature, instead of debating, reading, or following classical authority.[55] Replacing the homosocial sterility of past fealty to the ancients, the new enterprise, motivated by the collective desire ("on veut") for a new and embodied mode of study, promises a proliferation of new discoveries, offered up by "nature herself," if only scientists have the audacity to look and to probe. For the rest of this chapter, we will follow the ways the architecture of the Observatoire Royal articulated these aspirations but also incarnated the challenges and sometimes the failures of an observational science.[56]

Perrault designed the Observatoire Royal building symmetrically around the Paris meridian and sited the windows in the octagonal towers to optimize observations during solstices and equinoxes.[57] Glosses below prints of the building by a Pérelle (likely Gabriel) and published by Nicolas Langlois (figs. 15 and 16) note the building's "magnificence" and functional design.[58] The caption under the front view, for example, explains: "The OBSERVATOIRE is a Building that the King had started around the year 1667, in an eminent place, at the end of the Rue Saint-Jacques, to serve for Observations of the Orbit of Stars and for various Physical experiments. It has all the equipment and all the amenities [*commodités*] necessary to these two great Sciences," while the caption under the side view image reads: "This side of the OBSERVATORY faces the Faubourg Saint Jacques and does not yield to the other [façade] in magnificence and solidity, all parts of the building having the same advantages for working successfully at Astronomical and Physical Observations." The very existence of these prints, and the characteristics highlighted in their captions, demonstrates

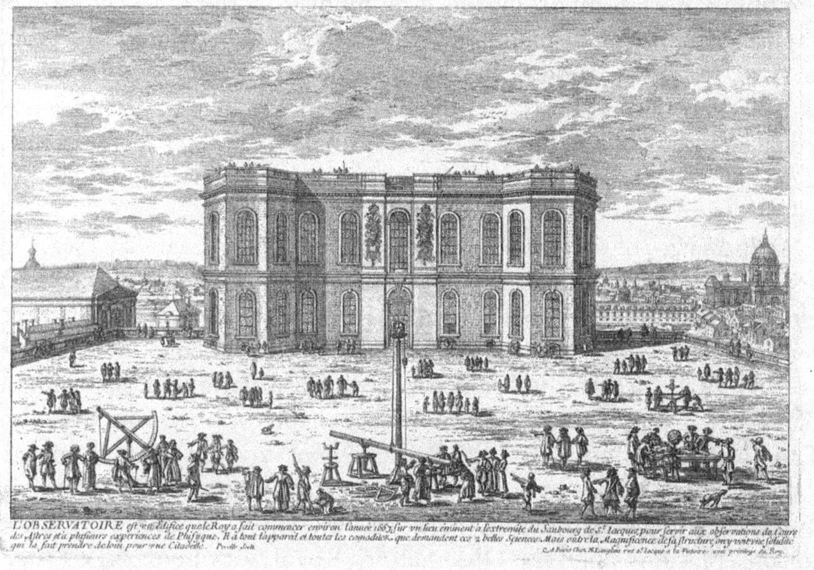

Fig. 15. Pérelle, Copperplate engraving of the Observatoire Royal façade. *Veues des belles maisons de France* (Paris: N. Langlois, ca. 1680), n.p. Bibliothèque nationale de France.

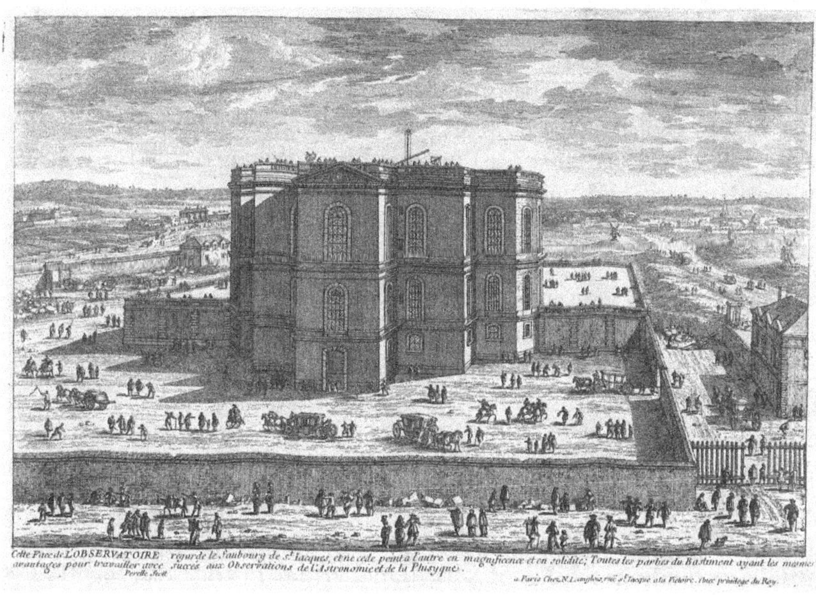

Fig. 16. Pérelle, Copperplate engraving of the Observatoire Royal from the side. *Veues des belles maisons de France* (Paris: N. Langlois, ca. 1680), n.p. Universitätsbibliothek Heidelberg.

the observatory's importance as a monument to the king's glory. The captions highlight the building's siting and layout, which, the prints declare, permit astronomers and other scientists the *commodités* to pursue their observations, as well as the building's external aspect and the visual impression it makes. The caption concludes: "In addition to the Magnificence of the structure, one notices a solidity that makes one take it, from a distance, for a Citadel."

Dictionaries of the period also reveal a building designed to serve as a setting for observations, a symbol of research, and a sign of the king's glory.[59] For Antoine Furetière, whose 1691 dictionary gives a nuanced sense of the contemporary meaning of words in late seventeenth-century French, Perrault's edifice is given as a paradigmatic example of a contemporary observatory: "A place made for observing the stars. The King had built near Paris a Royal *Observatory*, a castle for observation, which is a unique and magnificent building."[60] The key term "magnificent" and the definition of the observatory as a kind of "castle for observation" align the Observatoire Royal with Louis XIV's spectacular construction projects at the Louvre and Versailles and thus blur the building's vocation as a site where observation takes place with royal edifices built to be observed.

This alignment of priorities might illuminate the over 700,000-livre royal investment in the observatory project between 1667 and 1669—years when Louis XIV was at war, building Versailles, and spending heavily on myriad civil, institutional, and cultural reforms.[61] The project was sufficiently central to the regime's preoccupation that Charles Sevin de Quincy (1660–1728) included it in the entry on 1667 in his military history of Louis XIV's reign. "It was also in this year that the King, despite his great preoccupation related to the war, had the Observatory built for the Astronomers, as a monument to the care for the perfection of the sciences in his Kingdom [comme un monument des soins pour la perfection des sciences dans son Royaume]."[62]

As one of the signal accomplishments of Louis XIV's early reign, the construction of the Observatoire Royal merited inclusion in the grand series of medals, the *Histoire métallique*, which the Petite Académie designed to celebrate the great moments of Louis XIV's reign a as well as the richly illustrated folio *Médailles sur les principaux evenéments du règne de Louis Le Grand* (1702), which reproduces and glosses the medals, providing a kind of shorthand visual history (see fig. 17).[63] This intermedia portrait of the observatory—the printed representation of a numismatic representation of a building that represented the

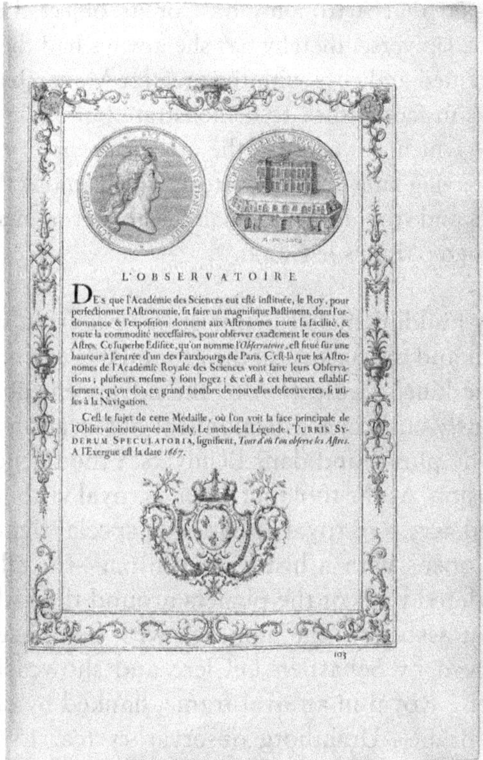

L'OBSERVATOIRE

DEs que l'Académie des Sciences eut esté instituée, le Roy, pour
perfectionner l'Astronomie, fit faire un magnifique Bastiment, dont l'or-
donnance & l'exposition donnent aux Astronomes toute la facilité, &
toute la commodité necessaires, pour observer exactement le cours des
Astres. Ce superbe Edifice, qu'on nomme l'*Observatoire*, est situé sur une
hauteur à l'entrée d'un des Fauxbourgs de Paris. C'est-là que les Astro-
nomes de l'Académie Royale des Sciences vont faire leurs Observa-
tions; plusieurs mesme y sont logez; & c'est à cet heureux establis-
sement, qu'on doit ce grand nombre de nouvelles descouvertes, si uti-
les à la Navigation.

C'est le sujet de cette Medaille, où l'on voit la face principale de
l'Observatoire tournée au Midy. Le mots de la Légende, TURRIS SY-
DERUM SPECULATORIA, signifient, *Tour d'où l'on observe les Astres.*
A l'Exergue est la date *1667*.

103

Fig. 17. Entry on the Observatoire Royal medal. *Médailles sur les principaux évènements du Règne de Louis Le Grand* (Paris: Imprimerie Royale, 1702), 103. Bibliothèque nationale de France.

crown's imprimatur over scientific observation—highlights the intricately imbricated stakes of visuality in the creation of Louis XIV's glory.[64]

While Furetière calls the Observatoire Royal a "castle," Sevin describes it as "a monument," and the Pérelle prints dub it a "Citadel" (figs. 15 and 16 above), Colbert sees the building as a kind of celestial arc de triomphe, one of the triumphal arches that were used to great effect in Louis XIV's public entrances and building projects.[65] In the opening passage of his 1680 *Voyage d'Uranibourg*, Jean Picard (1620–82) repeats this image, which he explicates to say that the Observatoire Royale will eternally glorify the future memory of France's king.

> It can be said that Astronomy has for its object the greatest things in the Universe: thereby has she always had the privilege of being granted audience with the greatest Monarchs; and His Majesty has in fact wished to demonstrate [*faire voir*] the particular care which He takes in the advancement of this noble Science, in having built an Observatory, which among the Triumphal Arches and trophies will remain as an eternal mark of the glorious Reign of Louis le Grand.[66]

The observatory building makes manifest (*faire voir*) in stone and mortar, in the present and the future, on the ground of Paris and in the pages of this book, the king's dedication to the advancement of astronomy and simultaneously enshrines astronomical observation as the pursuit of "ce qu'il y a de plus grand dans l'Univers": the science that reveals truth through vision. As the tool that extends royal visibility, astronomy enjoys unfettered access to royal attention, a special right to be seen.

Picard's text opens with a historiated initial—the ellipse of the O recalling the ordered path of the planets around the radiant visage of the sun king that associates the monarch with the frieze at the top of the page, engraved by Sébastien LeClerc and showcasing the newly built Observatoire Royal in an oval frame, flanked by circular medallions of Tycho Brahe's Uraniborg observatory (ca. 1576–80) on the left, and, on the right, the Rundetaarn of Copenhagen, an homage to the Tower of Babel built by Christian IV between 1637 and 1642 (see fig. 18). This vignette was reused by the Imprimerie Royale in an essay by Cassini, "De l'origine et du progress de l'astronomie dans la géographie et dans la Navigation" that opens a 1693 *Recueil d'observations faites en plusieurs voyages par ordre de sa Majesté pour perfectionner l'astronomie et la géographie: Avec divers traitez astronomiques par messieurs de l'Académie royale des sciences*.[67] For Berger, this frieze shows the observatory's architectural and institutional lineage while also declaring the building's innovation: "What distinguishes the Paris Observatoire from its predecessors is precisely that it is a building that no one could possibly mistake for a castle or a tower or anything else from traditional architecture, and in this it may be seen as embodying the contemporary theoretical demand for decorum in architecture. It is the first convincing monument of the scientific revolution, and it was designed by a man of science, Claude Perrault."[68] The term "decorum" that Berger uses is a key concept from Perrault's translation of Vitruvius. It is the principle holding that a building's design should be appropriate to its function. In other words, Perrault's Observatoire Royal

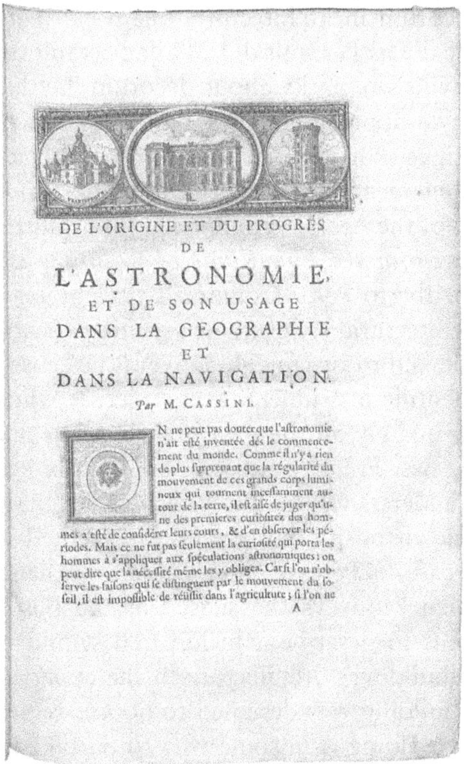

DE L'ORIGINE ET DU PROGRÉS
DE
L'ASTRONOMIE,
ET DE SON USAGE
DANS LA GEOGRAPHIE
ET
DANS LA NAVIGATION.
Par M. CASSINI.

ON ne peut pas douter que l'aſtronomie n'ait eſté inventée dés le commencement du monde. Comme il n'y a rien de plus ſurprenant que la régularité du mouvement de ces grands corps lumineux qui tournent inceſſamment autour de la terre, il eſt aiſé de juger qu'une des premieres curioſitez des hommes a eſté de conſiderer leurs cours, & d'en obſerver les périodes. Mais ce ne fut pas ſeulement la curioſité qui porta les hommes à s'appliquer aux ſpéculations aſtronomiques: on peut dire que la néceſſité méme les y obligea. Car ſi l'on n'obſerve les ſaiſons qui ſe diſtinguent par le mouvement du ſoleil, il eſt impoſſible de réüſſir dans l'agriculture; ſi l'on ne
A

Fig. 18. Jean Picard, *Voyage d'Uranibourg, ou Observations astronomiques faites en Dannemarck* (Paris: Imprimerie Royale, 1680). ETH Bibliothek Zürich, Rar 9475. https://doi.org/10.3931/e-rara-9308.

was designed to stylistically perform the functions it was designed to house. Calling the observatory the "first convincing monument of the scientific revolution," Berger echoes a contemporary appreciation of Perrault and the observatory published in the November 1688 edition of *Mercure galant*, following Perrault's death. "The construction, which is singular, demanded a genius with a perfect knowledge of the Sciences for which this Building was made."[69]

From the era of its construction, and continuing in modern assessments, the Observatoire has been considered an important example of Vitruvian decorum and thereby a powerful architectural image declaring the triumphant reign of observational science. In the estimation of Una Roman D'Elia, "Earlier commentators had echoed Vitruvius's discussion of decorum in architecture, but none had illustrated this point as dramatically as Perrault did. While Perrault did not comment directly on Vitruvius's remarks about decorum, he did show concern for creating a style appropriate to a building's function in his own architectural designs. For instance, the stark simplicity of the Observatoire, a building meant for scientific pursuits, contrasts with the dense decoration of the Arc de Triomphe."[70] In his introduction to Perrault's *Ordonnance for the Five Kinds of Columns after the Method of the Ancients*, Alberto Pérez-Gómez also celebrates the fit between the building's architectural style and its scientific vocation: "The building, practically devoid of ornament, conveys the sense of having been designed as a scientific instrument, as a structure whose sole purpose was to house adequately all the measuring devices and astronomical apparatus. In the best French tradition, but also in keeping with Perrault's scientific understanding of architecture, its character is defined by the stereotomic virtuosity of its precisely executed stone vaults and magnificent staircase."[71] Pérez-Gómez's assertion that the austere and symmetrical design "conveys the sense" of the observatory's scientific purpose seems to describe a fusion that would include exterior observers of the building's architecture in the experience of scientific observation the building was designed to house. To see the "scientific instrument" of a building is in some way to partake in the project of scientific observation for which the Observatoire Royal was created.

The Observatoire Royal in Print: A Symbol of the Power of Vision

The image of Perrault's sparsely decorated building appeared in prints and paintings that circulated in a wide range of contexts, as well as textually, in guidebooks and newspaper accounts and in books like Picard's and Cassini's, where a verbal description of the Observatoire Royal follows under Le Clerc's engraved frieze.

The Observatoire Royal appears just above the center of Sébastien Leclerc's frontispiece (see fig. 19) for the 1673 first edition of Perrault's translation of Vitruvius.[72] As Hilary Ballon observes: "The translation was regarded both as a masterpiece of classical scholarship and as a royal accomplishment in that it was sponsored by the crown and incorporated into the corpus of the academy. . . . The frontispiece of the translated treatise can be read as a condensed image of this common goal, the creation of a French classical architecture."[73] In

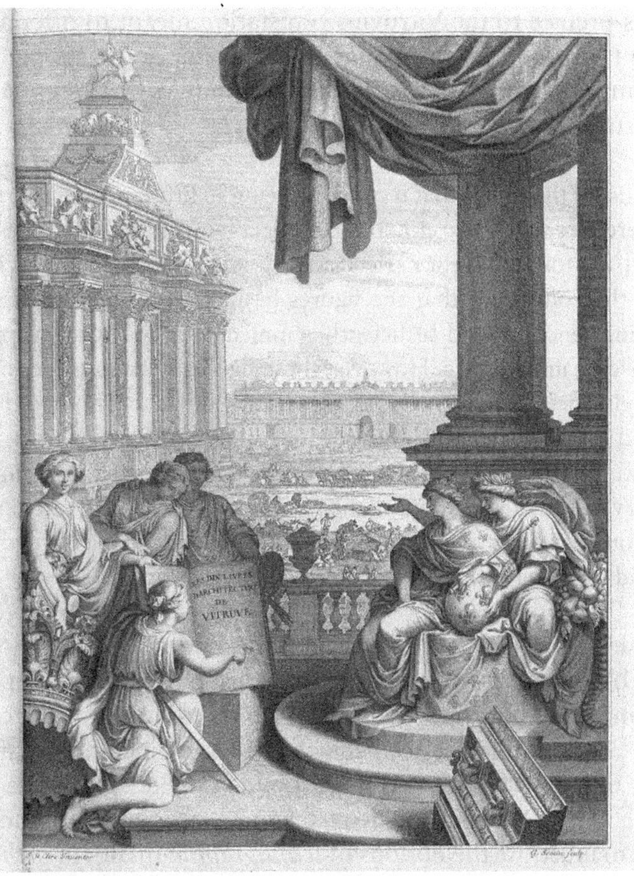

Fig. 19. Sébastien Leclerc, Frontispiece to Claude Perrault, *Les dix livres d'architecture de Vitruve . . .* (Paris: Jean-Baptiste Coignard, 1673), folio. Courtesy of Oberlin College Libraries Special Collections

the foreground, Architecture presents France with Vitruvius's works. Above are Claude Perrault's monuments to Louis XIV's reign, the realization of Vitruvian concepts, and the king's magnificence. The center horizon of the engraving features the stern colonnade of the Louvre, still under construction at the time of the engraving; to the left, the planned but never completed Arc de Triomphe for the Faubourg Saint-Antoine, topped by a statue of the king; and in the background at the center of the page, directly under theatrical drapery, the Observatoire Royal on the hill emerges from the clouds.

In his preface to the Vitruvius translation meant to offer a classical blueprint for Louis XIV's monuments, Perrault draws special attention to the importance of illustrations in communicating essential yet elusive Vitruvian concepts:

> Most of the things that it contains being so little expanded as they were, needed a clearer and more exact explanation than the text left us: for the Author tried less to make it clear than succinct, in the confidence that the figures he had added would explain things enough, and sufficiently supplemented what seems to be missing in language. However, these figures were lost due to the negligence of the first Copyists, who did not know how to draw, and who, incidentally, did not judge them altogether necessary because, the sight of these figures having instructed them in the very things talked about in the text, it seemed to them intelligible enough; just as one always manages to understand something said obscurely when the matter is already clear.[74]

The legibility of the text, he maintains, depended upon the inclusion of figures, but when they were lost by careless and ignorant copyists, the book's meaning became obscure. Perrault's edition of the architecture manual corrects for this accident of history by including a significant number of copperplate engravings to illustrate key principles of architecture.[75] The Observatoire Royal figures prominently among these images, to illustrate the notions of *ichnographia* (plan), of *orthographia* (front elevation), *skiagraphia* (cross section), and *scaenographia* (cavalier projection or perspective; see fig. 20). In Perrault's Vitruvius, the observatory he has just designed illustrates the ancient concepts the building was meant to reflect, occupying simultaneously the position of antique model and modern realization.

As already noted, the Observatoire Royal never fulfilled its intended purpose of serving as the headquarters and meeting place of the Académie royale des sciences, as the scientists found it to be simply too far from the city center. This did not, however, diminish the building's status as a symbol of the French pursuit of astronomical observation.[76] In an engraved portrait by Pierre Drevet, Cassini poses with a globe and telescope, gesturing with his right hand toward the observatory framed by the window (see fig. 21). Assistants look through various sized telescopes on the roof and the terrace in front of the observatory's south side. This is of course an imaginary view—Cassini normally would have been in the observatory, not looking at it, and any building

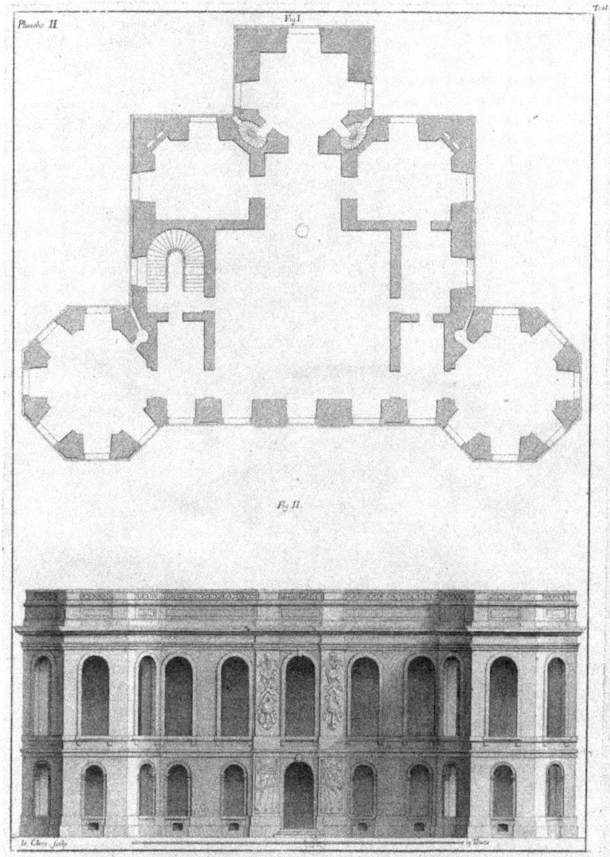

Fig. 20. Sébastien Leclerc, Plan and elevation of l'Observatoire Royal from the south side, plate 2. Claude Perrault, *Les dix livres d'architecture de Vitruve . . .* (Paris: Jean-Baptiste Coignard, 1673). Bibliothèque nationale de France.

with such a good view of the observatory from its window would have ruined the observatory's own views. Clearly, the Observatoire Royal here stands in as a kind of visual symbol of observation itself. This view had a long afterlife as the Louis Cossin/Drevet engraving seems to have inspired numerous copies in the seventeenth, eighteenth, and nineteenth centuries, including mirror-image versions with the Observatoire Royal on the other side, or framed in an oval aperture, and a nineteenth-century painted portrait used to this day by the Observatoire and other official entities to illustrate Cassini.[77] In the background

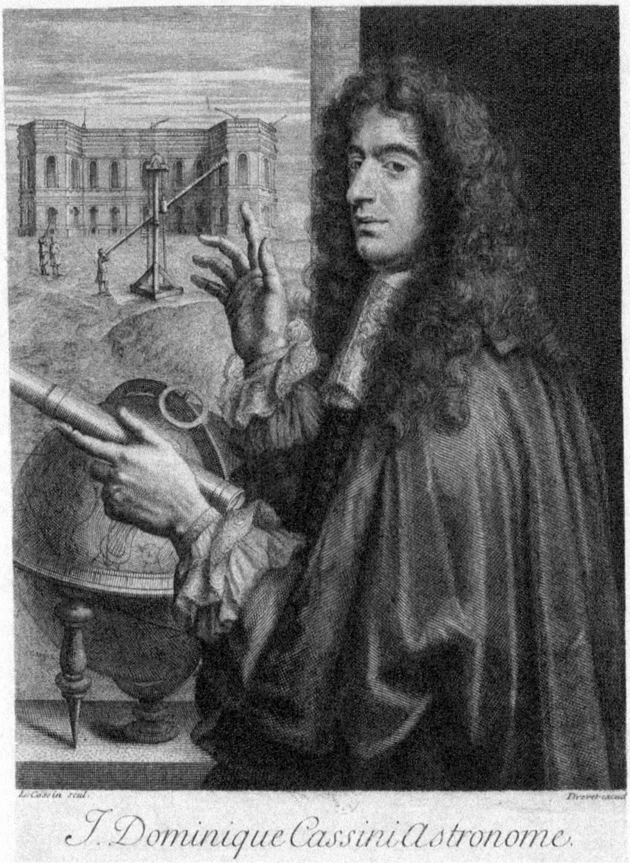

Fig. 21. Louis Cossin, scul., portrait of J. Dominique Cassini Astronome, Pierre Drevet excud. (seventeenth century). Bibliothèque nationale de France.

of these images, the observatory functions as the symbol of a certain kind of scientific observation.

The Observatoire Royal also served as a symbolic *toile de fond* for representations of the king that emphasized his support of observational science, his status as an observer, and the power of observation. Dramatically framed and just right of center, the observatory, still under construction, features prominently in an engraving by Sébastien Leclerc (see fig. 22) that served as frontispiece to the first volume of the 1671 Imprimerie Royale superfolio edition of Claude Perrault's 1671 *Mémoires pour servir à l'histoire naturelle des animaux* (Natural History of Animals) and several successive Imprimerie Royale works

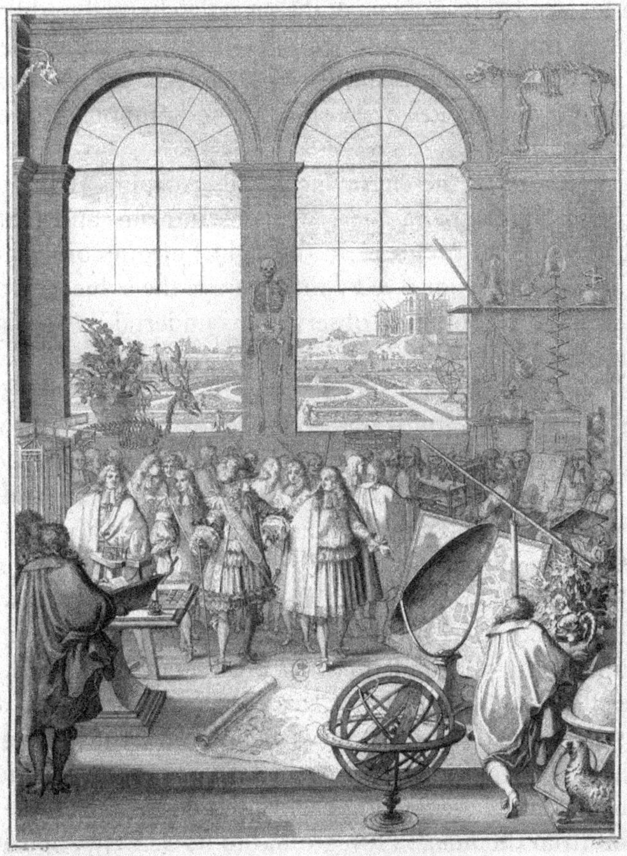

Fig. 22. Sébastien Leclerc, frontispiece to Claude Perrault, *Mémoires pour servir à l'histoire naturelle des animaux* (Paris: Imprimerie Royale, 1671), folio. Bibliothèque nationale de France.

by members of the Académie royale des sciences.[78] The scene imagines Louis XIV, wearing a sash and a feathered hat, addressing members of the Académie royale des sciences. Views of the Jardin du Roi and Observatoire Royal are framed by large windows. Although the king did not make an official visit to the Académie royale des sciences until a decade later in 1682, the engraving shows him at the center of the academy's activities. The scene's three registers offer a visual parable of Louis XIV's conquest of the world through scientific observation. Objects, apparatus, and discoveries related to scientific observation appear in the foreground, while the human contributors to science—the king and his academicians—pack the visually charged middle ground. Representing the worlds to be conquered by the well-equipped men in the bottom half of the engraving, the earth and the sky, the observatory and the Jardin des Plantes are visible in the luminous outdoor space framed by the windows.

The foreground of the engraving recalls Auzout's request in 1665 that the king supply researchers with instruments and materials to enable discovery and progress. In the right bottom corner an animal that might be a weasel or a red panda signals the subject of the book, the dissections and zoological observations undertaken by Claude Perrault and other members of the academy's physical section. A crouching figure, shown from the back, is caught in the act of moving a massive pot of showy flowers into or out of the tableau. The unruly masses of blooms, including bearded iris, jonquils, hyacinth, and peonies, might signal the domestication of the natural world by the Académie royale and also point to an initiative spearheaded by Claude Perrault to create a comprehensive natural history of plants.[79] A globe and a map signal the cartographical improvements that Auzout promised as the outcome of a well-equipped royally endowed observatory. A wigged and robed figure in the bottom left corner proffers an open book, signifying the results and discoveries this academy has achieved. An armillary sphere in the immediate foreground figures the crown's conquest of the sky, just as the globe signals cartographic precision and terrestrial dominion. A long telescope and a burning mirror make the pictorial connection from the image's foreground to its densely populated middle ground and, beyond that, to the Jardin des Plantes and observatory that occupy the full top half of the composition.[80] The telescope signals observation as the master trope of this image. The foreground also prominently features the large mirror purchased by Louis XIV in 1669 for the impressive sum of 7,000 livres. Turned toward this king, as if to reflect his light, the mirror draws an equivalency between Louis XIV and the sun, while, through the story of Archimedes, the object establishes a classical precursor and celebrates Louis XIV as a ruler who fosters scientific advancement in his realm. Furthermore, the mirror—by drawing attention to the brightly illuminated plan of military fortification (by Vauban) that it partially obscures—signals the military implications of investing in science.[81]

In the middle ground, surrounded by the machines, specimens, and specialists in natural philosophy, Louis appears in person to inspect the academy. Skeletons, an alembic, a vacuum machine, a diagram or print (of the kind to be found in Perrault's book, following this frontispiece)

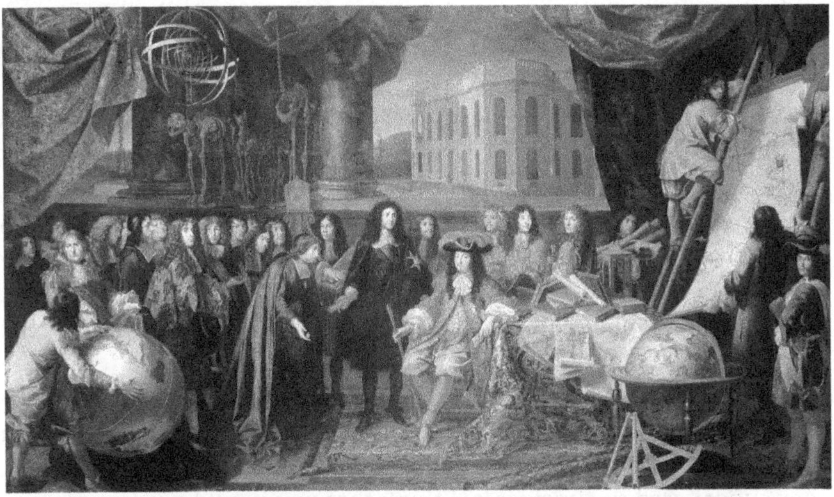

Fig. 23. Henri Testelin, *Colbert présente à Louis XIV les members de l'Académie royale des sciences*, seventeenth century. Château de Versailles.

from the dissection of an ibis or gazelle, and more flowers surround the academicians, signaling their work. With his head turned to his left toward the standing figure who gestures toward a burning mirror, Louis's profile turns in the direction of the observatory and his left arm, posed on his hip, and slightly turned body draw our eyes upward in a diagonal toward the building, which is dramatically framed in the bright bottom half of the right, eye-shaped window. The composition emphasizes the analogy between the standing figure of the king and the vertical mass of the building. The pleats of Louis's garment double the regular rhythm of Perrault's façade, and the two forms create a dramatic baroque diagonal. As the scene stages the spectacle of king as scientific spectator, everything refers our act of viewing to his.

Leclerc's engraving may have been loosely inspired by a cartoon, or preparatory drawing, prepared by Charles le Brun for an episode in the *Histoire du roi* tapestry cycle but never executed. This cartoon is credited as the inspiration for a painting by Henri Testelin showing a seated Louis XIV flanked by members of the Académie royale des sciences, their instruments, and collections (see fig. 23). The observatory, framed by drapes of crimson velvet, dominates the top third of the composition. The building's shape echoes the king's tricorn—crimson feathers redoubling the balustrade just as Louis's gold pantaloons echo the color of the stone. While Colbert and the academicians on the left (the king's right) direct their gazes to each other or at the king, the monarch and a courtier in the bottom right corner align with the building's eye-like arched windows and engage the spectator's gaze in a commanding performance of masterful looking.

Written descriptions of the Observatoire Royal likewise depicted the building in theatrical terms that highlight vision and observation, the conquest of knowledge over uncertainty. In volume 2 of his 1686, *Voyage des ambassadeurs de Siam en France*, Donneau de Visé gives an exhaustive description of the building, offering readers the opportunity to discover the observatory through the eyes of the visiting Siamese ambassadors.[82] The visit is scripted in phenomenological terms that emphasize the faculty of vision. After being greeted on the terrace by Cassini and members of the Académie royale des sciences, the ambassadors consider the exterior of the building, and the readers do so with them. "The Building having first caught the eye of the Ambassadors, they endeavored to consider it. I think that you would not be unhappy to learn many curious things about it."[83] Already in this initial presentation, several expressions emphasize the importance of sight in conjunction with the observatory. The building first comes into narrative focus when it has "frappé la vue" of the ambassadors, and they emphatically resolve to look at it, "s'attachèrent à le considérer." Even the final word of the phrase *regardent*, in this case meaning "regarding" or "about," evokes its homonym, meaning "to look at." Donneau de Visé nests a detailed introduction to the Observatoire Royal within the narrative space created as the Siamese ambassadors focus on looking at the building's exterior architecture, even when the narrator's account surpasses information visible from the building's exterior, for example, the size, layout, and scientific uses of the underground vaults and the collections of instruments and models.[84] He concludes the sixteen-page exterior description with the remark that Claude Perrault's "observations" in medicine and mathematics have given him unique knowledge, which he applied to the construction of the building: "What he knows of Medicine and Mathematics has given him the opportunity to observe things in the construction of this Building, that all of the other Architects are not obligated to know."[85] In this circular model of discovery, Perrault's discoveries are made manifest in the building, which we discover through (description of) observation.

While the exterior of the building communicates a great deal about the observatory, Donneau de Visé's description scripts a visit through the building's interior as well. "After the Ambassadors had considered this Building, of which the sole exterior view does not reveal all of the things for which it is useful, they entered in the first Room."[86] The first objects they encounter are "Instruments to observe the Stars" and a burning mirror five feet in diameter, which is used to set fire to a plank of wood and to instantaneously liquefy lead. This demonstration synecdochally demonstrates the way the observatory uses technology, design, and science to focus the light of the sun and the king, making visible to his guests and subjects its power to incinerate or change matter.[87]

As Donneau de Visé details the ambassadors' tour, describing the spaces as well as the instruments and their use, he emphasizes observation as the central value enshrined in the observatory. The ambassadors manipulate and use the instruments, including planispheres, which concretize the results of astronomical observation, and telescopes, through which even more precise observations can be made.[88] The massive world map painted and engraved on the west tower floor is the subject of a long ekphrastic passage.[89] The world, seen as if from the North Pole, also includes the southern hemisphere, an enhanced imaginary point of view that allows viewers "to see together, and all at once, all of the parts of the earth that we know to be habitable."[90] Readers follow the ambassador marking out the distance that has brought him here, to the west tower of the French king's Observatoire Royal, and this cartographic experience of seeing the world from its center and highest point. The ambassadors' effective illiteracy with respect to the French inscriptions accentuates the visual revelation of knowledge that occurs at this idealized encounter with the observatory. "Even though he had no knowledge of our letters or characters, [he] immediately recognized the Kingdom of Siam and the neighboring Kingdoms. He distinguished America, and several other Corners of the Earth, which he outlined. He described their journey all the way to Paris, tracing the route without hesitating or making a misstep."[91] In fact, the ambassador's virtual trip is smoother than the real one, for Cassini's celestial observations have allowed the rectification of distances and locations that appear distorted on "all the old Maps and Dutch Globes" like the ones that steered the Siamese delegation off track on their voyage to France.[92] The corrected map in the French king's observatory brings Siam a symbolic three hundred leagues closer to France, as does the shared activity of celestial observation, brought to Siam by the Jesuit fathers Fontaney and Tachard.[93]

The scene of territorial mastery from the map room is repeated outside from the viewing platform when the ambassadors are invited to use a telescope to look out over the surrounding landscape: "The First Ambassador having asked where the Château de Berny was located, in order to orient himself, recognized Vincennes, Montmartre, and Sceaux, which he had visited, and some of the most remarkable sites around Paris, which he had been shown."[94] The observatory and its telescope offer a comprehensive view that integrates the ambassadors' previous discreet encounters into a single scene. This vignette echoes a famous passage in Fontenelle's *Entretiens sur la pluralité des mondes*, published the same year as Donneau de Visé's description, but doubtless circulated earlier in manuscript.[95] A significant difference is that Fontenelle's bird's-eye view of Paris is summoned via the faculty of imagination, while Donneau de Visé's is a scene of (imagined) telescopic observation.[96]

The visit follows the ambassadors back inside, where they praise Cassini and manipulate various mechanical devices that replicate and predict the movements of the universe. For example, a clockwork planisphere made by the king's clockmaker allows the ambassador to discover heliocentrism, while enjoying the godlike experience of moving the planets and the stars and playing with time.[97] Over and over, the descriptive text emphasizes the mechanical and observationalist discoveries incarnated in this site. The ambassadors look through lenses and windows at mechanical curiosities and historical demonstrations, including machines depicted in Perrault's illustrations to his Vitruvius translation, an ancient catapult, and even commercial textile machines as the narrator describes in great detail the appearance and function of each one.[98] Toward the end of the visit, as the accumulated details get increasingly dense, the first ambassador is so transported into the pedagogy of observation that he cannot resist undertaking a kind of mechanical dissection to see inside the workings of one of the machines. "Curiosity pushed him to the point of taking apart another Machine to see its interior, in order to see if the essential parts look like he had imagined they would."[99] The visit's conclusion both touts and casts in doubt the primacy of vision in the scientific enterprise pursued at the Observatoire Royal. In a moment of reported indirect speech, the ambassador thanks Cassini for the visit with a strange locution that alligns *voir* and *expliquer*: "Next he expressed to Mr Cassini the satisfaction he had felt to see everything that [Cassini] had taken pains to explain to him" ("Il témoigna ensuite à Mr de Cassini la satisfaction qu'il avait eue de voir tout ce qu'il avait pris la peine de lui expliquer").[100] The terms used in parallel reveal a hesitation, hint toward a possible need to supplant visual demonstration with explication that mirrors in reverse Claude Perrault's commentary about the need for illustrations in the Vitruvian text.

"It's up to me to make these fools believe anything I want"

For all the pageantry of these images vaunting French investment in the enterprise of observation, upon closer scrutiny it seems that the scenes in Leclerc's engraving, le Brun's cartoon, Testelin's painting, and Donneau de Visé's account may have been most potent as theater, the king's neoclassical observatory more powerful as a mise-en-scène of the enterprise of looking than as a site from which to look. Whatever celebrators might say about Perrault's singular understanding of architecture and of science, from the moment Cassini arrived in France, the royal astronomer was displeased by the Observatoire Royal, already well under construction. Perrault had sited the building precisely on the meridian and designed three octagonal towers such that their windows would perfectly frame the two solstices and the equinoxes. Cassini demanded changes, which Perrault made, on the north-facing tower and on the height and disposition of the top floor, but the finished structure was a source of discontent to Cassini, his protégés, and generations of his progeny who succeeded him in the post of royal astronomer. In 1700, the Danish astronomer Ole Rømer, who worked with Cassini from 1672 to 1681, wrote scathingly to Gottfried Wilhelm Leibniz about the conditions afforded by Perrault's building: "I differ sharply from their opinion, who up to now have adorned observatories more for pomp than use, and have accommodated the instruments to the building rather than the building to the instruments."[101] The awkward setup at the observatory impeded observation and posed a hazard to precious instruments. In September 1682, Christiaan Huygens wrote home to The Hague describing a disastrous attempt to observe the current comet (now known as Halley's comet) with the long telescope. The observatory building was so awkward that he has to hastily assemble the apparatus and shove it out the attic window: "this lady Comet . . . was deadly for my good 21-foot lens . . . having passed the end out of the attic window which overlooks the garden, the lens fell from all of that height into the courtyard, so the tinplate piece was all bent and deformed. Luckily the lens remained whole because it was attached to a round of cardstock."[102] In his 1810 history of the Observatoire Royal, Cassini scion Jean-Dominique IV maintains that for all the "scientific"-looking gravitas of its architectural style, Claude Perrault's building was singularly unsuited for the actual practice of observation.[103] Cassini laments that "the architect who presided over the Observatory's layout had only a dim notion of the practice of observation, and that he barely consulted astronomers about the amenities he was supposed to provide them."[104] While the architect, according to Cassini, does a marvelous job translating into architectural form an esthetic *representation* of science, this esthetic vision comes into direct conflict with the exigencies of practicing scientists: "Doubtless he believed he had done everything for astronomy by building for it a vast, very high building, with a pleasing volume and in a style of architecture staid, severe and appropriate in form for science; but this was not at all what was needed. The elevation and expanse of the building were only an even more considerable inconvenience, for no matter where one stood, the mass of the building blocked most of the sky from view."[105] Cassini adumbrates a catalog of nuisances caused by the stately building. For certain observations astronomers were constrained to sit outside on a platform on the roof; for others they had to transport the telescope from one end of the building to another. "The massive vaults that crown all the rooms did not allow one to see the meridian from the zenith to the horizon from any spot in the Observatoire; finally, and this will seem quite surprising, not a single spot from which relative distances [parallax] could be measured without considerably disturbing the [telescope]."[106] The building *looks* like serious science, as it is built "d'un style d'architecture sage, sévère et convenable au genre de la science," but the good looks of the architecture hinder the astronomers' ability to see out of the building, and the building's careful orientation is only useful four times a year. In a cruel twist on the building's "decorum," decorum becomes a kind of performance—in the era's terms, a verisimilitude (*vraisemblance*)—that, with excellent dramatic irony, interferes with its ability to deliver on the claims of what it successfully (stylistically) signifies. Here it is important to point out the difference between decorum, where form follows the idea, and the modernist architectural dictum that "form follows function."[107] While in the case of Perrault's Observatoire, form comes in conflict with function, the building is still an example of good decorum. Form does in fact materialize idea, but unfortunately it does so in a way that directly interferes with the building's functionality.

The observatory was designed in 1667 in an era of unprecedented promise for astronomical observation, with converging breakthroughs in timekeeping, lenses and telescopes, and mathematical tables, all of which held the potential to revolutionize the ability to precisely map the position of celestial bodies. The kinds of instruments and observational practices Cassini was hoping to employ in his new position as French royal astronomer were only just emerging when Perrault started the building project on the outskirts of Paris. Images of the

Observatoire Royal, whether the views by Pérelle or those featured in Perrault's translation of Vitruvius, show telescopic observation and the rest of the astronomical action principally taking place on the terraces surrounding the Observatoire Royal. Taton concludes quite generously that while Perrault's building might not have contributed to the scientific advancements of his era, "its creation nevertheless occupied symbolic value."[108] When the next Great Comet preoccupied Europe in December 1680, Christiaan Huygens traced its motions from Perrault's observatory. He reports to his father Constantijn in The Hague that along with the "surprisingly big comet," first visible to Parisian viewers on December 26, the astronomers' activities at the observatory had also drawn significant public attention. "The weather is fine today, which will attract an infinity of people to the observatory because they think that they will see it much better there than from elsewhere and believe that astronomers should make sense of this phenomenon, and even what it means. I have already had inquiries about it from two or three places, it's up to me to make these fools believe anything I want."[109] A comet is a spectacle that activates the public's curiosity and credulity, but at least part of the spectacle is the astronomers observing the comet from the Observatoire Royal. And for the astronomer, the comet is an opportunity for fictions: "Il ne tiendra qu'à moi de faire accroire aux badauds tout ce que je voudrais."

Cassini observed the comets of 1680–81 and 1682 from Perrault's observatory. In his *Abrégé des observations et des réflexions sur la comète qui a paru au mois de décembre 1680 et aux mois de janvier, février, mars de cette année 1681*: *Présenté au Roy par Mr. Cassini*, he proports to make good on the promise made by Auzout to make the comets follow the king's laws, concluding that the recent comet follows the same path as the comet of 1577. However, twenty pages into his exposition, he hedges, admitting that comets are still unpredictable: "Several centuries' worth of observation were needed to resolve ("régler") the movement of the Planets. . . . That is why one cannot expect us to resolve instantly every Comet appearance."[110] The following year, in August 1682, while the next comet is still new in the sky, Cassini publishes what he calls "a first sketch" of his observations (see fig. 24). The dedication and historiated initial divert attention away from the comet toward stars on the ground; the king is represented in the sun behind the letter S, which connects the word "sire" with the depicted "soleil," and in his newborn grandson, Louis, Duke of Burgundy, as the astronomer poetically imagines the comet as a kind of echo in response to the éclat of the royal birth: "It seems as if the Sky

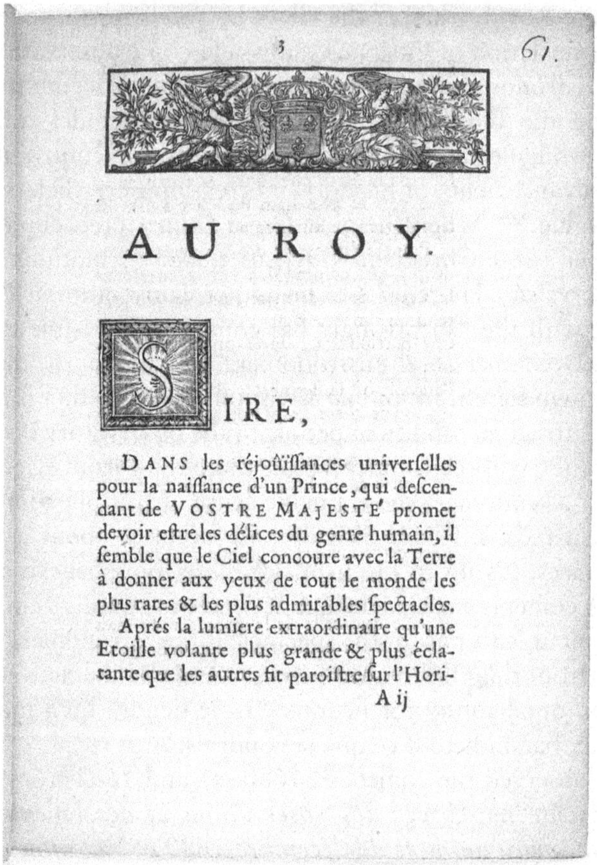

3 61.

AU ROY.

SIRE,

DANS les réjoüiſſances univerſelles pour la naiſſance d'un Prince, qui deſcendant de VOSTRE MAJESTE' promet devoir eſtre les délices du genre humain, il ſemble que le Ciel concoure avec la Terre à donner aux yeux de tout le monde les plus rares & les plus admirables ſpectacles.
Aprés la lumiere extraordinaire qu'une Etoille volante plus grande & plus éclatante que les autres fit paroiſtre ſur l'Hori-

A ij

Fig. 24. Jean-Dominique Cassini's *Premières observations de la comète . . .* (Paris: S. Mabre-Cramoisy, 1682) opens with a visual and textual nod to the sun king and his lineage. Bibliothèque nationale de France.

competes with the Earth to give everyone's eyes the most rare and most admirable spectacles."[111] After presenting initial observations, Cassini reports that it is still too early to make any definitive conclusions about the comet's path and position. However, Cassini declares in the pamphlet's conclusion that the very fact of the comet's observation from the Observatoire Royal is news: "This first sketch will demonstrate to YOUR MAJESTY my obedience in presenting him, following the command with which he honored me, the most rare and remarkable things discovered at his Royal Observatory."[112] What is most dramatically demonstrated by this pamphlet is the astronomer's fealty to the king and the king's creation of the observatory. Solely by royal command and locale, the observations become rare and remarkable signs of the astronomer's obedience. The comet is reduced to a supporting role in its own appearance.[113]

In 1699, the king moved the Académie royale des sciences from the building it still shared with his library on rue Vivienne, having never moved to Perrault's observatory, to more spacious quarters in the Louvre, and declared that twice a year its meetings should be open to the public. The May 1699 *Mercure galant* reports on the first such meeting, on April 29, emphatically signaling the highly visual nature of the experience: "His Majesty wanted the public to be able to judge by its own eyes the form and the utility of these Meetings, persuaded that the mysteries treated there would engender even more esteem if they were more exposed to the light of day."[114] Donneau de Visé reports on the "vast" new rooms "entirely packed with spectators" even including some curious women, discreetly peeking from behind blinds.[115] The abbé Bignon, presiding, explained that anyone who had come expecting eloquent speeches would be disappointed: "The Académie Française had as its domain the art of speaking with all of its niceties, while the Academy of Sciences only aspired to the truth, and often to the truth of the driest and most abstract sort. . . . [They] were satisfied with a useful truth, and gave it dispensation from having to be pleasing."[116] The abbé further explained that academicians made their presentations by reading written speeches out loud, for "these are subjects too deep and too precise to be presented otherwise."[117] The very first lecture, written by Cassini and read by his son (who was better at French), was "a text which supported the opinion of a Philosopher past—very unique and very brave at that time, especially more ingenious as the opinion was at that time less-founded [*moins fondée*]—that Comets, which appear to be newly created, or at least very irregular Stars, are [in fact] true Planets, as old as all the others, which return at regular intervals."[118] In this first paper read at the first public meeting of the Académie royale des sciences, Auzout's prediction and wish has come true: comets were declared no longer *fort irréguliers*. They had been subjected to law, their movements recognized as *réglés*.

However, the observation so emphatically referenced in the *Mercure galant* retelling of this inaugural public meeting of the academy founded in response to Auzout's request is the observation . . . of a rhetorical act. It is a textual account of people watching Cassini Jr. read his father's speech, with an emphasis on the approving crowd, filled with "Residents, Envoys, many Foreigners of distinction, and even some Princes."[119] Observation—and the abstraction of new mathematical systems—has progressed immeasurably in the thirty-four years since Auzout's publication, and in the twenty-eight years since the observatory's completion—yet as this newspaper report of a public conference (in format much like our academic conferences today) reveals, the spectacular encounter that ties subject to sovereign, far from being a limpid and immediate visual connection, is complicated, layered, and mediated. In other words, the king *has* become the sun; the once-precarious rule has succeeded in dazzling the present and seems sure to dazzle the future. But in the spectacular encounters that consolidate the subjectivization of subject to sovereign, sometimes, at least in part, les mots *sont* les choses, the things that are witnessed are words on a page.

In fact, if we had been looking carefully, the precariousness of the phantasm of a total regime of vision that would conquer comets and all of the unruly elements that they stood for was already visible in Auzout's letter of request. In his epistle, Auzout emphasizes the dire need for better instruments and a permanent facility as the decisive factors needed to submit the comet to the order of law. This focus on instruments and the nuts-and-bolts mechanics of taking measurements and observations accords with the strong experimentalist focus visible in the letters of Auzout and his closest collaborators in the early 1660s. It might seem curious, then, that the ephemerides Auzout published following this influential letter are not a monument to careful observation. His ephemerides are based on mathematical projection rather than on the fine-tuned mechanics of observation.

Scholars have argued that a rigorous focus on experimental processes and observations to the exclusion of philosophical questions or models was a way of depoliticizing learned collaborations by framing out potentially problematic issues such as heliocentrism versus Arisotelian geocentrism, or debates between proponents of diverging philosophical systems (Cartesianism, Gassendism, etc.).[120] Auzout employs this kind of strategic distortion of his work in the epistle when he argues that the linchpin of future French scientific glory will be a permanent facility and good instruments, even as his *Éphéméride* is actually speculative in nature and based on few observations. The audacious part is his mathematical projections of where he thinks the comet will travel. But he downplays this to focus in his textual framing exclusively on the technical issues of seeing and measuring (if you just give us a great building and tools, we will conquer the sky for you, make maps, and all the rest). Auzout downplays the abstract theoretical focus of comet research so much that to this day, histories of the observatory tend to repeat his distortion. However, looking closely at the Observatoire Royal's beginnings in Auzout's letter, we can discover the roots of a disjunction between the needs of science and the needs of political power. Observation of the comet seems, at the outset of this adventure, to present the ideal alignment of the exigencies of these two regimes. Following the building's realization, however, we see them diverge. The symbolic needs of politics impede the function of the science that was supposed to confirm political ideologies. Due to the math of Newton and the observations, financing, and initiative of Halley, comets did soon join the ballet of orderly celestial bodies orbiting the sun, but all of this took place outside the official landscape of royal observation: The theory of cometary motion that Halley proposed while in Paris during the *cométomanie* of 1681 earned him ridicule from French savants. He observed what would become his eponymous comet the following year while on his honeymoon. Newton recounts (perhaps spuriously) that he had squirreled away the draft pages of what would become his *Principia* in a forgotten drawer. Ultimately, the joint case study of Auzout's comet and Perrault's observatory reveals that the Observatoire Royal, where the staging of all-powerful observation was meant to align with the imperatives of Louis XIV's glory, instead marks the divide between the theatrical-political needs of a king and the inevitably more complex realities of developing science.

Part Three

Comètomanie, 1680–81

CHAPTER 4

Comets as Commercial Circulation in "la petite comédie de la comète"

In January 1681, only a few weeks after the initial public sightings of the Great Comet, the *Mercure galant* publicized the upcoming premiere of "la petite comédie de la comète." The notice for a one-act play about a new celestial phenomenon seems like exactly the kind of public attention-grabbing event the *Mercure galant* was made for. The small-format monthly, written as a letter addressed to a fictional female correspondent, included political, technological, and social news of the day, literary diversions, songs, fashion, and poetry with a special focus on all that was novel and popular.[1] The play was likewise a rapid-response occasional piece, capitalizing on the comet, which had, since its appearance, galvanized public attention, interest, and curiosity.

The magazine's editor/narrator presents the play's appearance among noteworthy happenings of the last month specifically related to the comet. He characterizes the play as a kind of philosophical demonstration in dramatic form: "People who want to cure themselves of Comet-phobia can go see the little comedy that the Royal Troupe of French Actors has started performing under the same name of *The Comet*, with great success. It demonstrates that there is no reason to be afraid of it, and shows, in a very playful manner, the opinion of the famous Descartes on this matter."[2] The *Mercure galant*, in emphasizing that the play shares the same name as the celestial occurrence, suggests the substitution of one kind of observation for another, replacing the observation of the natural world with its theatrically mediated representation on stage. If the spectacle of the comet itself might occasion fear, then going to see the other comet "represent[ée] sous le même titre" will cure comet phobia. At the same time, the periodical suggests, seeing the play also offers an enjoyable or playful (*enjouée*) substitute to reading Descartes's vortex theory of planetary motion, an easy titrate of Cartesian physics conveyed audiovisually in vernacular French, through theatrical "pleasure and instruction" rather than through scholarly textual exposition.

Like many occasional pieces, *La comète*, has been relegated to near critical oblivion. There are zero subject word references to the play in the MLA International Bibliography, and only one contemporary chapter on the play in French.[3] Its only lasting imprint has been as a footnote to the fugitive event that inspired it, and to more regarded chapters in the careers of the authors to whom it has been attributed: Jean Donneau de Visé (1638–1710), the creator and coeditor of the *Mercure galant*, and Bernard Le Bovier de Fontenelle (1657–1757), the young nephew of Donneau de Visé's official partner in the *Mercure galant*, Thomas Corneille.[4] Fontenelle went on to achieve great things in his century of life, but in January 1681 he was a twenty-three-year-old aspiring writer and *Mercure galant* contributor. While I cannot promise that *La comète* offers a viable substitute for both celestial observation and reading Descartes, in this chapter I propose to study the play as a means of accessing some of the cultural concerns and attitudes that the comet's passage occasioned and crystalized.

While the comedic one-act may not live up to the *Mercure galant*'s promise of a philosophical exposition *enjouée*, the play, I try to show, offers a distinctly worldly response to superstition that can be read productively in concert with the brilliant and epochal critique elaborated by Pierre Bayle in his *Pensées diverses*, which is the subject of this book's fifth and final chapter. The play, I will demonstrate, dismisses superstition outright—and science along with it—to privilege what I call "worldly commerce." For this reading, the play's connection with the *Mercure galant*, a publication obsessed equally with fashionability and solvency, self-promotion and reader engagement, is absolutely crucial.[5] The logical conclusion to such a reading would demonstrate the contributions of commercialized writing to the disenchantment of traditional modes of thinking. Against the backdrop of the comet's passage, *La comète* dramatizes the symbolic reorganization of family structures and the concomitant discrediting of received knowledge in a new regime of practicality and pragmatism, and a rapidly circulating marketplace of ideas.

"Crowds of revelers came from all the Neighborhoods of Paris"

La comète was written and debuted at an exciting moment in which theater in the French capital was being reshaped. *La comète* premièred on a billing with Molière's comedy *Le misanthrope* (1666) on Wednesday January 29, 1681, a few months after the official merging of the actors from the Hôtel de Guénégaud, the former troupe of late playwright/actor Jean-Baptiste Poquelin de Molière (1622–73), with the actors of the Hôtel de Bourgogne company to form the Comédiens-français.[6] Within the next two weeks, the newly combined troupe, sharing a newly combined repertoire, performed *La comète* six more times on a public stage.[7] On January 30, the day after the premiere, they performed *La comète* twice: once, in the theater with Racine's *Britannicus* (1669) and then at the Hôtel Sallé, with Racine's *Andromaque* (1667) for the Venetian ambassador, as part of a carnival celebration. The February *Mercure galant* reports: "After a grand dinner, they performed Mr Racine's *Andromaque* and then the little comedy of *La comete*. During this time crowds of masked revelers came from all neighborhoods of Paris. Then the ball started and the whole night was spent dancing."[8] "La petite comédie de la comète" was at the center of the capital's social swirl. For these two performances on January 30, *La comète*, shared the stage with Racine's serious tragedies drawn from Roman and Greek history. Most likely the extreme timeliness of a play about a comet made *La comète* a suitable choice, and perhaps also the comic one-act was a good fit for a carnival performance. The record of performances kept by La Grange, an actor from Molière's troupe, lists a brisk round of performances in early February that paired *La comète* with a series of farces and marriage comedies: Pierre Corneille's 1644 *Le menteur*, Jean Desmarets de Saint-Sorlin's *Les visionnaires* (1637), and Molière's *L'école des maris* (1661). However, after a February 10 billing with Thomas Corneille's Spanish-style comedy, *Le geôlier de soi-même* (1655), *La comète* disappeared from the troupe's performance records.

The "petite pièce" appeared in print almost immediately. *La comète* was approved for publication eleven days after its stage debut and appeared for sale shortly thereafter, published in Paris by printer-bookseller Claude Blageart in a simple, but nicely printed, octavo format book of sixty-six pages, without illustrations, which included neither the author's name nor a press date. Blageart's name and the distinctive zigzag typography of the word *comédie* identify *La comète* as part of the *Mercure galant* imprint (see fig. 25).[9] Riffaud emphasizes the expense of this custom typography, which Blageart's workshop reserved for the sole printing of the *Mercure galant* and associated titles. Using the unique and highly recognizable font and decorative elements on the title pages of *La comète* and selective other plays "shows to what extent the author demonstrated his desire to associate his name with the *Mercure galant* brand."[10] Between the title page and the list of characters, permission to publish from the powerful Lieutenant General of Police La Reynie is dated February 9.[11] Seventeenth-century playwrights often held their texts to be published only after their original theatrical run had finished in order to preserve the performing troupe's exclusivity. The fast timeline of *La comète*'s publication and the elements highlighted on the spare title page seem to indicate the priorities of its publisher. The title page does not highlight a connection to any author or troupe; however, attention had clearly been devoted to achieve the play's timely publication at the height of *cométomanie* and to emphasize the connection with the stylish and up-to-the-minute *Mercure galant* and the imprimatur of the powerful public safety official, appointed by the king and synonymous with his campaign to establish tighter control both of Paris and of the printing market.

Before looking closely at the text of the play, I would first like to examine how the critical reception of *La comète*, especially regarding the question of authorship, relates to perceptions about the comedy's critical stance toward the questions raised by comets. Assumptions about the play's authorship tend to refract readers' perceptions of how the play engages the philosophically important questions that the passage of the comet raised. How critics have resolved this open question of the play's authorship also highlights the extent to which the suggested reception of *La comète* (as a work of philosophy) is a red herring that distracts from the play's actual depiction of the same commercial mode of literary production of which it was a product. As comets drew attention to disordered motion and open questions, so too might *La comète*—a middlebrow, commercially circulating text and likely a product of collaborative authorship—help revise our understanding of seventeenth-century French literature and esthetics.

La comète and the Questions of Authorship and Authority

Critical practices of attribution in the face of the anonymous publication and uncertain authorship of *La comète* reflect and frame the

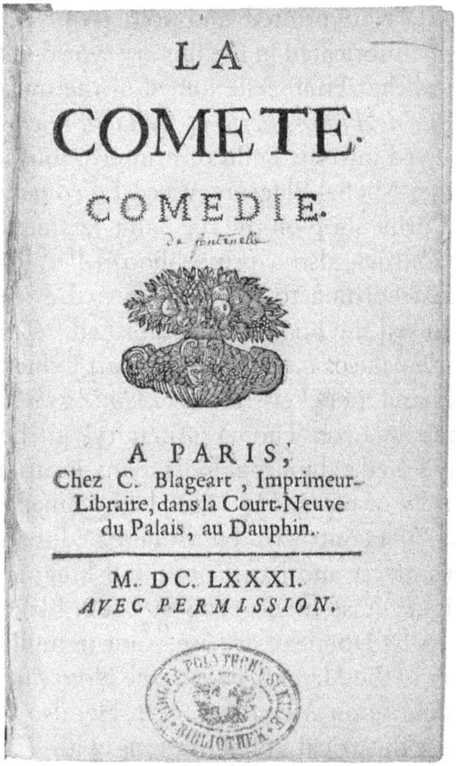

LA

COMETE.

COMEDIE.

A PARIS,

Chez C. Blageart, Imprimeur-Libraire, dans la Court-Neuve du Palais, au Dauphin.

M. DC. LXXXI.

AVEC PERMISSION.

Fig. 25. Title page from *La comète: Comédie* (Paris: C. Blageart, 1681). ETH-Bibliothek Zürich, Rar 6553. https://doi.org/10.3931/e-rara-25883.

ways we understand its presentation of the comet of 1680–81. The play has been variously attributed to two polymaths of the late seventeenth century who were working in close collaboration at the time *La comète* was written and performed. On the one hand, the comedy has been attributed to Fontenelle, the young writer who would go on to become a celebrated playwright, author, *salonier*, and secretary of the Académie royale des sciences. On the other hand, it has been ascribed to innovating editor and writer Donneau de Visé, whose diverse pioneering commercial literary projects had found a successful incarnation in the widely distributed monthly, the *Mercure galant*, which advertised the play in the January 1681 issue and mentioned it again the following month in a description of a lavish high-society carnival celebration.[12]

Claude Blageart, who printed and sold the 1681 edition of *La comète*, was deeply imbricated in the literary world in which Donneau de Visé moved and that Fontenelle joined in the mid-1670s. According to the *Histoire de l'imprimerie et de la librairie* (1689), Claude Blageart was received into the guild of printer-publishers in 1651 and succeeded his father Michel Blageart. When *La comète* was published, he was running a print shop on the rue Saint-Jacques, near his stepfather Christophe Journel, also a printer-bookseller. Together Blageart and Journel had, according to Riffaud, opened a shop in the Cour neuve du Palais to sell the books they published.[13] This is the address featured on the title page of *La comète*. Blageart published descriptions of court festivities and a pocket-sized *Académie galante* (1682, in-12), joined Houry (and in Lyon Thomas Amaulry) in the publication of Nicolas de Blégny's medical journal in its many incarnations, and published a wide variety of plays.[14] In 1666, he became the major printer of Molière's plays, first contracted by Molière's editor, Jean Ribou, and then after Ribou's arrest and ban from publishing, in direct contract with the playwright himself.[15] Starting in 1660, Blageart printed and sold a dozen works by Donneau de Visé, some in multiple editions. He became the printer of the *Mercure galant*, the *Nouveau Mercure galant*, and the *Extraordinaire du Mercure galant*. He also printed and sold Donneau de Visé's theatrical collaborations with Thomas Corneille, Fontenelle's uncle and from 1675 Donneau de Visé's official partner in the *Mercure galant*.[16] Blageart would also go on to produce Fontenelle's first major work, *Les nouveaux dialogues des morts* (1683). After Blageart's death in 1686, his widow, who succeeded him, printed Fontenelle's *Histoire des oracles* (1687).

Judging from marginalia in the copies of Blageart's *La comète* edition extant in libraries, the play's attribution to one or the other of these authors, Fontenelle or Donneau de Visé, was an open question even in the late seventeenth century.[17] In a catalog of books advertised by bookseller Michel Gerout in the back of the Paris edition August 1689 issue of the *Mercure galant* sold by the Palais Royal booksellers Gerout, Guillaume de Luyne, and Théodore Girard, *La comète* is listed on the page following a list appearing under the subheading "Oeuvres de Mr de Fontenelle" (see fig. 26). While Gerout's list shows that *La comète* was still in demand eight years after the comet's appearance (or at least he hoped to sell a few more copies), it is not clear that *La comète* is listed as part of Fontenelle's works. *La devineresse* appears on the same page as *La comète* (both priced at ten sols), but so do plenty of works by other authors, such as Ménestrier's *Blasons*

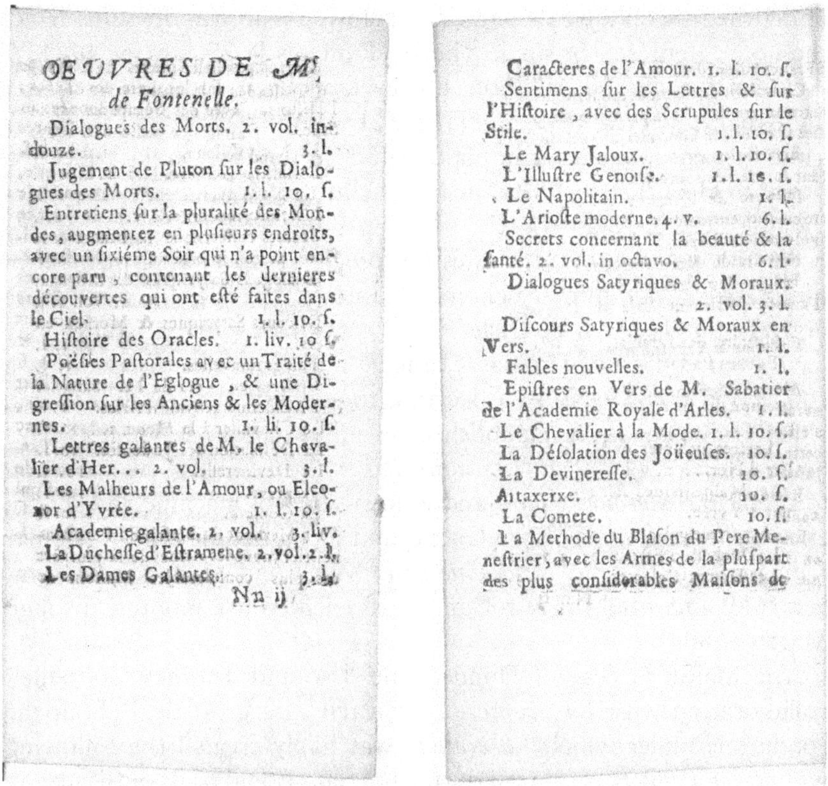

OEUVRES DE Mr de Fontenelle.
Dialogues des Morts, 2. vol. in-douze. 3. l.
Jugement de Pluton ſur les Dialogues des Morts. 1. l. 10. ſ.
Entretiens ſur la pluralité des Mondes, augmentez en pluſieurs endroits, avec un ſixiéme Soir qui n'a point encore paru, contenant les dernieres découvertes qui ont eſté faites dans le Ciel. 1. l. 10. ſ.
Hiſtoire des Oracles. 1. liv. 10 ſ.
Poëſies Paſtorales avec un Traité de la Nature de l'Eglogue, & une Digreſſion ſur les Anciens & les Modernes. 1. li. 10. ſ.
Lettres galantes de M. le Chevalier d'Her... 2. vol. 3. l.
Les Malheurs de l'Amour, ou Eleonor d'Yvrée. 1. l. 10. ſ.
Academie galante. 2. vol. 3. liv.
La Ducheſſe d'Eſtramene. 2. vol. 2. l.
Les Dames Galantes. 3. l.
Nn ij

Caracteres de l'Amour. 1. l. 10. ſ.
Sentimens ſur les Lettres & ſur l'Hiſtoire, avec des Scrupules ſur le Stile. 1. l. 10. ſ.
Le Mary Jaloux. 1. l. 10. ſ.
L'Illuſtre Genoiſe. 1. l. 10. ſ.
Le Napolitain. 1. l.
L'Arioſte moderne. 4. v. 6. l.
Secrets concernant la beauté & la ſanté. 2. vol. in octavo. 6. l.
Dialogues Satyriques & Moraux. 2. vol. 3. l.
Diſcours Satyriques & Moraux en Vers. 1. l.
Fables nouvelles. 1. l.
Epiſtres en Vers de M. Sabatier de l'Academie Royale d'Arles. 1. l.
Le Chevalier à la Mode. 1. l. 10. ſ.
La Déſolation des Joüeuſes. 10. ſ.
La Devinereſſe. 10. ſ.
Artaxerxe. 10. ſ.
La Comete. 10. ſ.
La Methode du Blaſon du Pere Meneſtrier, avec les Armes de la pluſpart des plus conſiderables Maiſons de

Fig. 26. "Oeuvres de Mr de Fontenelle," from an advertisement, "Catalogue des Livres nouveaux qui se débitent chez le Sieur Guerout, Court-neuve du Palais." *Mercure galant* (August 1689), n.p. Bibliothèque nationale de France.

and Sabatier's *Epîtres en vers*.[18] For this reason, the listing is not clear evidence for an attribution to Fontenelle but instead points to the commercial enterprise that connected Donneau De Visé, Thomas Corneille, and these book sellers.

In libraries, we see evidence that some seventeenth-century collectors considered the play to be the work of Donneau de Visé, as it is included in two contemporary *recueils factice* of plays by Donneau de Visé held by the BNF, one in the Richelieu theater collection and one at the Bibliothèque de l'Arsenal.[19] Seventeenth-century books tell an ambiguous story, however. A copy of the play in the BNF main collection, bound in leather with gilt tooling and inscribed in front "De la Reynie" (so possibly belonging to the collection of the superintendent of police), includes Donneau de Visé's name written carefully in ink under the title on the title page, and below, possibly in the same hand, the note "Représentée sur le Théâtre François le 29 Janvier 1681."[20] Next to the inked attribution, however, a pencil notation in a later hand marks "en réalité de FONTENELLE." Another copy of the play's same first edition, bound in plain leather in the rare books collection of the ETH Zürich, includes "de fontenelle" in pen in small hand under the title (see fig. 25).[21]

La Grange's register attributes *La comète* to yet another member of the *Mercure galant* circle, Thomas Corneille, referring to it as "Pièce nouvelle du fils de Mr de L'Isle."[22] It is easy to see why La Grange may have been confused. Thomas Corneille and Donneau de Visé had a connection going back two decades, to some of their earliest work, when Donneau de Visé published an appreciation of Thomas Corneille's *Stilicon* (1660) in *Les nouvelles nouvelles*.[23] The two went on to coedit the *Mercure galant* and write a series of machine and special effects plays for the Théâtre Guénégaud including the wildly popular *L'inconnu* (1675) and *La devineresse*, which had opened in November 1679 and would play for an unprecedented five months, through March 1680.[24]

The major success of Donneau de Visé and Thomas Corneille's collaboration with *La devineresse* can tell us a great deal about the conditions under which *La comète* was likely created the following winter. *La devineresse* played during the sensational trials of the *affaire des poisons*, which galvanized public attention even while it was subject to a complete press blackout. While the accused in the *affaire*—aristocrats, fortune tellers, midwives, and alchemists—appeared before the specially convened "Chambre ardente" on charges of purveying or obtaining poisons, abortions, black masses, and the like, *La devineresse* used theatrical illusions and a standard love plot to denounce charlatans and the credulity of those, like the aristocratic female lead and her amorous rival, who put their faith in soothsayers and witchcraft.[25]

In the January 1710 *Mercure galant*, a few months before his own death in July at age seventy-one, Donneau de Visé recounts in an "Éloge de Feu Mr Corneille" how he wrote the play with Thomas Corneille.[26] At the urging of the theater troupe to write a piece based on the exploits of the recently executed midwife-poisoner-abortionist Catherine Monvoisin, known as la Voisin, Donneau de Visé writes that he created "a great number of scenes which could have provided the material for three or four plays but which could not form a plot because it was too uniform." Thomas Corneille, according to Donneau de Visé, shaped the scenes into a proper play. "He chose a number [of these scenes] and composed a plot out of them with the most pleasing climax, which has been considered to be like a masterpiece."[27] Declaring *La devineresse* to be both critically acclaimed—"regardé comme un chef d'oeuvre"—and one of the most successful of the century or era ("Le succès de cette pièce a été un des plus prodigieux du siècle"), Donneau de Visé is clearly proud of the play and of the collaborative process by which it was written. Indeed, he emphasizes that the success of *La devineresse* stems from the complementary contributions of both writers. However, this collaboration, which seems so self-evident to an aged Donneau de Visé, became invisible to subsequent generations of readers and critics due to changing notions of authorship.

Donneau de Visé, Thomas Corneille, and Blageart deployed an extraordinary publicity apparatus around the multimedia theatrical and publishing event that was *La devineresse*.[28] In November, just as the play opened, Blageart published an eight-image almanac featuring illustrations of the play's dramatic scenes.[29] An announcement in that month's *Mercure galant* cross-publicizes the play and the almanac, two products that also promote each other. Philippe Cornuaille observes, "It's a communications two-for: the editor announces the publication of an almanac, which itself promotes a play that has just opened."[30] Cornuaille stresses that Blageart took the extraordinary step of obtaining a *privilège* and printing *La devineresse* during the high-profile trial, and while the play was still on stage.[31] Approximately one year later, this strategy of hasty publication would be repeated with *La comète*. It seems likely that during the *cométomanie*, the collaborators were hoping to reprise their success tapping into public interest in current events and superstition. They again tried a similar formula, mining similar material with a special effects–laden comic takedown of alchemy and Rosicrucianism, *La pierre philosophale*, which, however, closed after only two performances, in February 1681.[32] The text of this play, which was presented as *La pierre philosophale: Comédie mêlée de spectacles*, was also quickly published by Blageart, with addresses of both his print shop and his book shop, "chez C. Blageart, Imprimeur-Libraire, rue S. Jacques, à l'entrée de la rue du Plâtre, et dans la Cour-Neuve du Palais, au Dauphin."[33] The zigzag typeface of the *Mercure galant* appears on the all-capital line "Comédie mêlée de spectacles."

Despite Donneau de Visé's open attitude toward the collaborative approach he and his associates used to write plays, eighteenth-century critics and editors attempted to attribute *La comète* to a single playwright. What is perhaps most notable, considering the likelihood that Donneau and Fontenelle collaborated on *La comète* (probably with the participation of Thomas Corneille), is the ways in which editors and critics begin to separate the legacies of the two figures. Pierre-François Godart de Beauchamps in *Recherches sur les théâtres de France* and François and Claude Parfaict in *Dictionnaire des théâtres de Paris* attribute the comedy to Donneau de Visé alone.[34] In the entry on the play in his eighteenth-century theater dictionary, Antoine de Léris acknowledges the disputed authorship but gives sole credit to Fontenelle. He opines that *La comète* is "attributed to de Visé yet is by Fontenelle," although he also lists the title under the names of both authors, Donneau de Visé on page 537 and Fontenelle on page 443.[35] Clément and LaPorte's *Anecdotes dramatiques* echoes this assessment: "This play, written on the occasion of the Comet that appeared in 1680, was played under the name of de Visé. It was written by Fontenelle, who had many little Pieces in Verse and in Prose printed in the *Mercure galant*, which was at that time under the direction of his uncle Thomas Corneille and Visé."[36] The earliest editions of Fontenelle's collected works, notably those produced during the author's lifetime, do not include *La comète*.[37] However, a year after the author's death the comedy is included in the final volume of the ten-volume pocket-sized edition of the "new expanded" *Oeuvres de Monsieur de Fontenelle* published by B. Brunet in Paris.[38] There is no note of any other author or collaborator. Henceforth, *La comète* starts to appear reliably in editions of Fontenelle without mention of Donneau de Visé.[39]

This shifting of the play toward Fontenelle's corpus reflects the bias of literary history. Posterity has not treated the two authors equally. Fontenelle has been enshrined as an avatar of the early Enlightenment and as an important contributor to more prestigious and official cultural life. Donneau de Visé, by contrast, associated with a commercial context, has been until recently less studied and certainly has commanded less cultural and literary prestige.[40] In 1680, however, Fontenelle was just finding his footing in a literary world and book marketplace very much animated by the *Mercure galant*. Donneau de Visé had spent the previous twenty years publishing an astonishing range of titles in diverse formats among which the common thread was innovative marketing strategies and an attention to readership. His own career as a writer included some twenty-odd plays, some of which, although mostly forgotten today, numbered among the era's most popular.[41] According to Schuwey, Donneau de Visé's theatrical writing "distinguishes itself from a classical conception" of theater by its entrepreneurial logic, episodic construction, and the preoccupation with current events and trends that also structure Donneau de Visé's prose writing.[42] The breathless announcement of *La comète* from the *Mercure galant* that I cited at the start of this chapter strongly resembles the way Donneau de Visé promoted his other plays, using what Cornuaille calls the "worldly detachment—very useful—of a journalist who neglects to mention that he is the author of the play."[43]

Both Donneau de Visé and Fontenelle frequented the same literary milieu in this era; however, in 1680 the literary scene was more prominently marked by Donneau de Visé—and even by the now forgotten Nicolas de Blégny—than by Fontenelle, whose literary career had heretofore been mostly aspirational. Situating *La comète* within the literary and publishing scene that produced the comedy (and Blageart's edition of it) highlights *La comète*'s connection to the innovative, but commodified and feminized, realm of popular culture and literature, with its concerns of time and timeliness.[44] Best known as the editor and often writer of the *Mercure galant*, Donneau de Visé was born in Paris into a lineage distinguished by several generations in service to the royal family.[45] The youngest child, and thus destined by convention to a career in the church, he instead opted to pursue literature, and in particular what we might call trendy forms and topics.[46] His first published book, *Les nouvelles nouvelles*, received a *privilège* on February 28, 1662, when the author was only twenty-three years old. Published almost a full year later in February 1663, *Les nouvelles nouvelles* introduces many of the techniques, interests, and preoccupations that Donneau de Visé would explore over the course of his career, from the socially imbricated spread of stories, gossip, literature, and information to the reception of new literary forms and an interest in theater.[47] Topics and genres mix heterogeneously within a frame narrative that recounts how stories and news circulate and the reactions they inspire. Donneau de Visé's inventiveness as an author very much anticipates not only the *Mercure galant* but also *La comète*.[48]

At the time of *La comète*'s performance and publication, Fontenelle was just recovering from the dramatic flop of his first tragedy, *Aspar*, which played only three times in December 1680 and was immortalized in a satiric epigram by Racine. According to legend, Fontenelle burned the manuscript. Fontenelle, who was only twenty-three when the Great Comet appeared, had already established an active collaboration with both his uncle, Thomas Corneille, and his uncle's coauthor and friend, Donneau de Visé. Fontenelle had published poems in the *Mercure galant* in 1677 and 1678 and had worked with Corneille on the operas *Psyché* (1678) and *Bellérophon* (1679).[49] He went on to an extraordinarily long life of literary and cultural acclaim, dying in

1757, one month shy of his one hundredth birthday. Straddling the worlds of literature and science, an ardent opponent of superstition and a popularizer of Cartesian natural philosophy, he was the author of the phenomenally successful *Entretiens sur la pluralité des mondes* (1686)—a series of dialogues presenting cosmology in a seductive, accessible, and entertaining fashion—*Histoire des oracles* (1687), and multifarious other works.[50] Taking the side of the Moderns in the Quarrel of the Ancients and the Moderns, he was elected a member of the Académie française (1691) and associate of the Académie des inscriptions et belles lettres (1701). He was also a member of the Académie royale des sciences, of which he became the permanent secretary in 1697 and which he chronicled in the *Mémoires de l'Académie royale des sciences* for the years 1699–1740.

Despite his association with the Académie royale des sciences, Fontenelle is famous not so much as a popularizer of rigorous science as a talented writer whose work stylishly engaged with scientific ideas. Nicolas-Charles-Joseph Trublet's *Mémoires pour servir à l'histoire de la vie et des ouvrages de Mr de Fontenelle* captures this paradox perfectly, praising Fontenelle as one of the few authors whose work is both esthetic and substantive, "aussi bon qu'il est d'ailleurs beau et bien fait." In fact, Trublet emphasizes, Fontenelle's writing is so good that his book outlives the death of the Cartesian science it popularizes. "Even if the Physics of this Book was wrong in several regards, an excellent Philosophy reigns. Consequently, it is as good as it is beautiful and well-written, for that matter. Its fate is thus independent of the System which it presents; and Mr De Fontenelle will still have readers when Descartes no longer has spectators."[51]

The attribution and the presumed meanings of *La comète* are intertwined. The attribution of *La comète* to Fontenelle colors readers' understanding of how the play presents and interprets the comet, which they assimilate to a global understanding of Fontenelle's later work. By the time Claude-Nicolas Le Cat pronounced his "Éloge de Monsieur de Fontenelle" in front of the Académie royale des sciences, belles lettres et arts de Rouen in August 1757, Fontenelle's reputation and prestige had far surpassed those of Donneau de Visé. Consequently, Le Cat takes pains to attribute the play to Fontenelle and to write Donneau de Visé out of its conception. He focuses on the play's demystification of superstition, which he ascribes to Fontenelle's worldly wit: "The *Mercure*'s author had played, under his own name, a Comedy entitled *La comète*, a short Play by M. de Fontenelle, where, on the occasion of the famous comet of 1680, our Author agreeably turned to ridicule, in one sole Act, the same prejudices that Bayle combatted so seriously in three volumes."[52] The contrast Le Cat makes between the unscrupulous-sounding "auteur du *Mercure*" and "notre Auteur" disentangles the two legacies. One is an author-philosopher; the other is a peddler and a literary thief. As in the original *Mercure galant* notice of 1681, *La comète* is presented as a condensed theatrical translation of the kind of serious discourse presented in a long book; only here its connection to the *Mercure galant* is excised. In order to assure the play's serious posterity, Le Cat places Fontenelle in the more unimpeachable proto-Enlightenment company of Bayle.

While *La comète* does not appear in contemporaneous editions of Louis Moréri's *Grand dictionnaire historique*, the final mid-eighteenth-century edition (20th ed., 1759) likewise attributes *La comète* to a young Fontenelle.[53] The entry declares, "Cette petite pièce est très-philosophique, sans en être moins comique," thus offering a reading of *La comète* as a serious, albeit amusing, work that intervenes meaningfully in the philosophical questions that the comet's appearance occasioned. By the mid-eighteenth century, such a reading resonated with Fontenelle's weightier legacy so the editors also take pains to separate *La comète* and Fontenelle from Donneau de Visé and the *Mercure galant*: "The one-act, prose comedy entitled *La comète*, although played under the name of M. de Visé in 1681, is likewise [i.e., like the works discussed above] by M. de Fontenelle. This short play is very philosophical, without being less comedic. It was written on the occasion of the famous comet that had appeared the year before, and to combat the still rather widespread prejudice that comets are evil portents."[54] As in Le Cat's *Éloge* the dictionary's editors present the one-act play as a serious contribution to Enlightenment thought and frame its themes, preoccupations, and contributions as demystification, which is a very accurate description of Bayle's *Pensées diverses sur la comète* and not at all, as we will soon discuss, an accurate description of *La comète*.

This is very much the way *La comète*, if mentioned at all, continues to be referenced in contemporary scholarship. It is the reading Julie Boch offers, for example: "What Fontenelle invites the spectator to is a change in perspective, a reversal in point of view to which *La comète* is a prelude, *Nouveaux dialogues des morts* the continuation, and *Entretiens sur la pluralité des mondes* the culmination."[55] Boch is at pains to integrate *La comète* into Fontenelle's oeuvre as a prelude to his best-known work. Noted exceptions to this mode of critical reading are Riffaud and Schuwey, whose meticulous focus on the printing and publishing history of theatrical works and whose archeology of Donneau de Visé's entrepreneurial mode of authorship, respectively, make them more receptive to the elements of the comedy that do not correspond to a historical narrative of disenchantment. Neither of these scholars dwell at length on the play, however. It falls outside the time frame of Schuwey's study of Donneau de Visé and is proximate, yet not central, to Riffaud's interest in Jean Ribou and the publication of Molière's plays.

Rather than read *La comète* as a prelude to Fontenelle's elegant fictionalization of Cartesian science, the rest of this chapter proposes to look at the play with fresh eyes to see whether we can uncover what other issues or dynamics the comet may have evoked for its seventeenth-century spectators. Like the nouvelle "La comète" published in the *Mercure galant* and discussed in chapter 1, where curiosity about the comet and a series of misunderstandings conspire to marry off a penniless young noblewoman, the "petite comédie" seems to exploit contemporary fascination with the comet to peddle a stock marriage-plot drama. In the play, a modern young couple in love must triumph over the superstitions of the ingenue's astrologer father. If Bayle famously argued that a moral society need not believe in God, the stakes of this play at first glance seem much scanter: a nice young couple might use the distraction a comet provides to circumvent the *non du père*. I would like to argue, however, that while the play might not contribute significantly to the debate about the merits of astrology as a system of prognostication (indeed it did not need to, as astrology as depicted on the comedy's stage seems so bereft of authority as to necessitate no further dismantling), *La comète* stages a shift away from decisive authority figures and toward the dynamic collectivism of systems of circulation. While no one could decisively determine that comets followed a regular and predictable path, literature like *La comète* offered a place for social observation, and could describe the brisk circulation of desire, capital, and information within France's capital city.

Alternative Circulations

In its framing and its content, *La comète* meditates on the comet in very cometic ways. By his own account in his obituary for Thomas Corneille, Donneau de Visé had a talent for intriguing and topical bits and depended on a collaborator to tie them into a unified whole. *La comète* follows this pattern, introducing a series of characters variously reacting to the comet within the conventional frame of a marriage plot. While the *Mercure galant* advertised the one-act comedy as a dramatization of Descartes, *La comète* can much more accurately be characterized as a dramatization of the comet coverage in the *Mercure galant*, for many of the play's comic vignettes echo the January 1681 *Mercure galant*, like *pièces* put to life on stage.[56]

The comedy's opening scene introduces *La comète*'s concern with astrological phenomena within the commercial context of a competitive print marketplace. The maid and valet discuss the upcoming marriage of their young mistress, Florice, to her suitor, Monsieur de la Forest. The conversation, as it often will throughout the course of the play, pivots on financial matters. Maturin, the valet, comments on the wealth of the groom and their master's lucrative work as an astrologer. He has been trying to pick up astrology so he can make his fortune with an almanac of his own. "Our Master has also amassed a lot of money with his astrology. My goodness, Françoise, it's a good profession being an astrologer! I'm trying to learn it without anyone noticing; I carefully collect everything that our master says. . . . I am working on an almanac for the year 81, which we are about to start."[57] Here in the first act of a play performed in January and February 1681, Maturin inscribes the action in the very recent past of December 1680. This is a play about the present moment asking what the comet appearing in the sky during the very nights of *La comète*'s performance has to say about the year to come. Maturin tells Françoise he has finished his weather predictions through the upcoming December but he needs to fill out his volume. Should he include the lives of some great men, or instructions on planting cabbages? Françoise opines for the latter as more adapted to his expertise. She also suggests that to be a proper astrologer Maturin needs a portrait. "But above all don't forget to have yourself depicted at the beginning of your almanac, instruments in hand, gazing at the stars."[58] The audience may laugh easily at the maid's literalist understanding that a print representation staged with the right accessories guarantees credibility rather than the actual mastery of the astronomer's tools to which the portrait is meant only to testify. However, while the comedy pivots on Françoise's misprision, it is at the same time a substantial joke because she is pointing to something that is true: within a marketplace of ideas mediated by books rather than bodies, it becomes increasingly difficult to ascertain what is substance and what is simulacra. (Of course, hoaxes exist before there are books, but as we well know, new media and dissemination networks open up not only the possibility of connection but also the means to disseminate misinformation.) The scene ends with an exchange that contrasts the serious attention Maturin and Françoise give all matters of commodified textuality (almanacs, printed portraits) with the pair's complete lack of seriousness regarding predictive astrology. "And what did you do to predict good and bad weather?" asks Françoise. "Like everyone else. I don't really know stars; I used three dice."[59] As opposed to the planfulness (albeit half-cocked) of his commercial-literary aspirations, for the content of his book the valet resorts to a method based on contingency and randomness, one literally synonymous with chance.[60]

When the play's ingenue, the lovely young mistress Florice, enters, Maturin plies her with promises of marital happiness and beautiful children born under lucky stars. The triad seems comfortable in the shared understanding that astronomy, if it is *about* anything at all, is about commerce. Françoise objects to Maturin's posing as astrologically knowledgeable in his compliment: "Look at the great astrologer, promising beautiful children to a lovely woman!" To Florice, however, the predictions are only a sign of Maturin's desire to please and to curry favor. She sees through his posing and discerns financial motivation: "Never mind, Françoise; I see that Maturin means well, and it is my plan that on my wedding day you both should have reason to praise my generosity, and M. de la Forest's."[61] Before any mention of the play's titular subject, the comet, this framing presents star gazing as thoroughly imbricated with economic concerns. No one in these first two scenes is interested in the motion of the stars per se. If they are interested, it is in terms of economic opportunity—the chance to peddle an almanac, or flatter one's employer, which Florice rightly recognizes by promising money in exchange.

The play's third scene introduces Monsieur de la Forest, the romantic lead, joyfully announcing to Florice that their wishes are about to come true. His uncle has agreed to their union: "He will be coming here to sign the Contract with Monsieur your Father."[62] Florice jubilates that her father has also consented, despite what both lovers acknowledge as his difficult nature. "I admit that I trembled a thousand times for you," says Florice, to which Monsieur de la Forest responds, "It is true that People of his Profession are of a rather difficult humor, and the commerce they have with the Stars usually makes them less fit to have commerce with People."[63] Florice and her lover both use the language of astrology in their exchange, but just as Maturin and Françoise's discussion of astrology is undermined by their focus on commercial goals, the young couple's astrological references are all absorbed into a language of sentiment. They seem perfectly comfortable evoking the stars, but each time they do, the reference is turned into a literary trope. Florice tells Monsieur de la Forest that her father's reading of the stars matches the starstruck state of their hearts. "Apparently, my Father consulted the Stars on our Marriage. It seems they promise only good things, and we should be rather happy to see that the Celestial Influences match those which Love put in our hearts."[64] He replies in thc samc vcin, "For me, lovely Florice, it was my destiny to love you, and my inclination."[65] For Florice and her love, what is real is their feelings. The alignment of stars and the "destiny" celestial bodies were long ago thought to predict are useful as a means to express poetically the abstract but deeply felt reality of sentiment.

The framing of the first three expository scenes deflates in advance the play's central conflict: the question of whether the comet has any commerce with normal mortals and, consequently, whether it should cancel Florice's engagement with Monsieur de la Forest. The play's exposition thus seems to belie in advance the philosophical readings of the play presented in Donneau de Visé's own *Mercure galant* and echoed in later critical appreciations, such as Moréri's dictionary, which assimilated *La comète* to Fontenelle's oeuvre. The dramatic moment when the ingenue's *astrologue*-father "calls it all off," in the fourth scene (of seventeen) reveals the astrologer-father as a figure of broad humor. He bursts in, declaring, "What disasters for the coming Year! What storms, what famine, what plague, what war!"[66] Without so much as a name, he is a comic type whose dialogue is composed chiefly of concatenations of astrological terms and hyperbolic non sequitur exclamations of upset about the comet—a gibberish ecphonesis devoid of reasoned propositions. The scene is composed according to a classic comedic structure, placing in conversation three characters whose diction and reasoning occupy entirely different registers and systems of logic and social value. The agitated astrologer is flanked by de la Forest's appeal to reason and by the valet Maturin's financially motivated buffoonery.

> ASTROLOGER: What do you mean? This great Book of the Heavens, printed in Letters of fire, does it not contain every Man's Destiny?
>
> MR. DE LA FOREST: Permit me to reply that this great Book is not easy to decipher, and that with all of your Telescopes you still have great difficulty reading a few of its words.

ASTROLOGER: No, it is surely not for you that it is written. It is only for us, the Descendants of the famous Nostradamus, to understand these mysteries. What do you know about the Aspects of Stars, Sextiles, Trine and Quadrant? What about Conjunctions, Oppositions, Stations, Directions, Retrogrades?

MATURIN (*taking up his notes*): Let's put this down on our Tablets. Blimey, I'm going to be rich![67]

While Maturin feverishly takes notes on his *tablettes* about things to add to his almanac, the young suitor tries to convince the astrologer with common sense.[68] De la Forest argues that bad things happen often enough without comets and good things happen when comets are present as well: "Aren't there presently a thousand happy People, who have as much right to thank the Comet for their felicity, as Unhappy ones [who] have the right to blame it for their misfortune?"[69] He asks his prospective father-in-law to change his perspective and zoom out to consider the whole of humanity. "Are we such important People that we can imagine that the Heavens would take the trouble [lit: incur the expense] to make a Comet for us?" ("Sommes-nous des Gens si importants, que nous puissions nous imaginer que le Ciel fasse pour nous la dépense d'une Comète?"), he asks, returning the comet to the language of economics.[70]

Florice's unreasonable father and de la Forest's angry uncle trade insults over the astrologer's change of mind in scene 5, and all reason and common sense evacuate the stage as the young couple's plans for marriage are dashed on both sides. The young suitor tries repeatedly, with all his social grace, to mediate between his love's eccentric father and his own scandalized uncle, but it is Maturin and his financial concerns that have the last word in the scene. Seeing the wedding plans canceled, he laments the monetary loss this will entail for the servants. "Oh, my dear Françoise, farewell to our presents; look at all of our lovely hopes dashed."[71] Just as the valet Sganarelle's "mes gages, mes gages" in the final scene of Molière's *Dom Juan* (1665) is not only a final moment of comic relief but also points to his master's deepest trespass—abandoning an aristocratic-theological ethics of value for the market value of commodity exchange—Maturin's focus on money likewise highlights one of the play's most astute critiques of the *comé-tomanie* and all who would profit from it. If you would really like to know what the comet brings—fortune both good and bad—follow the money. If the comet brings disaster to the valet, it brings a flurry of clients to the master. However, where comets go and when they appear can never be predicted. Although they may circulate, comets destabilize order rather than prop it up.

Florice tries to repair the breach in the following scene. She chides her father for going back on his word and forcing her to go back on hers and tries to win him over to her way of seeing things, where comets and stars are (only) useful metaphors for the real organizing force behind the situation, which is her heart. "But Father, you let me give my word to him; and I am no longer in the position to obey you like I would have before. Concerning my marriage, do you need to consult the stars more than my heart? It has only promised me happiness, and its prediction must be just as certain as your comet."[72]

In the exchange between father and the young lovers, reason (de la Forest) and sentiment (Florice) emerge as the play's most explicit protagonists. This generational and hermeneutic conflict might seem to bear out the *Mercure galant*'s praise of the play, and Moréri's assertion in his *Dictionnaire* that the play is "très philosophique." However, these "reasonable" modern voices seem to withdraw from most of the important action on stage; they are absent for the majority of the drama. Florice emerges from the house in the beginning; speaks in scenes 2 and 3; merely witnesses (but for two short incomplete and totally disregarded interjections) the conflict between men in scenes 4 and 5; and tries to reason with her father in scene 6. She appears briefly on stage to hand her father a letter and then emerges near the end of the play to exit with her lover in an *enlèvement* scene utterly devoid of drama. Monsieur de la Forest speaks in scene 4 (with Florice) and scene 5, in which he tries to reason with the astrologer. In the crisis scene he is outshouted and totally ineffective. Thus accounted, the play's only confrontations between superstition and nascent *lumières* occur in scenes 4 and 6 and are so brief that they seem almost to be shorthand for an audience who already knows the terms of this debate. If we watch what happens on the stage for the duration of the play, a different ideological protagonist emerges: Maturin, the voice of commerce. Ultimately, the play advocates a regime of exchange value or *mondanité*. The world is changing so fast that the astronomer, following the comet from his garret observatory, becomes irrelevant to the action in his house and neighborhood. Rather than engaging superstition in a careful, reasoned debate, this play laughs it off. Yet it laughs off science as well. The real revolution is happening in the streets and in families, in bookshops, salons, and theaters.

If the play opens with the valet's comedic scheming, in *La comète*, the real ridicule lands squarely on the master, not the servant. In scene

7, Maturin talks to his master about his curiosity. He wants to borrow the telescope to see the twelve houses of the sun, wondering, hilariously, which one is the most habitable. When he also expresses an interest in seeing the moon, the *astrologue* assures him that it is covered with men and cities, a literal reading of the satirical voyage depicted by Savien de Cyrano de Bergerac (1619–55) in *Histoire comique des états et empires de la lune*, published posthumously in 1657.[73]

A comic exchange in scene 6 between father, daughter, and servants throws the play's focus on questions of value into sharp relief. Florice, having failed to convince her father through recourse to the values of society (his word and hers) or sentiment (her heart), throws herself at her father's feet and repeatedly begs him to reconsider, for all he holds dearest, "Par tout ce que vous avez de plus cher."[74] Françoise misconstrues the reference, clearly made to the astrologer's deeply held values, and perhaps his love for his daughter. The servant adds her plea to her mistress's: "In the name of your Telescopes, and all of your Instruments."[75] Maturin joins his voice to the entreaty: "By the conjunction of Venus with Saturn."[76] The servants' comic misunderstandings are in fact apt observations that serve to highlight the even greater folly of their master, who should know better than to prize instruments and observations above his family and his word.[77]

Other comic elements are female types, perhaps meant to resonate with the inscribed readers of the *Mercure galant* and echoing the witty *pièces* published in the January edition. The comtesse de Goustignan, panicked that her house seems to be directly in the comet's path, rushes to consult with the astrologer and remains with him for the duration of the play. *Salonnière* Madame Fraisier is not interested in the composition or trajectory of comets, which she dismisses as "rather vain and useless questions."[78] Rather, she asserts that "our question is more important. We made a large wager and agreed that we needed a professional to decide it."[79] She and her circle want to know whether the word "comet" is masculine or feminine. The exchange between the superstitious countess and the *salonnière* is revealing about the real conflict happening around the comet. When Fraisier assures Goustignan that "people of quality don't give a care to what the comet is, they only want to know the true name" ("les gens de qualité ne se mettent point en peine de ce que c'est que le comète [*sic*]; mais ils voudraient seulement en savoir le véritable nom"), the countess retorts by evoking her rank, "I want you to know that there are people of quality here" ("Je veux bien que vous sachiez qu'il y a ici des gens de qualité").[80] This conflict over the comet changes the subject from the cosmological questions, which have only been voiced in the most broad of comic tones, to a disagreement over who can claim the category of "gens de qualité"—whose questions about the comet express refinement or social status.

La comète undermines paternal, epistemological, religious, and even scientific authority. Once the countess appears, she doubles the astrologer in a classic example of buffoon comedy. Along with Maturin, she drinks in the astrologer's confused rantings, and dramatizes, with her well-timed misapprehensions, just what a bad observer and analyst the astrologer is. What is interesting—and what makes me diverge from classic appreciations of this work as a theatrical rendering of Descartes and company—is that the play's heavy-handed comedy tars "good" (modern, rational) natural philosophy with the same brush it does superstition. The astrologer goes up to his attic to use his up-to-date telescope, but the action that he is missing, the action that propels the play, isn't in the sky but in Madame Fraisier's salon, in the street, in his own house.

Scene 12 prepares the imminent crisis when the astrologer vacates the house and unwittingly leaves the terrain open for his daughter to leave. It is also a one-joke scene that dramatizes the fact that the subject of the play is not the comet itself but rather its mediatization. Letters arrive from Rome. The maid, countess, and valet watch in suspense as the astrologer opens and reads the news sent by, he says "one of my colleagues, a very skilled astrologer. They assuredly contain some curious observations of the comet."[81] The rapid dialogue that follows creates suspense as the observers note the astrologer's expression of surprise and urge him to explain a succession of expressions of wonder, each one more maddeningly imprecise than the last. With excellent comic timing, the astrologer reveals, finally, that the wonder seen in Rome is "Une comète dans un oeuf," the print hoax reported, and then updated, in the very *Mercure galant* issue that advertised the play.[82] The countess and Maturin both fret: if the nucleus of a comet is an egg, is it safe to eat an omelet? This printed hoax, not anything he has seen in the sky, is ultimately the news that motivates the astrologer, with the countess in tow, to grab his telescope and warm bearskin and leave his home for the superior height of his neighbor's rooftop. "Let us not waste any more time discussing; something supernatural is happening here. Let us hurry, my telescopes, my bearskin, my equipment, so that I may go observe the comet," he declares while making a dramatic exit.[83]

In the play's final scene, the astrologer and the countess, who has caught a horrible cold stargazing from the attic of the neighboring house (another comic comet trope familiar to readers of Loret's rhyming depiction of the comet gazers sixteen years prior), return from their observations, proclaiming the misery the comet portends. The agitated valet begins to announce, "Sir, you do not know what misfortunes . . ." ("Monsieur, vous ne savez pas quels malheurs . . ."). But before Maturin can announce Florice's abduction, the astrologer cuts him off: "Oh! I know them only too well; I read in the comet all of the misfortunes it portends. I already see war in Europe, plague in Asia, famine in Africa."[84] Perhaps there will be war, pestilence, and famine, but a certain *malheur* has transpired in the *astrologue*'s own home. Convinced by her fiancé, Florice has left the family house, right under her father's nose. The young man concocted a tale of abduction for the servants to tell so they and his young bride could save face (scene 15). Rewarded for their cooperation—with money, at last, to marry (scene 16)—the nervous servants try to break the news to their master, comically reaching for the comet when their tale of a dramatic armed abduction stalls.

> MATURIN (*low*): Nothing is left for America. Haven't you discovered anything else?
> THE ASTROLOGER: What, doesn't that suffice? Did a Comet ever announce more misfortunes?
> M: Sir, we know more than you, Françoise and I, even if we have not ever observed the Comet. Did you not happen to see among the misfortunes of Europe, Asia and Africa . . . Oh, the sadness leaves me speechless. Toinon, tell Monsieur . . .
> THE ASTROLOGER: What? What are you trying to tell me?
> FRANÇOISE: The Comet that revealed to you so many things, should have told you that Monsieur de la Forest . . . I was with Maturin. Your Daughter at this same moment . . . Oh, Monsieur, what a Comet!
> THE ASTROLOGER: You are not telling me anything. Why are you two so upset?
> MATURIN: Monsieur, it's that the Comet . . . You were observing it, we were here, and it abducted Monsieur de la Forest.[85]

After Maturin faithfully rehearses Monsieur de la Forest's fabricated tale of armed abduction, all present—servants and master and countess—agree that the comet, flying in to interrupt Françoise's story just before its awkward climax, must be the cause of this ultimate familial misfortune. As Maturin tries to pick up the story from a flustered Françoise, it comes out topsy-turvy: "Monsieur, c'est que la Comète . . . Vous l'observiez, nous étions ici, et elle a enlevé Monsieur de la Forest." In this carnival retelling, the female gendered comet comes out on top, having abducted the young hero. The astrologer proposes to pursue the audacious fiancé, but the countess reminds him that Florice is already beyond his power and his house; her marriageability is no longer in his purview: "Vous n'en êtes plus le Maitre."

Following the conventions of the genre, this comedy ends with a marriage. However, rather than being sanctioned and dictated by the father in a classical affirmation of paternal order (and all of the other social orders that it condenses), this marriage has shown the father's authority to be entirely irrelevant. The marriage takes place off stage in another house, leaving the paterfamilias forlorn to contemplate the emptiness of the house over which he was supposed to preside but which he foolishly abandoned, fueled by a fake news story and comet mania. The (near) rhyme of the last two lines of dialogue only serves to highlight the father's redundant *désoeuvrement* and comic dispossession:

> ASTROLOGER: Let us go, Madam. I wish to consult all my friends.
> COUNTESS: Let us rest first. Our colds need a good night.
>
> (A: Allons, Madame, j'en veux consulter tous mes amis.
> C: Songeons d'abord à nous reposer. Nos Rhumes ont besoin d'une bonne nuit.)[86]

The astrologer's final recourse will be gossip or commiseration, but only after a good night's sleep. Does *La comète*—a circumstantial piece that tried to bank on *cométomanie*—in the end damn its own chances at success by suggesting the inconsequence of comets as well? I would rather argue that the play's comic thesis is that the debate between superstition and rationalistic understandings of the comet misses what comets really have to tell us as signs of the novel, the new, the faddish, and the commercially successful. *La comète* focuses our attention on the ways that the forces associated with these economic and social registers were rapidly transforming life in late seventeenth-century France. The comet in this comedy sidelines authority, as, preoccupied with the sky above, the father and the countess leave the house and the street up for grabs. This absence opens a breach for the revolutionary mores of a new generation: marriage for love, daughters circulating on their own judgment, valets with publishing projects. At the same time, the comet draws people into public spaces and disrupts the normal social order, a kind of celestial carnivalesque. It galvanizes conversations and sparks the development of a new refined language and new social forms. The comet likewise provides the opportunity for class transgression: Maturin becomes rich enough to marry Françoise and establish an independent household. Who knows, maybe he will overcome Françoise's reticence and publish an almanac, and the comet will provide him an opportunity to enter into the print economy that made this comet mania possible, and that was made possible by this comet.

CHAPTER 5

"La comète appartient à tout le dessein du livre"

Genres of Disenchantment in Bayle's *Pensées diverses sur la comète*

Perhaps the most famous literary monument to the Great Comet of 1680–81—the comet seen one year before Halley would chart and, with the resources of Newton's *Principia*, solve the riddle of comets' celestial movements—was not actually written in response to a comet but will forever be remembered for the fiction that pretends it was. This paradox seems like an appropriate legacy for Pierre Bayle's *Pensées diverses sur la comète*, a book that purports to answer the question of comets' influence on life on earth but ranges in cometic ways to engage questions of superstition and belief, community, idolatry, and ethics.[1] Bayle's book broke new ground for its daring assertion that an atheist could be a moral person and that a society of atheists could be ethical. Beginning my inquiry with the fiction of the text's origin story, I will examine the prefaces and introductory paratexts, where Bayle most explicitly highlights and reflects on the modes and nature of his discourse. In these metatextual reflections, I will discover how Bayle employs fictionality and other tools of literature to explore serious ethical and theological issues.

Bayle was a theologian whose *Pensées diverses sur la comète* sought a higher standard for theology; the *philosophes* of the subsequent generation credited him with taking the first decisive steps toward the Enlightenment disenchantment of the world. Bayle also was a wide-ranging reader who from 1684 to 1687 served as the founding editor of *Nouvelles de la République des Lettres*, an Amsterdam-based French-language journal highlighting recent book publications.[2] In this chapter I will trace how, in prefaces and introductions to successive editions of his controversial bestseller, Bayle's fictionalized editor personas comment on textual dynamics that in turn offer a guide for reading his famously dense and digressive text. Bayles's peritexts deploy and call attention to techniques of literary address and writerly modes of fictionality that he discovered in the contemporary novels, critical conversations, and periodical press that he followed with great intensity. These often self-reflexive frames, I argue, point up literary modes of discourse that extend rich possibilities to readers for understanding the audacious, and in equal measure cautious, digressions and indirections of Bayle's *Pensées diverses*.[3]

In a treatise Bayle read and discussed with friends, Pierre Daniel Huet (1630–1721) defined the emerging and somewhat controversial genre called the *roman*, or novel, as a fictional adventure written in prose that brings readers pleasure and instruction: "Today . . . the books properly called Novels are fictional love stories, written in artful prose, for the pleasure and instruction of readers. I say fictions to distinguish them from true stories/histories [*histoires véritables*]."[4] Catherine Gallagher, in her influential essay "The Rise of Fictionality," also argues that the development of fictionality and the novel were inextricably linked. "The novel is not just one kind of fictional narrative among others; it is the kind in which and through which fictionality became manifest, explicit, widely understood and accepted. The historical connection between the terms novel and fiction is intimate; they were mutually constitutive."[5] Foundational to this pairing of novel and fiction that became popular in Bayle's formative years is the category called *vraisemblance*, or verisimilitude, debated and dissected by Huet and others.[6] For Huet, the *vraisemblable* defines the novel: "Verisimilitude, which is not always found in history, is essential to the Novel."[7] He sums up the concept in the following, frequently quoted way: "Novels are fictions of things that could have been, but never were" ("les Romans sont des fictions de choses qui ont pu être, et qui n'ont point été").[8] Novelistic fictionality cultivates a habit of mind in which readers discover something potentially or generally true through the mediation of the fictional.[9] This provisory epistemology that underpins novelistic reading becomes a linchpin for Bayle's innovative critique of superstition and for the uniquely modern ethical universe his work proposes, a world where ethical actions are not assured by unshakable and logically or mystically accessible foundations but rather by the ability of human beings to think, discern, and imagine their way through situations and quandaries in much the same ways readers of novels negotiate fictional worlds.[10]

While Huet's treatise contrasts novels and histories, delimits novels as paradigmatically restricted to the realm of love plots, and draws specifically on associations between novels and female readers, French-language novels in the seventeenth century were not separate from conversations in which learned pursuits were articulating themselves into distinct discourses.[11] Huet's research interests exemplify this. Ordained a priest and named a bishop in midlife, he was also deeply involved in pedagogy, philology, and philosophy as well as natural science, mathematics, astronomy, chemistry, and anatomy.[12] In a series of readings of cosmological texts, which straddle what we now perceive as a divide between the literary and the scientific, Fréderique Aït-Touati follows how in late seventeenth-century France, the modern novel and natural philosophy became differentiated into distinct discourses, arising from a shared if divergent history in which both define themselves in relation to what she calls "the poetics of the credible."[13] Like the pairs of texts Aït-Touati reads as she traces the entanglement and progressive disarticulation of literature and science, Bayle's *Pensées diverses sur la comète* addresses questions and objects that we now consider to be scholarly with means that we have come to associate with the world of pleasure reading. Unlike the cosmological texts by Kepler or Cyrano, Fontenelle or Huygens, Hooke or Cavendish that Aït-Touati analyzes, when Bayle's book dispels outdated interpretations and meanings ascribed to the comet, it does so with an eye toward theological rather than scientific questions. Yet as in some of the cosmological texts Aït-Touati explores, Bayle's *Pensées diverses* employs fictional conceits alongside narrative structures and rhetorical stagings and strategies characteristic of the novels of his contemporaries and thereby brings the toolbox of novelistic fictionality to bear on questions of faith, ethics, and epistemology.

Written in epistolary form, Bayle's is a digressive and multifarious text that promotes reason and audaciously denounces superstition and oracular thinking by refuting the belief that comets are portents. As part of this critique, Bayle created a seismic cultural shift by proposing, in contrast to both Catholic and Protestant apologetics, that atheism was a less serious error than idolatry.[14] The comet that fascinated Bayle's contemporaries serves as the opportunity for Bayle to demonstrate that an atheist could be virtuous and that an ethical society need not believe in God, while, conversely, individuals claiming strong religious faith may perpetrate heinous actions. In a series of short meditations on questions such as "The Authority of Tradition" (7), "Proof That Demons Prefer Atheism to Idolatry" (113), and "Reflection on the Diverse Way(s) in Which People Act against Vices and Errors" (199), Bayle's *Pensées diverses* imagines new ways of understanding human relations to divinity, community, power, and faith.

The thrust of my argument is that in *Pensées diverses sur la comète* Bayle employs the kinds of narrative techniques that were emerging in French prose literature and the related realm of the periodical press in the last decades of the seventeenth century in order to challenge the persistence of superstition and denounce its relationship to power structures. Introductions and prefaces to the editions of *Pensées diverses* published before Bayle's death (that is to say, the introductions and prefaces written by Bayle himself) announce these narrative strategies and prime readers to consider the text in this light. I believe that the affinity between Bayle's critique and contemporaneous literary forms extends beyond the use of formal and semantic elements shared with the developing modern novel and the periodical; Bayle's defense of rationality also eschews systematic thinking and organized and hierarchical structures to encourage individual readers to experiment speculatively with big ideas. In Bayle's textual universe, access to the divine is available through a process of imagination and faith resembling a novel's *vraisemblance* rather than a mathematician's geometrical proof. In other words, Bayle's *Pensées diverses* draws on the narrative conventions of fiction above all to mine their epistemological, ethical, and political potential.

In focusing on Bayle's writing style, and its imbrications with literature, notably the kinds of fictional techniques being developed in the emergent psychological novel, the periodical press, and the era's salon, I in no way want to belittle the serious stakes of Bayle's interventions.[15] The positions Bayle articulated in this book could have cost him his life and may in fact have led to the death in (French) captivity of his beloved brother. The literary stylings and novelistic elements of Bayle's text in no way soften, obscure, or undermine the book's profound and daring critique. On the contrary, I wish to highlight the serious stakes of the literary devices and fictionality that Bayle deployed, the fundamental role they played in his critique of authority. It is the discourse of literature that offers Bayle the resources to propose an alternative. For Lynn Festa, "the novel rescales and repartitions objects, redistributes power, and makes new aspects of familiar objects visible. In the process, it constitutes its reader in relation to the rules by which the world it depicts can be made intelligible."[16] In his *Pensées diverses*,

Bayle attempts just such a rescaling and redistribution. Bayle might not have been inspired to write his book by the comet, but the comet's passage becomes the narrative opportunity for Bayle to reexamine humans' relationship to the divine and to each other through distinctly literary means.

Bayle, I will show, leaves various clues that he intends us to read his work like a novel. In the prefaces (and likewise in key points in the body of the text, which I will allude to but not analyze in depth) the question of genre and voice is fundamental to Bayle's culture-changing critique. Simply put, Bayle's most central and innovative argument pivots on the *literary* categories of (human) reading and (divine and human) writing. Bayle famously asks, Wouldn't God (if there is a God) send an unambiguous legible sign if they wanted to address humans? Would God send a message so prone to misinterpretation that it reinforces pagans in their errors and despots in their dominion? This question, which provides the fulcrum of Bayle's refutation of superstition, conceives of faith as a kind of sensemaking, as an act analogous to reading a letter or a *roman à clef*.

One may wonder why (as I claim) Bayle might use novelistic tropes, and literary turns from within the toolbox of imaginative prose fiction, to effect a critique as serious as he undertakes in his *Pensées diverses sur la comète*. But, seen another way, the novel is perfectly apt for Bayle's letter, his diverse *pensées*. His book does not propose to observe or measure or rationally study the comet, but rather to make the comet speak to a sometimes maddening variety of issues and concerns, the underlying coherence of which is that they stand in some way or other against authority and for the sovereign autonomy of individuals and dialogic communities of readers. In this way, Bayle's text, with all its serious claims and ethical repercussions, draws on the models of epistemology speculatively explored in the pages of novels, and debated in salons, critical texts, and the pages of periodicals such as the *Mercure galant*. Reading his prefaces alongside contemporaneous works of literary criticism will shed new light on the importance of literary models and worldviews for Bayle's shocking and radical ethics.

Bayle uses the tools of literary writing boldly to imagine a world where meaning is not found, and ethics are not grounded, in messages from God, but rather from conversations between mortals. As we will soon see, Bayle's famous book might not have actually been inspired by seeing a comet as Bayle's authorly conceit would have it. This is, rather, a tale Bayle spins, a fiction that allows him to animate his critique in concrete terms in the imaginations of his readers. What is indubitable, however, is that his book itself was a kind of comet, luminous and breaking new ground outside the established orbits of the era's cultural meanings.

Pierre Bayle: Exile and a Bestseller

When the Great Comet appeared in December 1680, Bayle (1647–1706) was a thirty-three-year-old philosophy professor at the Académie de Sedan, a Protestant seminary in a formerly independent Protestant principality, which had been annexed to France in 1651. He had yet to publish any work. Born into modest means, the second of three sons of a Huguenot minister in southwestern France in the small community of Carla, Bayle started his studies at the Protestant Academy of Puylaurens in 1666, attending school only when his brother was done and the family could afford it.[17] Three years later, he converted to Catholicism in order to study at the Jesuit college in Toulouse. After defending his thesis in philosophy, however, Bayle returned to Calvinism, a decision that made him a relapse in the eyes of the Catholic Church and the French authorities and necessitated a move out of the country. Forced to leave behind his parents and brothers, whom he would never see again, Bayle fled to Geneva, where he began, and then abandoned, the study of theology. He subsequently took positions as a private tutor to a series of wealthy households, eventually making his way back to France, first to Rouen and then Paris, before in 1675 successfully postulating for the philosophy chair in Sedan, where he kept his past brushes with the Catholic Church to himself.

Bayle reports in the various prefaces to his *Pensées diverses sur la comète* and its expanding subsequent editions that he began writing the book during the winter of 1681, during his final months in his native France. This was a hard and bloody time for France's religious minorities: Louis XIV's dragonnades terrorized Protestant communities by billeting rowdy soldiers in their homes, a terrifying signal to Huguenots that if they refused to convert or emigrate, they would be subject to bodily harm and destruction of property. In July 1681, despite a mission by Bayle to Paris to argue in favor of his seminary, and the understanding that Sedan should retain some confessional autonomy, the crown declared the suppression of various Huguenot institutions, including the Académie de Sedan. Bayle left France as part of a large wave of Huguenots seeking refuge abroad in the years leading up to the 1685 Revocation of the Edict of Nantes. He accepted an invitation from Adriaan Paets to settle in Rotterdam and become a professor of history and philosophy at the newly formed École illustre de Rotterdam at the recommendation of his erstwhile supporter, and later bitter critic, the theologian Pierre Jurieu (1637–1713).

In March 1682, a few months after Bayle's arrival in Rotterdam, the anonymous first edition of his famous critique appeared under the title *Lettre à M.L.A.D.C., Docteur de Sorbonne. Où il est prouvé par plusieurs raisons tirées de la Philosophie, et de la Théologie, que les Comètes ne sont point le présage d'aucun malheur: Avec plusieurs Réflexions Morales et Politiques, et plusieurs Observations Historiques; Et la Réfutation de quelques erreurs populaires*. Written in epistolary form, in the voice of a fictionalized Catholic narrator, in reply to an equally fictional query from a Sorbonne theologian, the published "letters" are dated from January 1 to October 11, 1681, and they stage the period of their composition as beginning during the period of the comet's appearance. While at the time of publication, Bayle, in Rotterdam, was outside the reach of French authorities, his parents and brother remained in France, and vulnerable to reprisals. Due to the dangerous nature of the book's implicit critique of Catholic dogma as a kind of idolatry, and probably also in hopes of making his text more convincing to French (Catholic) readers, *Lettre à M.L.A.D.C.* was published anonymously by Reinier Leers under the widely used pseudonym for clandestine print runs, Pierre Marteau of Cologne.[18] The book became and remained an international publishing phenomenon, with a new edition only one year later.[19]

Eventually unmasked as the book's author when a reader recognized his handwriting on a manuscript response to a critique, Bayle published expanded editions and sequels under the acknowledged imprint of Leers, whose Rotterdam publishing house was strategically situated for international distribution of this and other controversial bestsellers.[20] In September 1683 Leers published an expanded two-volume duodecimo edition, entitled *Pensées diverses, écrites à un docteur de Sorbonne, à l'occasion de la comète qui parut au mois de décembre 1680*. For the next decade Bayle concentrated on other projects: the *Nouvelles de la République des Lettres* (1685–87), a series of polemical texts, and his teaching. In 1692 he published the prospectus for what would become his most enduring work, the *Dictionnaire historique et critique*, which would come out in 1696 (dated 1697) with Leers.[21] In March 1694, Bayle revisited the comet in response to polemical accusations by Jurieu, with *Addition aux Pensées diverses sur les comètes, ou Réponse à un libelle intitulé, Courte revüe des maximes de morale*

& des principes de religion de l'auteur des Pensées diverses sur les comètes, &c. This *Addition* was republished along with a third edition of *Pensées diverses* in 1699. In 1704 Bayle published a fourth edition of *Pensées diverses* and a few months later responded to critics in his *Continuation des pensées sur la comète*. Bayle died two years later from pneumonia at the age of fifty-nine.

Republications of Bayle's *Pensées diverses* continued into the middle of the eighteenth century in Trévoux and Amsterdam. In 1708 fellow Huguenot refugee Pierre Des Maizeaux (or Desmaizeaux; 1666–1745) released a posthumous English translation, published by John Morphew under the title *Miscellaneous Reflections, Occasion'd by the Comet Which Appear'd in December 1680. Chiefly to Explode Popular Superstitions. Written to a Doctor of the Sorbon, by Mr. Bayle. Translated from the French. To Which I Added the Author's Life.*[22] A German translation was published in Hamburg in 1741. After significant interest in the first three-quarters of a century following its publication, it was not until the twentieth century that a modern edition of *Pensées diverses sur la comète* was published in France, and not until the twenty-first that modern editions appeared in English and Dutch.[23]

To modern readers Bayle's *Pensées diverses sur la comète* can seem confounding, diffuse, and unfamiliar. Before turning to my reading of the prefaces, I want to situate Bayle as a reader attentive to the development of literary forms, someone who participated in conversations about the way things are written and how different kinds of writing can provoke different ways of thinking about things. Because Bayle was a serious and erudite figure, a never-married intellectual engaged in theological debates and formidably erudite scholarly endeavors, scholarship has attended less to his engagements with fiction. Yet reading his letters, it becomes clear that fiction is something he is keenly interested in and very actively seeking out.

"Dying of impatience to read it": Bayle as a Reader of Novels and Literary Theory

Bayle is a famously cryptic thinker whose writings explored some of his era's most pressing and toxic ideological issues. But Bayle was more apt to raise difficult questions than to resolve them. Scholars seeking the key to Bayle's own philosophical or theological orientations often scrutinize his extensive correspondence. Bayle's letters were an important part of his intellectual and personal life and as such can provide insight into the attitudes he held and his place within the intellectual landscape. Jonathan Israel remarks, "He has long been a perplexing figure, yet one of immense historical importance, for no one had greater impact on the early Enlightenment. The now completed edition of the correspondence reveals Bayle to be the central figure connecting, expanding, and even at moments orchestrating the era's Republic of Letters—more so than any other single contemporary."[24] Because Bayle is circumspect, even in his letters, the comments and critiques Bayle shared with his intimate circle regarding the philosophical and religious books he read voraciously is a particularly interesting source for those trying to understand Bayle's own closely guarded convictions.[25]

Reading Bayle's correspondence reveals that Bayle was not only a reader of learned literature, however. He was also a dedicated reader of the kind of popular literary works that scholars focusing on Bayle's theological references and antecedents have tended to frame out of their discussions: prose fiction and the periodicals that offered both a promotional springboard for these works and a participatory forum in which readers could discuss them. This is a proclivity that Joan DeJean picks up on in *Tender Geographies: Women and the Origins of the Novel in France*. She cites Bayle's definition of the novel in the *Dictionnaire historique et critique* to demonstrate that in the seventeenth century a certain segment of the intellectual reading public (given voice here by Bayle) recognized women's writing as central to the prose tradition being established in France.[26] According to his contemporary translator and biographer Des Maizeaux, Bayle kept a journal called "Jugements ou journal de littérature."[27] While this journal has been lost, it echoes in the astonishing density of citations in his *Dictionnaire*, where entries include numerous citations from published books and especially from the periodical press.[28]

Bayle's letters written in the years leading up to the publication of *Pensées diverses sur la comète* reveal his deep engagement with literature. After leaving home for Geneva, Bayle was never able to visit his family in southwest France again, but he maintained a close epistolary relationship with them.[29] Books and letters seem to be the two major expenses in Bayle's budget.[30] Letters sent home almost always prominently include discussion of books, as Bayle himself reflects in the closing of a letter to his brother Jacob in 1671: "I will not say goodbye without having yet spoken to you about some books."[31]

Bayle wrote constantly to his father, brothers, and friends about what he was reading and titles he was hoping to procure.[32] His letters testify to what McKenna calls "the obsessive attention with which he follows *l'actualité littéraire*"—literary news or literary innovation.[33] During his sojourn in Switzerland, Bayle became part of a close-knit group of friends, passionate readers of everything new, and after leaving Geneva for a succession of private teaching jobs and then positions in Sedan and Rotterdam, he continued to converse with them in letters, sharing news of books he had heard about and opinions on those he had read himself.[34]

Bayle's letters share information and titles gleaned from book reviews he had seen in the periodicals he also closely followed—everything from the *Journal des sçavans* and the *Mercure galant* to other new journals his friends passed on to him.[35] Bayle, as McKenna notes, goes so far as to copy out passages of the *Journals des sçavans* summarizing newly printed books in letters to his family. "The cultural obsession is so significant that even in the letters where he expresses his sincere and profound grief in learning about his mother's death in June 1675, emotion does not prevent him from continuing with literary and political news."[36]

In the years he began working on his first publication, *Pensées diverses sur la comète*, Bayle's letters testify to a serious interest in new genres of prose fiction that were sparking fierce debate among French-language readers. Bayle's correspondence mentions works of fiction by Catherine Bernard (1662–1712), Jean de La Fontaine (1621–95), Anne de La Roche-Guilhem (1644–1707), Madeleine de Scudéry, and Marie-Catherine Desjardins, known as Madame de Villedieu (1640–83), among many others. For example, in a letter to his brother Jacob in September 1671, Bayle describes Villedieu in enthusiastic terms, detailing her latest books and remarking that she "regularly publishes works of admirable elegance." Bayle praises the very recent publication of Villedieu's *Les amours des grands hommes*, a collection of novellas that imagine important figures from classical antiquity as protagonists in counterfactual love stories: "The best part is that alongside the fictions of her invention, she retains the most remarkable elements of these figures' histories/stories, which is doubly useful to the reader."[37] Bayle highlights equally Villedieu's mastery of historical detail and her fictional invention, which he calls "les fictions de son esprit," and he informs his brother of the author's other publications, two volumes of *Annales galantes* and a *Journal amoureux*, appearing in installments, as was common for longer, episodic texts. Villedieu's novellas offer "une double utilité au lecteur," according to Bayle, because in addition to depicting important historical figures, they at the same time imaginatively plumb the psychological depths of various highly charged amorous scenarios.[38]

Bayle's *Dictionnaire historique et critique*, written at the end of his career, echoes this favorable appraisal of Villedieu. The language he uses in this description is so full of key terms from seventeenth-century culture that I am going to fully cite the original French:

> She did not hide her talent; on the contrary, having created a new fashion for Novelistic Narrations, she published a great number of them and succeeded admirably in it. She brought into fashion these little *Galant Stories*, which show the success or failure of affection, and replaced the vast, long books of Adventures about heroes, war, and love, which used to earn so much money for the Printers of Cassandra, Cleopatra, Cyrus and Clélie, etc. The new taste she created persists and even if this type of Work quickly loses the delight of novelty, we still read with Pleasure the first Novels she composed following her novel concept.
>
> Elle n'enfouit son talent; car au contraire s'étant fait un nouveau goût de Narrations Romanesques, elle en publia un fort grand nombre, et y réussit très-heureusement. Elle mit à la mode ces petites *Historiettes Galantes*, qui font voir bientôt le mauvais ou le bon succès de la tendresse, et fit tomber ces longs et vastes récits d'Aventures héroïques, guerrières, et amoureuses, qui avaient fait gagner tant d'argent aux Imprimeurs de Cassandre, de Cléopâtre, de Cyrus, et de Clélie, etc. Le nouveau goût qu'elle créa subsiste encore: et quoi que cette espèce d'Ouvrage perde promptement la grâce de la nouveauté, on lit encore avec Plaisir les premiers Romans qu'elle composa selon sa nouvelle idée.[39]

Twice repeating the phrase *nouveau goût*, Bayle cites Villedieu as a literary innovator, a writer who has turned the page on the long multivolume works of her predecessors (with a nod to Madeleine de Scudéry's *Cyrus* and *Clélie*) to establish and popularize a new style that "elle mit à la mode." Bayle emphasizes the author's talent and successes, and the pleasure readers continue to take in this style that he lavishly credits her for ("selon sa nouvelle idée"). Bayle's sense of literary history anticipates late twentieth-century scholarship that rehabilitated the *galant* esthetic and Villedieu and other women writers of the late seventeenth century as important avatars of the modern French novel.[40] The authors Bayle cites championed the modern worldview, valuing progress and the individual over tradition and hierarchy, and participated in salon culture and an esthetic style called *galant* that diverged from the esthetics and politics of classicism and the values of hierarchy, patriarchy, and tradition promoted by the Ancients.[41] We can see some of the key words in his description of Villedieu: *tendresse*, *grâce*, *plaisir*, and novelty in all of its forms—*nouveau*, *nouveauté*, *nouvelle*. According to Faith Beasley, this new literature translated "the values associated with the salon milieu and especially the women writers working within this milieu—*bon sens*, what is '*naturel*,' *vraisemblance*, authorial invention and liberty."[42]

In her 1939 Columbia dissertation, "The Early Reading of Pierre Bayle," Ruth Elizabeth Cowdrick, writing a half century before Beasley, Micheline Cuénin, DeJean, Myriam Dufour-Maître and others reclaimed women writers and placed their fictions in the history of the French novel, contextualizes and apologizes for Bayle's reading proclivities, which tend to prose more often than to drama and veer from what at that time still obtained as the canons of French literary classicism.[43]

> He was well informed on almost all branches of current literary production and . . . he saw the literature of the classic period more in its totality than we do today. He was moderate and tolerant in his judgments and could see some benefit to be derived from reading mediocre or even poor books. Ideas had great appeal for the thoughtful Bayle, whatever their source. Neither in practice nor in theory did he adhere to the classic ideal in literature; his preference was for a spontaneous and imaginative literary style.[44]

When Cowdrick was writing her dissertation at Columbia in the 1930s, the "classical ideal in literature" established more than three centuries earlier by Louis XIV's royal culture machine, and reinforced by successive centuries of disciplinary and esthetic norms, still endured. Even as she goes far to accommodate Bayle's taste for books outside the recognized canon of seventeenth-century greatness ("mediocre or even poor books"), she recognizes "his preference . . . for a spontaneous and imaginative literary style." The terms that Cowdrick employs are the very vocabulary we now appreciate as the hallmarks of the *galant*, often salon-associated, fiction that posed a kind of esthetic counterweight to the official regime of classicism, with its emphasis on singularity, order, discipline, and strict adherence to esthetic rules.[45] And it is this type of fiction that I contend provided Bayle the literary tools to construct his refutation of superstition in *Pensées diverses sur la comète*.

In addition to the contemporary novels Bayle read and discussed with his epistolary interlocutors, his correspondence also tracks critical texts that conveyed to far-flung readers the debates about the merits and defects of this new kind of literature that was animating Parisian salons.[46] The debates taking place in salons and even sometimes at court gained traction with a geographically diffuse reading public as they were disseminated in letters and in print. These debates were not "merely" esthetic debates but rather vied to define which forms and modes of expression were appropriate, allowable, and valuable. The conflict over literature—how to represent human perception, thoughts, and emotions—opened onto questions of what human experience, agency, and community could, should, or might look like, as Louis XIV and his administration tried to redefine France and the relationship between sovereign and subjects.[47]

Bayle's letters show a sustained interest in these critical conversations. His letters mention Dominique Bouhours (1628–1702), René Rapin (1621–87), and Huet, whose *Traité sur l'origine des romans* was published as the preface to *Zayde: Histoire espagnole*, a novel by Marie-Madeleine Pioche de Lavergne, comtesse de Lafayette (1634–93) but published under the name of the author's friend and collaborator Jean Regnault de Segrais (1624–1701).[48] He writes to Minutoli in October 1672, desperate to get a copy of Huet's treatise, writing in passionate terms about his desire to read the essay: "People say such positive things about this piece, and mention so many excellent things it contains that I am vexed at myself at not having read it yet, and I am dying of impatience to read it [je meurs d'impatience de le lire]. As I know you have it, Sir, I take the liberty of asking you for it and assure you I will devour it with such hunger [je le lirai avec une avidité si dévorante] that I will be ready the very next day to send it back to your study."[49] By the following summer, Bayle has gotten his hands on the treatise and writes of it admiringly in a long letter to his father. "I recently read a letter from Mr Huet to Mr de Segrais about the origin of novels, which really is the most *savant* letter anyone can read. . . . This piece is thoroughly fascinating."[50] Judging the essay "tout à fait curieuse" and "bien la plus savante qui se puisse lire," Bayle expresses serious regard for the literary discussion that he was so desperate to access nine months prior.

Bayle's appreciation of Huet's treatise remains short on details about the text that he praises so fulsomely to his father. However, we will see in Bayle's various prefaces to the *Pensées diverses* echoes of Huet's meditation on verisimilitude and the pedagogical power of novelistic fictionality.

Fictionalized from the Start: The Complicated Story of Pierre Bayle's Comet

Bayle's letters attest to the passionate interest he took in prose fiction and in theoretical discussions about it. Although his praise of the books he lauds lacks detail, we can discern Bayle's engagement with the emerging French novel in the way he stages his own first publication, the theological-philosophical critique that would become a controversial bestseller just like the books Bayle followed in the pages of the periodical press.

Bayle's *Pensées diverses sur la comète* draws on a range of narrative techniques developing in the literary writing of his contemporaries. The most obvious way Bayle's text engages with fiction is through the conceit that Bayle uses to explain its existence, which would eventually become an important part of the book's evolving title. In the prefaces to various editions of *Pensées diverses sur la comète*, Bayle tells related origin stories for his book. Most often cited is his claim, from the first edition, that he wrote it after having been beset by panicked inquiries about what the comet might portend. He maintains this fiction even after he was unmasked as the text's author, adding to it the assertion that he wrote in the voice of a Catholic, and began the work as a letter, which he planned to submit to the *Mercure galant*. Des Maizeaux repeats these stories in *The Life of Mr. Bayle*, an expanded version of the biographical sketch published with his 1708 English translation and later as an independent volume. "As People had an extraordinary Opinion of his Knowledge and Judgment, he found himself continually teased with Questions concerning it, proceeding either from Curiosity or Fear. He did all he could to quiet those who were disturbed at this pretended evil Presage, but gained very little on 'em by Philosophical Reasonings. Still they replied that God exposed these strange Phenomena o'purpose to give Sinners time to prevent their impending Ruin by Repentance."[51] A French version of this expanded *Vie de Mr. Bayle*, published in 1730 appended to Bayle's published works before eventually appearing in 1732 as a stand-alone volume, fills in precise dates for the book's composition ("Il commença à y travailler le 11 de Janvier de l'année 1681") to make clear that the dates on the letters reflect the text's actual composition and that *Lettre à M.L.A.D.C.* was completed during the period of the comet's appearance and before Bayle's departure for Rotterdam.[52]

Although perpetuated by Bayle and his admirers even after Bayle's death, this story that the titular comet of *Pensées diverses sur la comète* is either the sole subject of or inspiration for Bayle's book is almost universally disputed by scholars. Scholars note that Bayle's treatise meditates on philosophical and theological issues that had preoccupied him in the preceding months and years. "The passage of the comet is nothing but a pretext," according to Hubert Bost.[53] In their introduction to a modern edition of *Pensées diverses*, Joyce Bost and Hubert Bost note, following Élisabeth Labrousse's emphasis on Bayle's Protestant perspective, that Bayle was raised in the Calvinist tradition that condemned astrology as one of many forms of idolatry, Catholicism principal among them.[54]

Bayle's correspondence roundly confirms that the book that would become known as *Pensées diverses sur la comète* was not strictly "occasioned" by the comet (to use the language of the work's English title) and that the book's staging is fictional. Well before the comet of 1680–81, Bayle's letters betray his interest in the themes that he would pursue in *Pensées diverses*. In 1677 Bayle wrote to his brother Joseph about a treatise on comets published in 1665 by Gabriel Vandages de Malapeyre, and two years later he discussed two recently published critiques of superstition and oracular thinking in a letter to a friend in Geneva.[55]

The comet that did not in fact inspire Bayle's book provides our first insight into the important role fictionality plays in *Pensées diverses sur la comète*. The comet's appearance offered a precipitating event and also a kind of narrative flashpoint around which to consider thorny abstract questions: How, when, and where can human beings perceive the divine? What are the causes and inspirations of human ethical behavior? Where can people find the foundations for religious faith? Des Maizeaux's biography highlights the way the comet offered a dramatic example, which Bayle felt he needed to demolish what he saw as a pervasive system of idolatrous beliefs: "He thought t'would be in vain to reason any longer, unless he could produce some Argument to shew that the Divine Attributes permitted not the destining Comets to such an end. He applied himself to meditate upon it, and quickly lit on this Theological Argument, *That if Comets were Presages of Evil, God had worked Miracles to confirm Idolatry*."[56] The comet offered Bayle the kind of narrative focus described by Huet as necessary to the emerging novel: "They must be written with art, and according to certain rules, without which it would be a confused jumble, without order or beauty" ("Il faut qu'elles soient écrites avec art, et sous certaines règles, sans cela ce sera un amas confus, sans ordre et sans beauté.")[57] Largely repeating the stories Bayle tells in his prefaces (which we will soon examine in detail), Des Maizeaux portrays Bayle reaching beyond the limits of reasoning ("il serait très inutile de raisonner davantage") to arrive at a kind of spectacular demonstration. This vivid argument or "argument qui fit voir" could refer both to the comet as a kind of vivid and attention-grabbing example, as well as to Bayle's argument about it. The phrase, moreover, aptly describes the power of fiction, such as the fiction of the book's conceit, to focus and spark the imagination in order to make certain kinds of ideas, feelings, and emotions visible to human imagination.

In his *Entretiens sur la pluralité des mondes*, published in 1686, a few years after Bayle's own bestseller, Fontenelle employs a fictional frame and literary modes (a nighttime conversation between a man and a lovely marquise, in which he repeatedly asks her to picture various scenes) to help readers imagine cosmology. "It is precisely by means of the *Entretiens*' practices of fictionality that Fontenelle attempts to move scientific inquiry into unexplored territories," Juliette Cherbuliez asserts.[58] Bayle's *Pensées diverses* likewise uses the practices of fictionality to move human understanding into unexplored territories. We might expect a book about a comet to pursue scientific, observational knowledge regarding the circulation of celestial bodies as Fontenelle did. However, Bayle's *Pensées diverses* uses human attention on the sky as a means to help readers ponder the relationship of humans to each other, their world, and the divine. In other words, Bayle uses fictionality to break new ground in ethics and theology.

In his reading of Huet's *Traité de l'origine des romans*, Bayle encountered the proposition that the pleasure offered by reading fiction entices the mind to master difficult concepts. For Huet, imagination "works more easily . . . than reason [*entendement*], whose workings are more laborious"; the mind must be enticed to learn via reason "motivated by benefit, or hoping for a distant Pleasure, or by necessity."[59] In contrast, "the knowledge that attracts and flatters it [*l'âme*] is the kind it acquires easily, where the imagination acts almost alone, and on things similar to those that lie normally within our senses, and particularly if this knowledge excites our passions, which are the greatest motivators of our lives' actions. This is what novels do."[60] For Huet, novels facilitate knowledge and fiction's function is ultimately epistemological. Novels, he explains, bring *connaissances* easily, stoked by the passions, through the pleasurable evocation of realistic and familiar objects of the phenomenal realm ("matières semblables à celles qui tombent d'ordinaire sous nos sens") and through the faculty of the imagination.[61] The fictionality of these novels, Huet insists, does not engender falsehood, but, in the best examples, is "ingenious, mysterious, instructive" ("ingénieuse, mystérieuse, et instructive"), that is, intellectually productive.[62] Indeed, he insists that the mode of apprehension offered by novels may even reveal higher truths: "And Saint Augustine said somewhere that these meaningful falsehoods [*ces faussetés qui sont significatives*], covering a hidden meaning, are not lies, but figures of truth, which the sagest and holiest people, and even our Lord, employed."[63]

Bayle's correspondence reveals that he regarded fictionality and the *Pensées diverses*'s fictional conceit in much the same terms as Huet. The fictional for Bayle represented a crucial tool to engage readers in the innovative thinking his book proposed. In a letter to his brother Jacob in January 1684, following the public attribution of the anonymously published text, Bayle shows a particular concern with maintaining the fiction of the book's genesis:

> I should admit that all of the details about places, conversations, people named and designated are little tricks to entertain the readers even more. I tell you this in confidence for to keep the reader more agreeably engaged, one has to let him think that the circumstances are true, which is so necessary that people would have almost no pleasure in reading novels if well-guarded verisimilitude did not prevent the reader from reflecting on the fact that it is pure fiction [*des pures fables*]. In other words, one has to persuade oneself that the adventures really happened, otherwise one is not moved.[64]

Bayle perceives his book's ability to influence readers to think in new ways as depending on the same understanding of fictionality Huet somewhat disingenuously attributes to Aquinas: "Ces faussetés qui sont significatives, en enveloppent un sens cache, ne pas des mensonges, mais des figures de la verité." Rather than focusing his writerly commentary, for example, on the need to construct an impeccable logical argument, he evinces concern for the reader's experience and frame of mind. He turns to the novel as a model of the kind of engagement he sought to cultivate with the fictional framing of *Pensées diverses*. Like a novel, he explains, his book succeeds through pleasure, affect, and immersive imaginative projection, deploying characters, conversations, and settings ("des lieux, des entretiens, des personnages"). He calls these elements "des petites adresses pour divertir davantage les lecteurs" constructed "pour tenir le lecteur plus agréablement attaché." Bayle meditates on the pact and pleasures that engage readers of novels with the fictional universe of the text in order to help them imagine and motivate them to invest in what they are reading: "Il faut lui laisser penser que les circonstances sont véritables." This readerly investment in the novel has epistemological stakes in the way it extends the fictional illusion. Bayle draws on Huet's concept of novelistic fictionality to conceive a mode of writing to not only express novel ethical and theological ideas but also to frame his text in a way that he believes will help readers imagine their ways into new conceptions of ethics and the divine. The phrase "le vraisemblable bien gardé" draws a further clear line from Bayle's thinking about his *Pensées diverses sur la comète* to the intense debate around novels that Bayle was following. The critical debate continued in the wake of the 1678 publication of an anonymous literary bestseller.

Huet's treatise, which Bayle was "dying of impatience to read" in 1672, was published as a preface to Lafayette's *Zayde*. During the very period when, according to Bost and others, Bayle began contemplating the ideas he would explore in his *Pensées diverses sur la comète*, Bayle's letters demonstrate that he was closely following the literary news as French readers were clamoring over another work of fiction written by Lafayette and also not published under her name: *La Princesse de Clèves* (1678), a work often hailed as the first modern novel of the French tradition.[65] The active critical debates the work fomented centered around the concept of *vraisemblance*, that is, what various readers deemed realistic in the eponymous heroine's decisions and actions.[66] Bayle mentions Lafayette's controversial bestseller in at least five letters from the year of its publication, first noting in April 1678 that he has read about the recently published book in the *Journal des sçavans*.[67] Through the summer and fall, we can see him closely follow the lively exchange between defenders and critics of the text that was taking place in salons across Europe, in dueling critical publications, and in the pages of the *Mercure galant*.[68] In the postscript to a short letter dated December 15, 1678, he writes to Minutoli appreciatively of several new works of fiction, including a historical novel ("une espèce de roman") he has just read, as well as the much talked about *La Princesse de Clèves* and the critical conversation around it.[69] The next month he writes his older brother, "When I find a Novel like *La Princesse de Clèves* and *Le Prince de Condé*, or a historical novella like *Dom Carlos*, *Comte de Ulfelt*, or *Memoirs from Holland* (these last two works are rather new), I read them with pleasure."[70]

When Bayle writes to Minutoli in mid-December 1678 to share his appreciation of *La Princesse de Clèves*, he emphasizes not only that he has read this new novel but also that he was closely following the critical debate around the text from his provincial post in Sedan. He has read the anonymous *Lettres à Mme la marquise *** sur le sujet de la Princesse de Clèves* published in Paris by Sébastien Marbre-Cramoisy.[71] In mid-December 1678, Bayle has already gotten wind of the impending publication of *Conversations sur la critique de La Princesse de Clèves* by the abbé Jean-Antoine de Charnes (1641–1728), which would be published, also anonymously, by Claude Barbin in February 1679.[72]

Huet's and Jean Baptiste Henri Du Trousset de Valincour's critiques were part of a lively debate, also waged in the pages of the *Mercure galant*, about the potentialities and perceived flaws of new kinds of prose fiction. Critics were especially concerned about the behavior of the title character of *La Princesse de Clèves*, whether a confession the princess makes to her husband about a secret and unconsummated passion is *vraisemblable* or not. For Beasley, this critical debate taking place in the realm of late seventeenth-century literature reveals a literary and social world in transformation, and a wholesale reconsideration of what it means to read literature.[73] "As Charnes states, *La Princesse de Clèves* was composed for 'les honnêtes gens et les gens raisonnables' (cultivated and reasonable people)."[74] For Bayle, whose book replaces the kind of God who might speak to people in signs and wonders with a theology firmly rooted in the perspectives and intellectual and ethical agency of individuals and communities, these types of questions model modes of discovery and authority made possible through literary discernment, imagination, and community-constituting dialogue and critical debate.

In his January 1684 letter to Jacob, Bayle casts his *Pensées diverses* in terms that evoke the critical debate over *Clèves*: "This [recourse to fiction] occurs several places in the letter [*Pensées diverses*], and saves people from the bother of having to imagine a hundred different things, and having to seek out paths to reflection [*voies d'application*]. There is no other mystery in all of that than the intent to vary the subjects and present a more beautiful spectacle to the reader's eyes."[75] Bayle again centers the reader's experience in terms that are very evocative of Huet. In crafting his theological opus, he explains to his brother that he is concerned above all about how readers will experience the book, revealing a theory of mind that valorizes the way the fictional can inspire thoughts and ideas and how vivid fictional scenarios can galvanize reflection, offer new perspectives, and change points of view. He crafts his *Pensées diverses* to be varied and a "beau spectacle," in accordance with the esthetics of *galant* literature. The modern discourse that can supplant the logic of superstition, which is false, corrosive, and open to abusive power, and even more radically, the most apt discourse for a modern *theological* critique (to use the word Bayle uses in the "letter" and its prefaces) is the discourse of the novel. While so many of his contemporaries labored to accommodate the comet's orbit to rules regulating the functioning of a clockwork universe, Bayle eschews a unified rhetoric of rationality. He turns to literature to highlight the social inscription of new regimes of knowledge, a heteroglosia susceptible to being framed out of more purely—or narrowly—scientific or rational discourse.

This recourse to fiction and fictionality as a kind of provisory epistemology is visible elsewhere in Bayle's letters to his brother, which offer some of his most candid declarations of his ideas. In a May 1681 letter to Jacob, Pierre Bayle professes that he regards the philosophical systems of his day not as potentially true explanations of the lived world, but rather as heuristic narratives, more or less apposite, more or less pleasing, for various situations and states of mind. He asks Jacob and another friend to assure potential employers that he judges Cartisianism to be "but an ingenious hypothesis" ("simplement comme une hypothèse ingénieuse") and asserts that "the more I study philosophy, the more I find uncertainty."[76] This letter, written ostensibly during the time Bayle was drafting his *Pensées diverses sur la comète*, concludes, "I am a philosopher without obstinacy who considers Aristotle, Epicurious, Descartes like inventors of conjectures, which one follows or abandons depending on whether one is looking for this or that amusement of the mind [*esprit*]." Bayle subverts the traditional hierarchy of textual authority as he recasts philosophers as authors of conjectures that amuse the mind—much like inventors of fictions—while he empowers readers to voyage among them according to their own textual and imaginative pleasure.

Pensées diverses as an Epistolary Novel

Bayle's preface to the first published edition of his comet book, entitled *Lettre à M.L.A.D.C.*, employs the standard narrative staging and vocabulary of one of the novelistic subgenres that was becoming popular in late seventeenth-century France: the epistolary novel. The title, *Lettre à M.L.A.D.C. Docteur de Sorbonne*, draws attention to the text's epistolarity, and the preface, likewise, presents the entire text as a found letter. The book opens with an "Avis au lecteur" addressed to the reader in the first-person voice of a (fictional) editor. "It would be useless for me to explain how this letter fell into my hands," writes the "editor." "I will only say that after having read it with much attention, I thought that it was not undeserving of the public's curiosity, and that people would find in it something ineffably [*je ne sais quoi*] novel that would be powerfully apt for entirely disabusing those who persist in imagining that comets presage great misfortunes" ("Il me serait inutile d'exposer comment cette lettre m'est tombée entre les mains. . . . Je dirai seulement qu'après l'avoir lue avec beaucoup d'attention, j'ai cru qu'elle n'était pas indigne de la curiosité du public, et qu'on y trouverait je ne sais quoi de nouveau qui serait fort propre à désabuser entièrement ceux qui persistent à s'imaginer que les comètes présagent de grands malheurs").[77] This preface is printed in a roomy italic font meant to evoke handwriting and thus provide an intimate personal connection between the reader addressed in the page's header and the editor, who is made present by the repeated use of the first-person pronoun. The phrase "m'est tombée entre les mains" at the close of the first sentence conjures the materiality of a stack of manuscript pages and the hand of the editor who holds them, who in turn writes to address us in particularly human and embodied terms.

The conceit of the mysterious found letter, whose origins remain secret but which promises to pique our curiosity and show us something new—("j'ai cru qu'elle n'était pas indigne de la curiosité du public, et qu'on y trouverait je ne sais quoi de nouveau")—follows the formula of the epistolary novel that was becoming increasingly popular in the later seventeenth century.[78] For example, the anonymous bestseller, *Lettres portugaises traduites en françois*, published by Barbin in 1669, spawned much debate about whether the text was a real collection of letters, written by a heartbroken Portuguese nun to her faithless lover, or whether it was a novel. Valincour's critique of Lafayette was likewise styled in letter form as *Lettres à Mme la marquise* ***. Bayle's anonymously published letter-styled text is a philosophical-theological treatise rather than a love story or a critique of one. However, appealing to "the public's curiosity" and claiming that the letters contain "je ne sais quoi de nouveau," in words that echo the 1674 translation by Nicolas Boileau (1636–1711) of Longinus's *Traité du sublime* and also a subtle pun on the name of the *nouvelle* genre, the preface's "editor" describes the missives in terms that for contemporary readers would evoke the new kinds fictional texts Bayle discussed with his family and friends. These novelistic tropes of wonder that the editor uses to characterize the text could just as easily be used to describe the book's subject (or ostensible subject)—the comet—as the letters about it. In this way, the preface to the first edition of Bayle's *Pensées diverses* presents both the comet and the text (the "diverse thoughts") within the frame of a contemporary corpus of prose fiction.

The preface to *Lettre à M.L.A.D.C.* demonstrates the intimate connection between Bayle's investments in epistolarity and in literary style. For Bayle, a voracious reader and an engaged correspondent, epistolarity was a place to discover and discuss literature as well as a place to cultivate literary style that he would employ in the many and diverse texts he wrote and published from his refuge in Rotterdam. Said another way, for Bayle, the social inscriptedness of letter writing—writing as dialogue or conversation—becomes the paradigm of writing *tout court*. In *Salons, History, and the Creation of Seventeenth-Century France*, Beasley tells the story of the novel as multifariously inscribed in the social practice of the salon. Due to the circumstances of his life—the poverty and rural setting of his roots, his exile from France as a Protestant and a "relapse," and the rigor of his studies, teaching, and publishing enterprises—Bayle never participated in this Parisian sphere. However, through his epistolary relationships, Bayle discovered and discussed the literature being produced in salons, and perhaps even more importantly, he developed the kind of voice and pursued the kind of epistemological concerns that he found in novels. These are the voice and the concerns that he brings with him to his first published book.

For Bayle, letters offered a way of staying connected with the intellectual community despite his peripatetic career, and eventually a medium for literary expression, both in the epistolary form of his first published book and, later, in the correspondence network he drew on for his *Nouvelles de la République des Lettres*.[79] Epistolarity was also the arena in which Bayle became a writer. McKenna notes that Bayle's first attempts at writing were epistolary. He refined his literary style in early correspondence with friends, asking for critique and reflecting on style: "One feels that in his witty repartee, Bayle is seeking a style and tries on various models from his era: these exchanges indeed confirm the importance of the struggle for literary preeminence between [Jean-Louis Guez de] Balzac and [Vincent] Voiture, and later [Paul] Pellisson."[80] Here McKenna flags some of the figures associated with *galant* literary style and epistolary writing as models for Bayle, although he signals only male writers, not including any of the women writers whom Bayle readily cites in letters to his friends and family, and later in his *Dictionnaire*.[81] McKenna asserts that in the years leading up to 1680, Bayle developed a style of writing influenced by the literary voices he read. "Little by little, Bayle finds the tone which suits him, the interplay between eloquent and familiar style, between philosophical abstraction and burlesque humor, all of it seasoned with agreeable irony; what one could call—and what was called at the time—'the conversation of *honnêtes gens*.' "[82] It is impossible to know whether McKenna is conscious that his characterization of Bayle's writerly voice resonates so strongly with Charnes's discussion of *La Princesse de Clèves*.

Preface as Literary Theory

Bayle's correspondence reveals that he avidly followed not only literature but also the latest writing about literature—texts where the stakes, claims, remit, and techniques of fiction were being actively battled out. These issues figure prominently in the introductory paratexts he crafted to introduce readers to his essay. In addition to providing the "effet du réel" that frames the *Lettre à M.L.A.D.C.* in the conventions of a found-letter novel, the "editor's" preface closely echoes the works of literary theory that Bayle followed with such interest. Like the books Bayle talks about, in his letters to close friends and family, of hungering for and devouring—Valincour's *Lettres à Mme la marquise****, Charnes's *Conversations sur la critique de La Princesse de Clèves*, or Huet's *Traité sur l'origine des romans*—Bayle's preface to the *Lettre à M.L.A.D.C.* examines questions such as, What form should writing take? What can writing do? How might the books we read help us conceive of novel ideas? How should we interpret what we see in the world? Where do meaning and understanding come from? In his *Pensées diverses sur la comète*, the answers that Bayle proposes to these questions very much resonate with the modes of believing and knowing associated with the nascent novel.

The fictional editor spends the major part of the preface discussing the text's stylings and its placement in a typology of genres, emphasizing its generic innovation even in the face of a surfeit of textual responses to the comet: "People have worked this subject so much, and from so many perspectives, that it did not seem possible to give it a new twist [*un nouveau tour*]."[83] The preface's fictional "editor" opens by citing a witty remark published in February 1665 by Denis de Sallo (1626–69), who had founded the revolutionizing *Journal des sçavans* just a month earlier: comets generate as many kinds of texts as there are kinds of readers.[84] From *physiciens* to *beaux esprits*, to *philosophes* to *rieurs*, to *astrologues*, to the frightened *peuple*, for each kind of person, there is a different kind of text written about the comet: "Some are written for people who like astronomy; and some are also written for people who scarcely take the care to observe the sky, and who have nevertheless curiosity for new things that happen there [les nouveautés qui s'y passent]."[85] Like fiction, or letters, comets are the ultimate news (*nouveautés*), and like the *nouvels* that they resemble, comets pique curiosity and inspire all manner of literary production by writers of all humors. Astrologers prognosticate and scientific spirits measure; *beaux esprits* "made incredible efforts to enliven the material and turn it gallantly. . . . They tried hard to amuse and instruct at the same time [de plaire et d'instruire en même temps]."[86] Even the interventions of comedy, "which boasts being the sovereign remedy for maladies of the intellect [*esprit*]," could not dispel superstitions. After all that spilled ink, "it is nonetheless true that the greatest feat still needed to be accomplished, and this is what the Author of this Dissertation attempted."[87] Neither comedy, nor astrology, nor philosophy, nor witticism, nor even scientific experimentalism can uproot superstition: "I say the same to the other huge entrenchment of popular opinion, in other words experimentation, which is often glorified."[88] Bayle's emphatically negative formulation, "that comets in no way portend any disaster" ("que les comètes ne sont point le présage d'aucun malheur"), emphasizes the kind of muscular discourse needed to definitively debunk superstition.[89]

As the editor figure posits how various kinds of writing might be employed to try to debunk superstitions about the comet, he stages a critical conversation that raises the kinds of objections that people would be apt to voice against each available genre's capacity to dispel superstition, whether the discourses of theater, philosophy, or observation. He inserts the contrary voices in roman type and then counters in the italic type of the rest of the preface how the "author" sidesteps these objections by employing a novel kind of theological reasoning together with a new kind of writing to make his argument against superstition. After making his way through the range of conceivable genres, suggesting and then discarding options from the academic to the popular, from the literary to the pedantic to the worldly, he asserts that a new kind of writing is needed to answer the fundamental question that his correspondent had addressed to him and that the comet seemed to pose to its human viewers. "I have read a good many books; but I had not yet seen one shrewdly attack popular errors at the jugular [*le jugulum causae*] and the true means to bring this controversy to an end."[90] The editor explains that the author uses theology itself to demonstrate to the scared comet watchers that their superstitious interpretation of comets as messages are incommensurable with the very essence of God. In doing so the writer turns theology in on itself, using the discourses that heretofore justified superstition to demolish it.

As the fictitious editor praises the fictionalized author, he dramatizes the process by which this new type of writing becomes, through the editor's own intervention, a new type of book. "These ways of proceeding [*manières*] inspired me to conceive a good opinion of the Author, and I easily thought that a man who could so satisfactorily find perspective and the heart of a problem, well deserved that his work be published."[91] Much like a novelist, this *auteur* works with "le point de vue" to resolve "le noeud d'une difficulté," a phrase that recalls the climax of a drama, the central moment of a plot's tension before conflicts are resolved.

Even as he signals the novelty and worthiness of the letters he is publishing, the fictional editor of *Lettre à M.L.A.D.C.* flags the fictional author's digressive style, pointing out the fluidity and diffuseness as flaws of an otherwise meritorious text. "If I had the honor of knowing him, I would have taken the liberty of giving him some advice before printing his book. I would have urged him to edit his Dissertation, to permit himself fewer diversions, to tighten his style a bit, and his thoughts, for even he recognizes that he takes a great deal of liberty because he only writes for a Friend."[92] The editor's appreciation of the work evokes many key words associated with *galant* novelistic style, with references to liberty, friendly exchange, and diversion. Here the editor emphasizes that what follows is not a professional scholarly work or a proper "Dissertation," but a conversation between friends that has by chance ended up in the hands of an editor, and thus only quite by accident become a book. The editor reminds us that his (professional) hands have had no commerce with the (entirely nonprofessional) text as he concludes that "not knowing to whom to address myself, I could not make any suggestions." This book, the editor emphasizes, has grown from the kind of collaborative process that flourished in the salons from which novels grew. Despite the editor figure's best intentions of tightening up the text, it has remained untouched by professional hands.

At first appearing strict about the mystery author's digressiveness and lack of discipline, the editor models for readers a critical posture in which his opinions and optics change and develop as a result of what he has read. As the editor reflects upon the experience of reading, his critique veers into a commentary on textual pleasure.

> In the end, I decided to publish this Letter after having carefully considered that all of the author's digressions are instructive, curious, and entertaining. . . . Apart from some geometrical souls, for whom this work was not intended, readers are not chagrined to be strolled from place to place, provided that, following the example of this author, they are instructed along the way and that they are brought back to the place from whence they had been diverted. How many intelligent people are bored by reading a work that confines their imagination, keeping it constantly focused on a single subject? Who does not fancy diversity? What greater charm than a well-turned episode?[93]

In describing the text as a kind of pleasure promenade, the preface's fictitious editor obliquely references both a commonly enjoyed society pastime and one of the early subgenres, frames, and topoi of the novelistic tradition that was emerging in France in these very years, for example, Madeleine de Scudéry's 1669 *Promenade de Versailles*, which recounts, within a promenade frame narrative, the adventures of Célanire, or Jean de La Fontaine's *Amours de Psyché et de Cupidon* of the same year, a retelling of Apuleius's myth framed by a visit to Versailles by a group of friends whose digressions into questions of style and literary technique frequently interrupt the framed tale. The editor character's description likewise accumulates a series of terms and images that associate Bayle's text with contemporaneous novelistic production. *Imagination*, *diversité*, *charme*, and *épisode* stand opposed to the *esprit géomètre*, who would stay on the circumscribed straight path.[94] The critical evaluation of the *Pensées diverses* offered by the book's fictional editor strongly evokes the *esprit de finesse* of Blaise Pascal's *Pensées*, which the fictional editor also cites at the end of the preface.[95]

The novelistic experience described by the editor character is not a formal dissertation but a promenade that allows readers to explore and find ideas detours and episodes. In this passage about digression, the editor figure furthermore provides a powerful mise-en-scène of the epistemological power of literature. He recounts that he started this text armed with a set of values and expectations. He thought he knew what thinking should look like—clear, concise, orderly—but the promenade the text has taken him on has changed his mind. This self-reflexive assessment voiced by a fictional character depicts how people come to know things through literary reading and how fiction can transform the way we value and hierarchize what we know. Here we see how the literary mode Bayle employs in *Pensées diverses sur la comète* surpasses solely esthetic or generic concerns. The novelistic voicing expresses an explicit part of his project and asserts the ways that literature and literary ways of exploring and knowing offer a political alternative to authoritarian regimes of knowledge.

In Bayle's new comet genre, detail and decoration take precedence over the central and the essential. In what Hubert Bost calls "a key to reading the work," the fictional editor of the "Avis au lecteur" evokes the beautifully detailed paintings of the ancient Greek painter Protogenes, in which ancillary details (a beautifully rendered partridge) completely overtake attention from the intended subject (a satyr).[96]

> I wonder if this work might even have a destiny similar to that of the satyr and the partridge of Protogenes. The satyr was properly speaking what the painter had in view, and the partridge but an accessory detail [*un accessoire*]; however, connoisseurs dwelt so intently on the partridge that they hardly looked at the satyr. It could well also happen that those who read this *Letter*, finding in the digressions something ineffably livelier, freer, more singular, might only care about the work because of what is beside the point.[97]
>
> Je ne sais même si cet ouvrage n'aura pas une destinée semblable à celle du satyre et de la perdrix de Protogène. Le satyre était proprement ce que le peintre avait eu en vue, la perdrix n'était qu'un accessoire: cependant, les connaisseurs s'arrêtaient si fort à la perdrix qu'ils ne regardaient presque point le satyre. Il pourra bien arriver aussi que ceux qui liront cette *Lettre*, trouvant dans les digressions je ne sais quoi de plus vif, de plus libre, de plus singulier, ne feront pas de l'ouvrage qu'à cause de ce qui y est hors d'oeuvre.

This passage recalls the language with which the marquise in Charnes's *Conversations* defended the historical frame that opens Lafayette's *Princesse de Clèves* from the often-voiced charge that it is long and disconnected from the story that follows: "On the contrary, I found it very well placed, and extremely pleasing. I saw it as a magnificent decoration" ("Au contraire, je l'ai trouvée fort bien placée et extrêmement agréable. Je l'ai regardée comme une magnifique décoration").[98]

Like Charnes's fictional marquise, Bayle's fictional editor celebrates the "accessoire" and the "hors d'oeuvre," the decorative and the pleasing as elements fundamental to the enjoyment and even the full meaning of the text and refutes critical voices that would condemn these interludes as beside the point. Here again Bayle's editor employs many of the key terms associated with fictionality. He characterizes the esthetic experience in terms that echo the novelistic ethic and esthetic as modeled by Lafayette and defended by Charnes and Huet when he gestures to "je ne sais quoi de plus vif, de plus libre, de plus singulier." Moreover, he consistently emphasizes readers' estimation of the text rather than any kind of inherent textual authority and highlights the natural and conversational style of the *Pensées diverses*. In contrast to Bayle's friend and contemporary Christiaan Huygens, whose writing, Aït-Touati demonstrates, works actively to disentangle scientific writing from fiction and the realm of the literary, Bayle's fictional editor emphasizes that this *Lettre* employs none of the scholarly citation or complicated astrological figures and charts that would please posturing "demi-savants" or discourage female readers.[99] As for "those who write about comets as astronomers . . . their books are so difficult and so full of circles and other figures that they scare those who are not part of the profession. All of these thorns were avoided in this letter, and there is hardly anything that ladies cannot rather easily understand."[100] Although simple, the *Lettre* will also appeal to *savants* with "an agreeable diversity, capable to instruct, or to touch, or to inspire new ideas."[101] This praise directly mirrors the ways critics such as Charnes characterized the new genre of the novel. It proceeds through diversity, emotion, and pleasure. It instructs, inspires, and pleases through an esthetic cultivating novelty and spontaneity, and valuing the pursuit of new ways of thinking.

The editor describes the textual voice of *Pensées diverses sur la comète* as candid and intimate rather than official and public. He praises "this free style people use when writing to a friend, not when people want to get published" and shares his own readerly pleasure as a model for the reader's, emphasizing the natural style that was both highly prized by his contemporaries and associated specifically with women and the emerging genre of the novel: "I was secretly thrilled to bring before the eyes of the public a work that naïvely represented its author's sentiments. It is rare to have one of this nature. Those who write in the hopes of publishing their thoughts accommodate themselves to the times and betray in a thousand ways their own judgement of things."[102] The editor reads as a private individual, experiencing a kind of pleasure that is personal and authentic ("une secrete joie"), as is the "rare" human honesty of the writer, who remains an anonymous individual, not a public figure wishing to garner professional esteem or royal sponsorship. This private honesty contrasts with the venal interests of author sycophants, who a thousand times over "betray" their own discernment.

Drawing on a citation from Pascal, the editor persona lands a final blow on the false and inflated style employed by "authors who had pensions from the State or who worked to have them, and who had published several beautiful panegyrics to the government and ministers."[103] From the start of Louis XIV's reign the crown consolidated control over every aspect of cultural production, including the writing of literature and, through an active program of sponsorship, encouraged certain styles, modes, and genres of writing very much to the exclusion of the novel and the salon-centered world with which it was associated.[104] With an approving nod to the collection of lapidary meditations published posthumously in 1669 as *Pensées de M. Pascal sur la religion, et sur quelques autres sujets*, Bayle's fictional editor heaps scorn on writers of officially sanctioned literature and positions the anonymous creator of the *Lettre* in opposition to their pompous and self-serving insincerity. While writers in the employ of the court are denatured or dehumanized by their fealty to their royal employer ("instead of finding a man as one expected, one was very surprised to find an author"), the editor asserts that the writer of this *Lettre* is just the opposite. "This is why, for the rarity of this fact, I did not want to miss out on the opportunity to publish a book where people talk like they think."[105] Bayle abdicates the position of the author and the authority it imposes on a text. Despite his own very visible professional and commercial interests, the fictional editor has come round to celebrating literature as a means by which humans, not state-sanctioned authors, converse and think together. In fact, although the editor offers the *Lettre* to the public, he wants to demonstrate that this book is not one, written as it was to travel in an epistolary exchange rather than a market economy.[106] Here especially we see the multivalent implications of Bayle's politics of novelistic literariness. The fictionalized voicing of his preface depicts and celebrates literary exploration, the exchange of ideas, and free roaming digressions, wherever thoughts take one, in dialogue with other free, independent people. In this way, it sketches out an esthetic, epistemological, and political alterative to centralized power and authority.

Talking, Speaking, Writing in the 1683 *Pensées diverses*

In September 1683, only a year after the first edition was published as *Lettre à M.L.A.D.C.*, Leers republished a two-volume duodecimo edition of Bayle's book under the title *Pensées diverses, écrites à un docteur de Sorbonne, à l'occasion de la comète qui parut au mois de decembre 1680.* This edition includes a new introductory text in which yet another fictional intermediary, here the "bookseller," addresses the reader. With this staging of multiple liminal voices, Bayle models his text's openness to multiple readers and readings. These paratexts also displace authority away from the "author" to instead authorize readers. Dated September 1683, a section entitled "Le libraire au lecteur" highlights the book's construction in terms that privilege digression, pleasure, and accessibility. The *libraire* highlights "the care that was taken to divide the Work into many more Sections than it was before in hopes that readers . . . will be able to rest where they wish and start in where they wish."[107] Unlike an academic syllogism, this book is designed to let readers dip in at will. Likewise, the Latin citations have been translated into French to expand the book's accessibility to the literate nonspecialist public. The bookseller uses the same term Charnes used to designate the reading audience for novels: "In this way, we believe we have made the work more agreeable to an infinity of *honnêtes gens* and witty people [*personnes d'esprit*]."[108] It is not possible to determine the gender of the readers Bayle's libraire designates with the feminine-gendered nominal phrase *personnes d'esprit*, but what is clear is that this formulation does not rule out female readers for the *Pensées diverses* but rather seems to welcome them, just as the translation of Latin passages into French makes the book more accessible to women, a majority of whom did not read Latin.

The attention Bayle pays to his readers' experience of the text through the various fictional intermediaries of his prefaces seems to have garnered enthusiastic reactions from contemporary readers. Writing to Bayle in 1686, a fellow Huguenot refugee named Hullin writes from The Hague to praise the *Pensées diverses* in terms that focus principally on Bayle's writing and the literary quality of the text: "I read your *Pensées diverses* with such concentrated application and attention, so many times, in so many different states of mind that this one volume occupied me for more than an entire month. . . . I praise and admire the purity, elegance, the facility; the abundance of style and the stylistic ornaments."[109] Hullin signals his absorption in the text and that he was able to engage in recursive and multiperspectival reading.

This is a book for reading and rereading, a book that prompts the reader to consider things from a multitude of angles and states of mind. The quality that Roland Barthes called *plaisir du texte* appears integral to the way Hullin evaluates Bayle's *Pensées diverses* for the kind of reading experience it affords, "l'abondance et les agréments de style."

When Hullin describes the insight he has gained from the month concentrating on Bayle's *Pensées diverses sur la comète*, he does so in deeply human terms: "I feel that I have discovered in the book a great understanding of the world, an admirable penetration into the heart of man—the capacity and art of uncovering the specious veils he covers it with to hide from others and from himself, the secret of revealing the diverse causes that animate him and showing the true and hidden motives that make him act."[110] Bayle's reader discovers in the *Pensées diverses* not a map to discover the rational systems of the universe, but a way to see and imagine the secret hearts of people, "une pénétration admirable dans le coeur de l'homme." In other words, the treatise provides the same kind of psychological insight that might be gleaned by reading—and rereading—a novel. Hullin goes on to praise the digressions in ways that both echo the literary preface and cast Bayle's book as a far-ranging adventure story: "What an infinite variety of things it presents to the imagination! And to how many places and peoples does the author transport his reader." He concludes that the book's detractors misunderstand its actual topic to be the comet: "La comète appartient à tout le dessein du livre et non le dessein du livre à la comète." This is not simply a book about a comet, it is a cometic book.[111]

Truth and Fictions: The *Avertissement au lecteur* from the 1699 and 1704 Editions

The preface added to the third and fourth editions of *Pensées diverses sur la comète* published during Bayle's lifetime differs from those in previous editions in that the author now addresses readers without the cover of a fictional persona. Yet fiction, I would contend, still plays an important role in the conceit of the *avertissement au lecteur* included in the 1699 and 1704 editions. Even as the author addresses readers in a straightforward first-person voice to peel back (or so it seems) the various layers of fiction maintained in the previous two editions, Bayle continues to use the tools of literary practice as a fundamental element of his textual construction.

Avertissement means foreword, but it also means warning, from *avertir*, to warn, inform, or advise someone of something. In his *avertissement*, Bayle opens without apparent stylistic flourish in order to take a straightforward approach to explaining some of his readers' possible questions. To explain why he writes in the voice of a Catholic he offers "the story I am going to tell about the origin of this work."[112] In this origin story, Bayle lays out an episode from which his text arose. It was December 1680. He was a professor in Sedan, beset by inquiries about the comet, unable to convince his interlocutors ("je ne gagnais peu de choses par les raisonnements philosophiques") that the comet was not a warning from God to inspire sinners to repent.[113] Bayle mulls over how to explain to his correspondents why it would be anathema to the nature of God to send such unclear and potentially misleading messages as comets.

Although now he is writing as the avowed author of the text, Bayle's new account of the book's origin echoes the earlier prefaces in his emphasis on the work's innovation as opposed to drawing on any kind of classical authority. "I thought about it, and I soon came up with the theological argument that you can read in this work. I did not remember having read it in any book, nor having heard anyone talking about it; that made me discover a new idea, which inspired me to write a letter on this subject to be placed in the *Mercure galant*."[114] The origin story Bayle offers for his *Pensées diverses* starts with conversations, questions, and novelty, and a reference to an innovative intertext that championed all three of these things: Donneau de Visé's *Mercure galant*.

Attempting to explain why he writes from a Catholic perspective, the Protestant philosophy professor now safely ensconced in Holland, and no longer fearing for the well-being of his family in France (they have tragically died), asserts that he began the work as a letter to be included in the *Mercure galant*. Depicting the *Pensées diverses* as a part of a publication staged as an active conversation, something to be *insérée* rather than *publiée* like a book reiterates the critiques of authority leveled by Bayle's fictional avatars in the two earlier paratexts. Even though his reflections quickly exceed the scope of a magazine article, Bayle says, "I never lost sight of Mr. de Visé." He continued to aspire to the style of Donneau de Visé's publication ("imiter le langage et les éloges de M. de Visé sur les affaires d'État") and hoped that Donneau de Visé would help him make a connection to a printer and obtain a royal *privilège* from the powerful lieutenant general of the police: "I resolved to send him my Letter and to ask him to give it to his printer, and to obtain permission from M. de la La Reynie if it would suffice for the printing of my work, as it had sufficed for the printing of some other treatises on comets, or the Royal Privilege, if it had to come to that."[115]

Whether this is a true account, or whether it is, like the other frames, a fictionalized presentation, Bayle clearly wants his *Pensées diverses sur la comète* to be read with the *Mercure galant* in mind. Bayle is either fabricating the story or underlining it, but either way, he emphatically situates his theological and epistemological critique in this popular, circulating site of literary discourse, a printing phenomenon that Schuwey characterizes both as a locus for following and participating in historiography and as a "bibliothèque de la galanterie" that invited readers to participate in "a veritable paper salon" ("un veritable salon de papier").[116] By evoking the *Mercure galant* as his model of writing in his preface Bayle conjures the magazine's particularly engaging model of apostrophe. With its epistolary address and interactive features such as the popular enigmas, the *Mercure galant* positions readers as writers and writers as readers to create a dialogic cultural institution, a kind of public sphere beyond the direct authority of the state.[117]

We can see from Bayle's letters that a few years before writing his own epistolary work, he was meditating about the way addressing a text to a fictional addressee transforms the way readers engage with it. In 1678, in the course of several letters to his brothers Joseph and Jacob, Bayle had puzzled over the identity of the *Mercure galant*'s author and also "had asked at least five or six Parisians . . . whether the lady to whom he writes is not an invention, to give his *Mercure* the appearance of a letter."[118] He does not mock or speak over the female-coded readers to whom the *Mercure*'s narrator-editor addressed *nouvelles* about everything from the cut of dress styles to artistic, political, and scientific news. Rather he laments that the counterfeit editions available in his far-flung province are missing the engravings. The model of discourse and readership exemplified by the *Mercure galant* serves not just as a neutral vessel for Bayle's critique but is in fact integral for his project. When comets no longer speak to us for God and about kings, Bayle entrusts epistemology to a forum where people attempt to grapple with meanings and make sense of the world through reading, exchange, and conversation.

Just as the question of form plays an important ideological role in the astronomers' mathematical modeling of comets, so is it also the bedrock of Bayle's groundbreaking treatise. Where astronomers appealed to regularity and rules in order to demystify the notion that comets are signs, to domesticate their erratic travels and make them bend, in Auzout's words, to the "Law," Bayle eschewed the spirit of geometry, preferring to demystify by indirection, through the techniques of prose fiction. Bayle does not chart, measure, and model; rather, he creates a densely layered discursive universe that deploys digression, irony, voice, imagination—in short, novelistic address. Yet as digressive and unruly as it was, Bayle's text was also inscribed in a logic of circulation: not the mathematical circulation of the astronomers' projections, but the circulation of letters and information between worldly readers and correspondents in the Republic of Letters, of which Bayle would serve as conduit in *Nouvelles de la République des Lettres*, which he edited from the journal's inception in 1684 until 1687.

For Bayle, the appropriate place for reflection on the 1681 comet—and the differences in epistemology, theology, political theory, and cosmology that its passage focalized—is a periodical, a form of writing in temporal flux. One thinks of the apt German word for periodical, *Zeitschrift*. It is always on the move, in the mail, to be continued. And if it cannot speak to or for God, if it lacks a singular position of authoritative knowledge, it incites reactions, responses, conversations, and letters between readers. If the comet no longer represents a singular and historically specific heavenly message, but rather—as everyone from Cassini to Halley to Newton was actively trying to prove—can be understood as a predictable rule-governed body in space, the celestial body equal to the multifarious demands of modern critique is *galant* Mercury, *Mercure galant*, traveling the firmament to share *nouvelles* and questions.

Legacies of Bayle's *Pensées diverses*: A Critical Enigma and a "Literary spice dealer"

Bayle's *Pensées diverses* fascinated readers in his own day and has been enshrined in intellectual history as a decisive step in what Paul Hazard called (in a 1935 book of the same title) "La crise de la conscience européenne." "If there was one man who, more clearly and more cogently than any of his predecessors, insisted that there was no necessary connection between morals and religion, that man was Pierre Bayle. . . . It was in his *Pensées sur la comète* that, taking his time, bringing all his powers into play, with great heat, yet with great lucidity, he drew up the *Magna Charta* of the Separation."[119] Although Bayle died in 1706, the *philosophes* consider his work to be an important first articulation of Enlightenment critique. "In the 18th century . . . Bayle's admirers included Frederick the Great, Thomas Jefferson, Benjamin Franklin and Voltaire. They revered him for his defence of religious liberty and his genius for undermining conventional ideas. Voltaire said that the 'immortal' Bayle was the greatest reasoner who ever set pen to paper."[120] Jefferson included Bayle's massive encyclopedic *Dictionnaire historique et critique* in the original collection of the Library of Congress, and Diderot credits this work as an inspiration for the *Encyclopédie*.[121] *Nouvelles de la République des Lettres*, the journal Bayle edited from 1684 to February 1687, was widely lauded for forging an expansive community of Enlightenment readers and critics.[122]

If Bayle is so important, then why is he so seldom read or cited today? Bayle's legacy, and the legacy of *Pensées diverses sur la comète*, his first published book, and the one that made his international reputation, are both inextricably tied to, and often hindered by, the intricate and literary qualities of his writing. While Bayle's books have long been acknowledged to have indelibly changed the philosophical landscape of European thought, starting with *Pensées diverses*, critics and readers have often qualified their praise of his work by noting the complexity and inscrutability of Bayle's writing style. From the era of its publication on, Bayle's writing has often been highlighted, often in culinary terms, as if the style were a kind of seasoning, separate and distracting from the more essential "meat" of the argument. Unfortunately, as readers and critics separate Bayle's style from argument in an attempt to find clarity in the intricate text, dispensing with the allegedly superfluous, they make a fundamental mistake in reading. In framing out the question of literary construction, this critical practice frames out a mode of understanding that elucidates, rather than merely decorates, Bayle's *Pensées diverses sur la comète*, for Bayle's book employs literary style and literary construction as a philosophical method.

In his history of *Le siècle de Louis XIV* (1751), Voltaire recognized *Pensées diverses sur la comète* as a first and foundational attack against superstition. Voltaire celebrates the blow to superstition delivered by the *Pensées diverses* in his chapter on *les sciences* in which he follows the nascent philosophical spirit of the age, or in his words, how "the spirit of wisdom and discernment, which spread among friends and associates, gradually destroyed superstition."[123] Bayle, for Voltaire, represents one of the avatars and authors of this changing mindset: "Superstitious notions were so rooted among men, that people were frightened by a comet so late as 1680; and scarcely any one dared to combat this popular fear. . . . It was then necessary that Bayle should write against these vulgar prejudices a book famous at that time, which the progress of human reason has since rendered less shocking [*piquant*] than it was at the time."[124] Voltaire assures his eighteenth-century readers that Bayle's *Pensées diverses* was scandalous at the time of its publication because, during that era, Enlightenment had yet to dispel the errors of superstition. Even as he calls Bayle necessary, and praises him for a decisive step forward, however, Voltaire's supersessionist narrative implies that whatever was brilliant about Bayle's book existed in its own evanescent historical context but now has faded and thus become inaccessible to current readers. It is *moins piquant*, less spicy, shocking, original, or biting—or in the words of the English translator "useless."[125]

Why does Voltaire find it so hard to access the *piquant* of Bayle's *livre fameux*? The answer might be the book's writing style. Seventeenth-century readers found *Pensées diverses sur la comète* not only piquant in its attack on superstition and the abuses of monarchical and clerical power, but also spicy in its execution. In his 1708 English translation of *Pensées diverses*, for example, Des Maizeaux draws attention to the way the book's writing defies the expectations set by its title and subject matter: "So that I'm only to acquaint the Reader, that he is not from the Title of the following Book to fancy it a dry Philosophical Dissertation on the Nature of Comets; far from it: Comets are the least of Mr. *Bayle's* care, farther than as they afford him occasions of very uncommon and surprising Reflections, and of the most exquisite Satyr and Ridicule on Things and Persons."[126] For Des Maizeaux, Bayle's translator and close contemporary, the book's juiciness and complexity are an "exquisite" enticement to read, not yet the alienating, faded flavor that Voltaire bemoaned. In November 1683, the *Mercure galant* announced the publication of Bayle's *Nouvelles de la République des Lettres*, praising in particular, and again using a metaphor of seasoning, Bayle's witty editorial voice: "The author is esteemed, the work is well received; it/he is witty [*il a beaucoup de sel*], which contributes more than a little to his/its success."[127]

Bayle's work had faded to relative obscurity by the nineteenth century. Those who sought to renew popular interest in Bayle evoked anew, although now in pejorative or apologetic terms, Bayle's piquant style, and the diversity of Bayle's "uncommon and surprising Reflections," which, for the eighteenth-century translator served as an enticement to read the book. Young Hegelian philosopher Ludwig Feuerbach (1804–72) saw Bayle as a sympathetic precursor to his own attempt to rewrite theology in anthropological terms, and God as a kind of projection of collective humanity. Feuerbach thus sought in 1844 to frame Bayle as something more than the "literary spice trader [*literarische Spezereihändler*] for the Republic of Letters" and to rescue him, rather, as a solid thinker who contributed to the history of philosophy and humanity.[128] In 1896, Louis-Paul Betz claimed that "no one in the literary tradition of the last two centuries has been unfairly neglected to the same extent as Pierre Bayle."[129] More recently, Anthony Gottleib mused along the same lines: "The year 1680 is a notable one in philosophy because of an occurrence that did not amount to anything much and a book about this non-event by a philosopher of whom few have now heard."[130]

One of the reasons Bayle's *Pensées diverses sur la comète* has remained on the margins of the philosophical or literary canon of its era, despite the great success of the book in the late seventeenth and early eighteenth centuries, is the text's richness and playfulness—the very salt and spice that made Bayle's text "far from . . . a dry Philosophical Dissertation on the Nature of Comets." For example, Edmond Lacoste, writing in 1928, seems to lament Bayle's unruly diversity in the very esthetic terms that Bayle celebrates in his various prefaces. "Our author's dominant passion, from his very earliest works, is curiosity. History, anecdotes, topography, mythology, prosopography, genealogy, heraldry, these are his game. He thinks himself ill-suited to long, orderly discussions. He recognizes his fickle intellect [son instabilité d'esprit]. Neither order, nor plan, this is consequently his motto."[131]

The digressive style Bayle adopts—flaunts even—in his first publication would become his hallmark in the important works to follow. Richard Popkin observes that "Pierre Bayle, working in Rotterdam in the 1690's, was able to wander hither and yonder through the world of man's intellectual and moral thought, from the beginning of written history to yesterday's newspapers and café gossip, and could portray enough of it from A to Z to encompass all that his age had to offer."[132] For Popkin, Bayle's masterpiece is a "wondrous suite of themes and variations."[133] Yet Gottlieb laments this same writing style in a 2016 article meant to restore attention to "the forgotten figure of the Enlightenment": "One reason for Bayle's eclipse is that his ideas no longer seem novel or shocking; moreover, his writings digress uncontrollably. This is not a winning combination."[134]

As Eric Jorink observes, "The *Pensées diverses* is a fascinating, complex book, which touches many different themes. As a consequence, many, sometimes conflicting and sometimes overlapping interpretations have been offered."[135] Bayle's writing is complex, engaging, and often enigmatic. He wrote and published in circumstances and about subjects not amenable to candor, and died young, without leaving a last word or work to clarify his oeuvre. Consequently, Bayle's work has inspired an extremely disparate critical reception. Although in the last half century, a dynamic scholarly conversation has arisen around Bayle's theological philosophical writing following important books by Labrousse and Popkin, and more recently in scholarship by Bost, Gottlieb, McKenna, Gianluca Mori, Gianni Paganini, and others, the understandings of Bayle in critical work remains highly varied.[136] In *Reading Bayle*, Thomas Lennon muses, "To take just the twentieth-century literature, the suggestions are that Bayle was positivist, an atheist, a deist, a sceptic, a fideist, a Socinian, a liberal Calvinist, a conservative Calvinist, a libertine, a Judaizing Christian, a Judaeo-Christian, or even a secret Jew, a Manichean, an existentialist."[137] Unlike those seeking to untangle the multifarious and often thorny whole of Bayle's oeuvre, or Bayle's true opinions of the many controversial issues his work raised, I do not offer a new totalizing interpretation even of this particular text, the sheer copiousness and diversity of which has made it available to a range of astonishingly divergent readings. Rather, I am interested in the diversity itself, the ways Bayle deploys and crafts this diversity that has been so perplexing to philosophers, historians, and theologians. It is a cultivated result, I argue, of Bayle's deep engagement with the literary experimentation that was transforming publishing and the field of letters during the same years he was taking a wrecking ball to the period's deep-seated ethical and theological ideas.

If Feuerbach wanted to rescue Bayle for posterity from the petty exchanges of the literary spice trade, or Voltaire noted apologetically that the text's piquancy had faded with age, I have concentrated on precisely the aspects of Bayle's *Pensées diverses* that have frustrated scholars wishing to square this unruly text with a serious theological or philosophical reading. The diversity and complex voicing performed in Bayle's various peritexts to *Pensées diverses* are relevant to the way Bayle's work takes the comet both as an occasion to decenter authority and as a kind of model for doing so, against the backdrop of the increasingly tightly orchestrated centralized France. By reading Bayle's prefaces—the spaces in which he adopts various fictional personae to comment on the text—alongside contemporary theoretical texts about the novels that Bayle read and commented on to his friends and family, we have seen the ways Bayle's writing engages in literary trade and exchange and generates pleasure, interest, and stimulation not by dispelling disorder and diffuseness but rather by reveling in it.

Voltaire celebrates what he sees as Bayle's modernity and Enlightenment spirit, the way he overturns an oracular epistemology that understood the comet as a celestial sign or portent. Contra Voltaire's reading, however, the epistemological regime Bayle puts in place is not simply a precursor to Enlightenment rationalism, or a mathematician's or natural philosopher's rationalizing gaze, but rather a decidedly literary form and voice very much characteristic of the literary experimenters of his time. In other words, Bayle's *Pensées diverses sur la comète* destabilizes unitary and hegemonic regimes of knowledge-power not only by way of the critique that it advances but also via the very literary forms through which Bayle's *pensées* proceed and that he cites and signals in his prologues. The comet that Bayle hails as the inspiration for his text seemed to challenge the sun's orderly organization of a stable cosmos by appearing, brilliant, in the sky without a scrutable shape or pattern or fixed abode. What the comet does, as Bayle notes, is draw people together to marvel, debate, argue, and attempt to decipher. Like the comet, then, Bayle's *Pensées diverses*, his diverse thoughts, model a form of ethics that is digressive, dialogic, socially mediated, disruptive, and profoundly based on community.

Bayle's argument against the oracular status of comets, and against the necessity of God as guarantor of ethical community, does not concern itself in the least with observation, mapping, or the philosophical-mathematic concerns or techniques that we may now consider the hallmarks of an emerging episteme of scientific reasoning. His heralded first step into the Enlightenment has little to do with analytical rationality. Bayle's response to the comet, rather, demystifies the universe by inscribing meaning in circulation that is social and textual. For one, Bayle relies on discursive reasoning, on the question of what it means to be a sign, what it means to signify or address. Furthermore, Bayle's "thick," digressive, emphatically textual narrative—and, especially, the fictional stagings of his prefaces—stake out a new role for literature in a world in which comets have ceased to hail readers as a legible sign from God. For Bayle, ways of knowing and of establishing values are negotiated collectively and socially. Truth is not antecedent and from above, but something more like the *vraisemblance* that practitioners and theorists of the novel were exploring in a new kind of prose fiction and in new literary venues such as the *Mercure galant*. In his *Pensées diverses*, Bayle uses the comet and the novel to explore radically new relations between people and heavenly bodies. In this universe, it is something like literary circulation—the interplay of narrator and narratee, text and a reading community—that can establish ethical modes of coexistence, stabilize meaning and ways of knowing, and cement connections between people.

Cometic Conclusions

When people looked at a comet in late seventeenth-century France, they looked at something they were not able to discern as a shape. This book has been an attempt to focus our critical attention on the ways people engaged with objects they did not have the wherewithal to apprehend and how these curious, uncertain interactions ramified in texts, in other words, to seek out what we can discover when we read the French late seventeenth century according to comet logic. Attending to objects, concepts, writings, and events that seem to pull us away from the center does not converge on a single unitary thesis or conclusion; instead, multiple connections and forms emerge, some unexpected and inhabitual. People and events appear that might otherwise recede, genres that we do not call genres come to our attention, and coherent images of authors and oeuvres become refracted. Proceeding in this way also leaves loose ends, which invite further research. By way of (in)conclusion, I want to explore three examples of such loose ends.

The Comet's Swerves: *Esquives*

Like comets themselves were thought to do, the texts I examine in this book all perform a kind of *esquive*—a swerve or dodge in response to the kinds of big questions that observation or the spectacular is commonly understood to pose in this period, questions about political subjectivity, epistemology, theology, and social organization.[1] Swerving, dodging, and feinting, these comet texts respond with provisory answers rooted in aesthetic or social practice. In chapter 2, ballets attempt to render the errant path of a comet into some kind of harmonious order. One of them succeeds by subjecting comets to the sun, whereas the other stages a carnivalesque celebration of the comet's disruption. In the case of the Observatoire Royal in chapter 3, the question raised by the comet is epistemological: What can people know by looking? Yet the response offered in the form of institutions and the building itself is aesthetic: the stylistic *performance* of observational mastery rather than real functional purchase on astronomical observation. In chapter 4, the play *La comète* purports to offer a satisfactory philosophical response to the comet for theatrical spectators but ends up instead diverting attention, comet-like, to the dynamics of terrestrial forms of social, erotic, and economic circulation. In the final chapter on Bayle's reflections, the questions posed by the comet move beyond its signifying capacities toward the intertwined questions of hermeneutics and belief: If there is a God, how does God communicate? How can humans recognize what is moral? The answers to these questions in the *Pensées diverses* are indirect and literary. Rather than offering readers firm ground for discerning God's will, Bayle's text engages them in the dynamics of reading, literary community, and textual circulation.

The Comet's Path: Tracing Inscrutable Networks

Reading with comets also reorganizes our perception of the social. The stories in this book were both made possible by and extend intricate and diffuse networks of connection. Sévigné writes to friends and associates of the disgraced former royal financier Nicolas Fouquet, whose trial coincided with the first *cométomanie* traced in this book and whose disgrace sent his former printer Asseline to distant Soissons, where he printed the program of the *Ballet de la comète*. Paris streets were full of heterogenous crowds speculating about Fouquet's trial and the comet at the same time. Both phenomena spurred social mixing (of the kind, for example, depicted in the play *La comète*) and spawned mini publishing booms of highly ephemeral texts.

Fontenelle is himself a nexus of heterogenous discourses and relationships. In chapter 3 he figures as the estimable figure of secretary of the Académie royale des sciences, while in chapter 4 he appears as the junior writer of a one-act comedy with commercial aspirations. Here reading with a comet unsettles classical notions of authorship and the orderly delineation of an author's oeuvre, highlighting instead commercially oriented collaborative literary production. Numerous other webs and networks traverse the readings: the Jesuit Ménestrier who wrote about both comets and ballets, Bayle's voluminous epistolary network, the Protestant diaspora, and the salon cultivated by Marguerite La Sablière, who conversed and corresponded with many of the figures who appear in this book. She had a telescope on her roof, and she even visited the newly constructed observatory.[2]

Following comets in the archive also unearths texts and people who might not come to our attention through other approaches. Following the confusing trace of Galileo, we find Nicolas de Blégny and his journal printed by the same press as the Paris edition of the *Mercure galant*, Molière's plays, and the one-act comedy *La comète* but largely forgotten by posterity. Looking closely at the pamphlet *Balet des cometes. Pour la tragédie d'Irlande* reveals a record of France's global aspirations embodied in the cast of dancers by two young *américains* and a family member of a missionary and prominent investors in Nouvelle France.

The Comet's Tail: the Globes that Never Arrived at Their Destination

Reading the late seventeenth century via comets brings to light some stories that normally do not get told, stories that do not lead to important cultural or historical developments or offer up a neatly usable past that can be tapped to mirror our present day. Coronelli's globes, which I described at the very beginning of this book, represented one such failure (literally a six-ton failure). They represented a huge investment and great intertwined hopes. Yet no matter the beauty of their execution and the drama of their size, the globes' cartography and scientific knowledge were out of date before they were finished. Their proponent was dead and the gift giver otherwise occupied. They were displayed for only one decade of Louis XIV's long rule and spent the twentieth century in storage. They never made it to Versailles.

Monique Pelletier theorizes that it may have been on Colbert's advisement that d'Estrées commissioned the imposing globes.[3] Colbert and d'Estrées each had their own motivations for the massive royal gift. Colbert, who had founded the Compagnie des Indes orientales in 1664, may have seen the globes as a means to encourage Louis XIV to take a greater interest in French overseas trade and expansion, which feature prominently on the terrestrial globe. For his part, d'Estrées, representing the king's interest in Rome, where his brother served as the French ambassador, hoped to advance his career. (D'Estrées aspired to become prime minister, a position that had been suppressed after Mazarin's death.[4]) The globes were intended to be displayed at Versailles, but Coronelli and his team were still working on them when the court took up residence at the palace in 1682. The commission was completed in November 1683 and the globes were officially presented to the king, but d'Estrées was absent on a diplomatic mission in Rome. Colbert had died two months prior so there was no longer a strong voice advocating to make space for them in the palace. Furthermore,

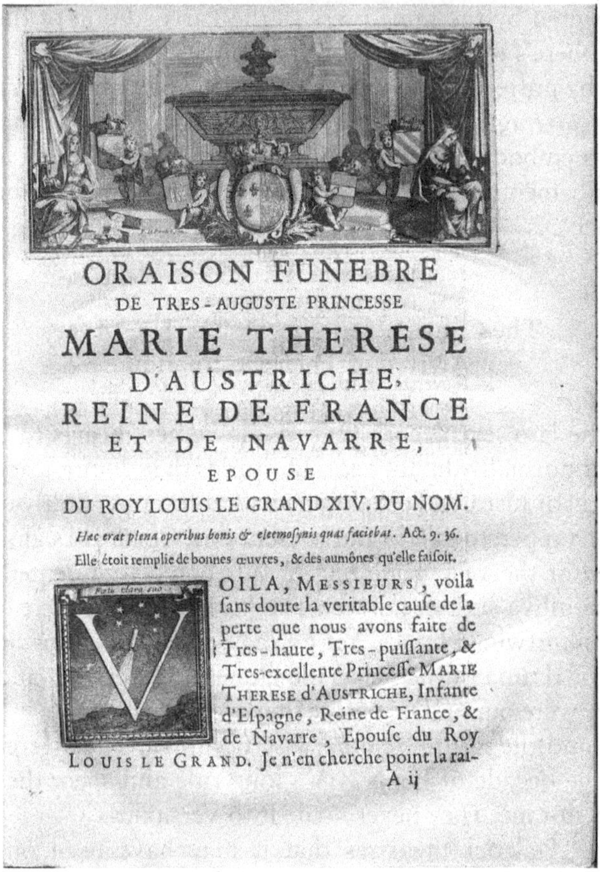

ORAISON FUNEBRE
DE TRES-AUGUSTE PRINCESSE
MARIE THERESE
D'AUSTRICHE,
REINE DE FRANCE
ET DE NAVARRE,
EPOUSE
DU ROY LOUIS LE GRAND XIV. DU NOM.

Hæc erat plena operibus bonis & eleemoſynis quas faciebat. Act. 9. 36.
Elle étoit remplie de bonnes œuvres, & des aumônes qu'elle faiſoit.

VOILA, MESSIEURS, voila ſans doute la veritable cauſe de la perte que nous avons faite de Tres-haute, Tres-puiſſante, & Tres-excellente Princeſſe MARIE THERESE d'AUSTRICHE, Infante d'Eſpagne, Reine de France, & de Navarre, Epouſe du Roy LOUIS LE GRAND. Je n'en cherche point la rai-
A ij

Fig. 27. At the start of a funeral oration for Queen Marie-Thérèse, the initial V, engraved by Pierre Le Pautre, depicts a comet and the inscription "Foetu[s] Clara Suo," shining with its own offspring. Antoine Gallois, *Oraison funèbre* (Paris: Guillaume De Luynes, 1683), p. 10. Bibliothèque de l'Institut National d'Histoire de l'Art, collections Jacques Doucet.

the queen's death the previous summer had spurred another round of renovations to the royal apartments (see fig. 27).[5] Not until mid-1687 would the *Mercure galant* report Coronelli's return to Paris to deliver the globes to the hotel de Lionne, recently acquired by the d'Estrées family, in preparation for their eventual installation at Versailles.

At this juncture Coronelli's globes seem to have drawn considerable interest in both France and Italy. A company was established to finance the planned publication of an engraved series that would "scale down these grand Globes into a volume convenient for Libraries and the Curious."[6] The *Mercure galant* published a description of the company's project and a lengthy list of subscribers.[7] Donneau de Visé enthusiastically declared that Coronelli's pair of globes would be "more beautiful, more exact, more correct, and more fascinating [*curieux*] than anything until now has ever been. This is thus written in the Privilege the king gave to Monsieur Coronelli. The Privilege even states that *these Globes will not be one of the lesser ornaments of Versailles*" (original emphasis).[8]

The globes' reputation and aura of importance seem to have been very much tied to the fact that they had been designed for display at Versailles. Coronelli's atlas *La Francia divisa in XII. prefetture* mistakenly or perhaps aspirationally lists the globes as on display in the Grande Gallerie at Versailles, and his 1706 *Teatro della Guerra* includes an illustration erroneously depicting them there.[9] However, the globes were never installed at Versailles. A never-ending series of additions and design changes ensued at the palace, and the globes were kept in storage until they eventually were moved in 1703 to Marly, the retreat of the king's sunset years. After Louis XIV's death in 1715 they were transferred to the royal library, where, as Catherine Hofmann traces, they were regarded as a treasure of the collection until "the globes gradually came to be seen as encumbering objects."[10]

Coronelli's globes spent most of the twentieth century in storage. Now they are prominently installed in a gallery built for their display. The Hall of Globes at France's new national library building, the BNF-François Mitterrand, displays the two globes in all of their glory as "Les Globes du Roi Soleil."[11] The globes' imposing presence in this edifice, the symbol and epicenter of the nation's literary tradition, gives the impression that they have always been central to French cultural history. This impression, however, belies the globes' circuitous trajectory. Instead, I wonder whether the tiny comets on the celestial globe, out of time and out of order, might offer an unexpectedly apt figure for the globes themselves, for their erratic path through French cultural history, and for the era that launched them on their highly contingent course.

NOTES

Introduction

1. Monique Pelletier and Hélène Richard both emphasize that Coronelli's globes were more an artistic and symbolic tour de force than cutting-edge science. Coronelli drew on previously established maps and travel narratives, emphasizing French strategic priorities, such as La Salle's navigation of the Mississippi. Monique Pelletier, "Les Globes de Louis XIV: Les sources françaises de l'oeuvre de Coronelli," *Imago Mundi* 34 (1982): 82. For Dennis Cosgrove, "Coronelli's globes are vast emblems, stores of empirical and allegorical information that serve equally Louis XIV's geopolitical and symbolic purposes." *Apollo's Eye: A Cartographic Genealogy of the Earth in the Western Imagination* (Baltimore: Johns Hopkins University Press, 2001), 167.

2. Like all such flattering presents to kings, the globes conjure a glorified vision of the sovereign's dominion, to be contemplated by the sovereign himself and by his subjects, foreign visitors, and posterity. This conjuring is what Hall Bjørnstad calls the dream of absolutism. See Hall Bjørnstad, *The Dream of Absolutism* (Chicago: University of Chicago Press, 2021).

3. For Lucia Ayala, the widespread print dissemination of the king's natal star chart, with the sun as his midheaven sign, bolstered the sun king mythology as well as the heliocentrism that reinforced this political ideology. "Cosmology after Copernicus: Decentralisation of the Sun and the Plurality of Worlds in French Engravings," in *The Making of Copernicus: Early Modern Transformations of a Scientist and His Science*, ed. Wolfgang Neuber, Claus Zittel, and Thomas Rahn (Leiden: Brill, 2015), 207.

4. The June 1687 *Mercure galant* republished and translated a short Latin poem written by Cardinal Ranuzzi and inscribed on the horizon of the terrestrial globe. The poem plays off of the symbolic meaning of the globes with the hyperbolic claim that the universe is not immense enough to contain the king's glory. There are not enough days in the year to contain all of the king's victories, and the universe will have to expand to include them all. *Mercure galant, dédié à Monseigneur le Dauphin* (Paris: Juin 1687), 1:14–17, here 16. Two manuscript books by François le Large, now at the BNF, transcribe the globes' multimedia narrative into codex form. *Explication des figures qui sont sur le globe terrestre de Marly* and *Recueil des inscriptions des remarques historiques, et géographiques qui sont sur le globe terrestre de Marly, executé par le P. Coronelli.*

5. During this period, comets were what Lorraine Daston and Katharine Park would categorize as a wonder. "Wonder and wonders hovered at the edges of scientific inquiry. Indeed, they defined those edges, both objectively and subjectively. Wonders as objects marked the outermost limits of the natural. Wonder as a passion registered the line between the known and the unknown." *Wonders and the Order of Nature, 1150–1750* (New York: Zone Books, 2001), 13.

6. Louis XIV came to be known as the *roi soleil*, or sun king, in retrospect; however, I will use this title by which he is now commonly recognized.

7. Alexandre-Gui Pingré, *Cométographie ou traité historique et théorique des comètes*, 2 vols. (Paris: Imprimerie Royale, 1783–84). Comets were only one of Pingré's obsessions. He got into trouble for Jansenism, participated in a scientific voyage to the Indian Ocean, and helped create the Revolutionary calendar.

8. All translations are mine unless otherwise indicated. I have often retained original capitalizations and some punctuation and syntax to give a sense of the original language, while making changes to facilitate legibility. In French citations I have modernized spellings, whereas I have left original spelling in book titles. Sévigné, *Correspondance*, ed. Roger Duchêne (Paris: Gallimard, 1972), 1:75. Pomponne was exiled at this time due to his friendship with disgraced financier Nicolas Fouquet, whose high-profile trial is the main subject of Sévigné's letter. According to Duchêne, de Foix may be either Gaston-Jean-Baptiste de Foix et de Candale, duc de Randan, or his younger brother Henri, who would inherit the titles after his brother's death the following year. Sévigné, *Correspondance*, 1:913n8.

9. Donald Yeomans, *Comets: A Chronological History of Observation, Science, Myth and Folklore* (New York: Wiley, 1991).

10. Kepler's tables synthesize three important discoveries: a set of measurements of the stars and planets unsurpassed in its precision that had been Tycho Brahe's life's work, a heliocentric model of the solar system, and John Napier's tables of logarithms, a method of simplified calculation that significantly reduced the (formerly staggering) risk of calculational error in complex problems.

11. Although Kepler used Tycho's observational data, Tycho himself subscribed to geoheliocentrism.

12. Yeomans summarizes some of the leading voices in the debates about these comets. *Comets*, 69–109. Luciano Boschiero examines Giovanni Borelli's modeling of the 1664–65 comets, arguing that Borelli's proposed trajectory engages with a number of natural philosophical issues rather than with the hypothesis of heliocentrism. "Giovanni Borelli and the Comets of 1664–1665," *Journal for the History of Astronomy* 40 (2009): 11–30.

13. Pingré emphasizes that comets remained a source of scholarly debate, especially between Cartesian and Newtonian cosmographies, until at least the 1740s, when thinkers coalesced around the Newtonian explanation of comets following the publications of Pierre-Louis Maupertuis's *Lettre sur la comète* and Pierre-Charles Le Monnier's *Théorie des comètes*. Pingré, *Cométographie*, 1:178–79.

14. Stillman Drake and C. D. O'Malley published a collection of translations of major texts from the ensuing controversy between Galileo and the Jesuits regarding the nature of comets, including Galileo's famous polemical essay, "Il Saggiatore" (The assayer). Stillman Drake and C. D. O'Malley, trans. *The Controversy on the Comets of 1618: Galileo Galilei, Horatio Grassi, Mario Guiducci, Johan Kepler* (Philadelphia: University of Pennsylvania Press, 1960).

15. The first volume of Gary W. Kronk's *Cometography* documents these comets and scholarly responses to them. *Ancient–1700*, vol. 1 of *Cometography: A Catalogue of Comets* (Cambridge: Cambridge University Press, 1999). See Ofer Gal and Raz Chen-Morris, *Baroque Science* (Chicago: University of Chicago Press, 2013), 81–89; and Tabitta van Nouhuys, *The Age of the Two-Faced Janus: The Comets of 1577 and 1618 and the Decline of the Aristotelian World View in the Netherlands* (Leiden: Brill, 1998).

16. Sophie Roux explains how the 1664–65 comets raised multifarious ontological and epistemological issues for natural scientists. Sophie Roux, "The Two Comets of 1664–1665: A Dispersive Prism for French Natural Philosophical Principles," in *The Idea of Principles in Early Modern Thought. Interdisciplinary Perspectives*, ed. Peter R. Anstey (New York: Routledge Studies, 2019), 103.

17. André Félibien famously explained the solar imagery prevalent at Versailles: "Il est bon de remarquer d'abord que comme le Soleil est la Devise du Roy, et que les Poètes confondent le Soleil et Apollon. Il n'y a rien dans cette superbe Maison qui n'ait rapport à cette divinité; Aussi toutes les figures et les ornements qu'on y voit n'étant point placés au hasard, ils ont relation, ou au Soleil, ou aux lieux particuliers où ils sont mis." *Description sommaire du chasteau de Versailles* (Paris: Guillaume Desprez, 1674): 11–12.

18. The royal image was overseen by the Petite académie, later the Académie des inscriptions, a group dedicated to overseeing and coordinating the symbols, allegories, and inscriptions used in the regime's multifarious productions, from buildings and sculptures to fountains and festivals. The bibliography on Versailles and Louis XIV's image conveyed in the arts is voluminous, including important works by Michel Baridon, Philippe Beaussant, Robert Berger, Peter Burke, Marie-Claude Canova-Green, Claire Constans, Louis Hautecoeur, Kathryn Hoffmann, Nicolas Milovanovic, Chandra Mukerji, Jean-Pierre Néraudau, Gérard Sabatier, Beatrix Saule, Pierre Verlet, and Guy Walton.

19. Sabatier emphasizes that the order of the planetary suite is not Copernican nor does it follow any earlier cosmological system. He argues that the order of the planets follows from the rooms' functions. Gérard Sabatier, *Versailles ou La disgrâce d'Apollon* (Rennes: Presses universitaires de Rennes, 2017), 40.

20. Louis's *Mémoires* for the year 1662 recount: "Ceux qui me voyaient gouverner avec assez de facilité et sans être embarrassé de rien dans ce nombre de soins que la royauté exige me persuadèrent d'ajouter le globe de la terre, et pour âme *nec pluribus impar*: par où ils entendaient ce qui flattait agréablement l'ambition d'un jeune roi, que suffisant seul à tant de choses, je suffirais sans doute encore à gouverner d'autres empires, comme le soleil à éclairer d'autres mondes s'ils étaient également exposés à ses rayons." Louis XIV, *Mémoires et réflexions* (Paris: Communication et tradition, 1995), 78.

21. In the sixth *entretien*, Ariste explains the king's *devise* to Eugène: "A la vérité, répondit Ariste, ce grand Prince ne pouvait prendre un symbole plus illustre, ni plus digne de lui que le soleil; ce bel astre est son véritable portrait." Dominique Bouhours, *Entretiens d'Ariste et d'Eugène* (Paris: S. Mabre Cramoisy, 1671), 259.

22. "La Devise est . . . une métaphore *de proportion*, qui représente un objet par un autre avec lequel il a de la ressemblance: de sorte que pour exprimer en langage de devise, par exemple, que notre sage Monarque est capable de gouverner lui seul tout le monde, il faut chercher une image étrangère qui mette cela devant les yeux, et qui donne lieu à une comparaison juste, comme serait un Soleil éclairant le globe de la terre avec ce Mot, *Mihi sufficit unus*." Bouhours, *Entretiens*, 259.

23. In the BNF copy of Nicolas de Fer's *Atlas curieux*, the first text entry in "Des principaux's sistemes du monde," follows an engraved depiction of an armillary sphere and two globes (1:n.p.). Although we cannot be sure whether this placement at the front of the collection reflects authorial or collector priorities (the atlas was published between 1700 and 1705 in yearly installments), it certainly presents the cosmos (and the question of its shape and mapping) as the point of departure for discovering France, its colonies, and principal sites. De Fer follows standard practice by presenting heliocentrism as one of four main hypotheses about the shape of the cosmos. He lends the model credibility by emphasizing its classical roots: "Le système de Copernic n'est qu'une ancienne opinion rétablie et renouvelée mais éclaircie et enrichie par tant d'observations Nouvelles, et par des remarques si particulières en faveur de ce system qu'il en peut passer l'Inventeur même." Nicolas de Fer, *L'Atlas curieux, ou le Monde représenté dans les cartes générales et particuliéres du ciel et de la terre*. (Paris: l'auteur, 1705), 1:n.p.

24. For Keith Hutchinson, the inverse dynamic was also true. The movement toward political centralization likewise supported the adoption of heliocentrism: "It was . . . the growing notion that society ought to be tightly organized around a central authority, instead of being loosely dominated by a dispersed nobility, that made heliostatic astronomy succeed." "Toward a Political Iconology of the Copernican Revolution," in *Astrology, Science and Society*, ed. Patrick Currey (Woodbridge, UK: Boydell & Brewer, 1987), 95.

25. Sarah R. Cohen argues that the gardens and château of Versailles rendered permanent shapes first explored in Louis XIV's ballets in the 1660s: "The role of Sun, which Louis XIV had by [1670] made his principal performance identity, moreover, assumed in the complex of Versailles an unprecedented physical agency, shifting from emblem, image, and analogy into a generative, even cosmic, force." *Art, Dance, and the Body in French Culture of the Ancien Régime* (Cambridge: Cambridge University Press, 2000), 79. As Rose Pruiksma emphasizes, court ballets depicting the revolution of the planets around the sun did not entirely reflect Copernican cosmology; however, the sun's placement at the center around which the planets orbited nevertheless emphasizes the element most important to the sun king mythology. "Embodying Cosmological Order and Motion: Celestial Bodies, Royalty, and Mythology in French Court Ballets," Renaissance Society of America, Annual Meeting, Boston, MA, 2016.

26. Hélène Himelfarb, "Versailles, functions and legends," in *The State*, vol. 1 of *Les lieux de mémoire*, ed. Pierre Nora (Chicago: University of Chicago Press, 2001), 2:287–88. Sabatier, *Versailles ou La disgrâce d'Apollon*, 9.

27. According to Kronk, the comet was first observed by telescope in Spain in November 1664 and was visible to the naked eye from December 1664 through February 18, 1665. It was visible through telescope until March 20. *Comets*, 10. Muhammad Zaman, "A Night-Time Gathering," folio from the Metropolitan Museum of Art Theodore M. Davis Collection, Bequest of Theodore M. Davis, 1915. Accession number 30.95.174, https://www.metmuseum.org/art/collection/search/448460.

28. Johannes Hevelius, *Prodromus cometicus, quo historia, cometae anno 1664: Exorti cursum, faciesq[ue] diversas capitis ac caudae accurate delineatas complectens nec non dissertatio, de cometarum omnium motu, generatione, variisque phaenomenis, exhibetur. Ad . . . Dn J. Bapt. Colbert, Regis Christianissimi a Sanctioribus Consiliis* (Gdansk: Auctoris typis, & sumptibus, imprimebat Simon Reiniger, 1665). The following year he republished the description with the prodromus as a mantissa in a volume dedicated to Leopold of Toscany: *Descriptio cometae anno Aerae Christ: 1665 exorti, cum genuinis observationibus, tam nudis, quam enodatis, Mense Aprili habitis Gedani; Cui addita est mantissa prodromi cometici, observationes omnes prioris cometae 1664, ex iisque genuinum motum accuratè deductum, cum notis, & animadversionibus exhibens . . .* (Gdansk: Auctoris typis, & sumptibus, imprimebat Simon Reiniger, 1666). Hevelius's comet observations drew criticism from French colleagues like Auzout, whom we will meet in chapter 2, although his methods drew praise from Halley, who visited his observatory in Gdansk. Hevelius owned his own printing press and made his own copperplate engravings illustrations to guard against piracy. See Inga Elmqvist Söderlund, "Taking Possession of Astronomy: Frontispieces and Illustrated Title Pages in 17th Century Books on Astronomy," PhD diss., Stockholm University, Department of Art History, Stockholm, Sweden, 83–84. Colbert organized royal patronage of sciences to further his mercantile and administrative vision for France. See Alice Stroup, *Botany, Patronage, and Community at the Seventeenth-Century Parisian Royal Academy of Sciences* (Berkeley: University of California Press, 1990), 29–33; and Jacob Soll, *The Information Master: Jean-Baptiste Colbert's Secret State Intelligence System* (Ann Arbor: University of Michigan Press, 2009).

29. The comet is called Kirch, after Gottfried Kirch, who first sighted it on November 14, 1680, from Coburg.

30. Jacques Gallot, "La comète." D-Lem ms.II.6.14 f.52v–54r (Leipzig Städtische Bibliotheken, Musikbibliothek). See Jacques Gallot, Alexandre Gallot, and Pierre Gallot, *Oeuvres des Gallot*, ed. and transcribed by Monique Rollin (Paris: Éditions du centre national de la recherche scientifique, 1987). Douglas Boyce's "La comète" partita for solo guitar (2023) was composed starting with a theme from Gallot's chaconne..

31. Writing of the Galileo-Grassi controversy around the comet of 1618, Gal and Chen-Morris stress that the comet debates held significant implications for cosmology: "The idea that comets were superlunary did not lose its revolutionary cosmological significance: the sharp dichotomy between the realms under and above the moon, which the heavenly position of comets undermined, was not an arcane scholarly conviction: it was well entrenched in religion and common sense." *Baroque Science*, 82.

32. See van Nouhuys, *The Age*.

33. We will meet Ménestrier in chapter 1 as a Jesuit theorist of ballets. Claude-François Ménestrier, *Lettre d'un gentil-homme de province à une dame de qualité sur le suject de LA COMÈTE* (Paris: Estienne Michallet, 1681), 5.

34. Roger Ariew, "Theory of Comets at Paris during the 17th Century," *Journal of the History of Ideas* 53 (1992): 355. Gal and Chen-Morris emphasize that Galileo's position is Aristotelian. *Baroque Science*, 81–113.

35. Andrea Gualandi, *Teorie sulle Comete da Galileo a Newton* (Milan: Franco Angeli, 2009). Roux's rich and fascinating chapter studies "the books that were published on these comets in France as a dispersive prism to enable us to see the ongoing controversies about which principles should be accepted in natural philosophy in the mid-sixties." "The Two Comets," 100.

36. Roux, "The Two Comets," 101.

37. Roux, "The Two Comets," 128.

38. Adam Mosely examines catalogs of comets from the sixteenth and seventeenth centuries to index changing notions about comets as portents and also as natural phenomena. "Past Portents Predict: Cometary *Historiae* and Catalogues in the Sixteenth and Seventeenth Centuries," in *Celestial Novelties on the Eve of the Scientific Revolution, 1540–1630*, ed. Dario Tessicini and Patrick Boner (Florence: Leo S. Olschki Editore/Museo Galileo and Istituto e Museo di Storia della Scienza, 2013).

39. Paul Hazard, *The Crisis of the European Mind 1680–1715*, trans. J. Lewis May and intro. Anthony Grafton (New York: New York Review Books, 2013); Voltaire, *Le siècle de Louis XIV, édition de 1751*, Préparée et présentée par Ulla Kölving (Centre international d'étude du XVIIIe siècle, 2018) http://c18.net/vo/vo_textes_siecle.php?div1=29.

40. Sevigné relays Mazarin's deathbed *bon mot* about comets; see chapter 3.

41. Sara Schechner Genuth, *Comets, Popular Culture, and the Birth of Modern Cosmology* (Princeton, NJ: Princeton University Press, 1997), 5.

42. Translated by Leon J. Richardson in *Sir Isaac Newton's Mathematical Principles of Natural Philosophy and His System of the World*, vol. 1, ed. Florian Cajori (Berkeley: University of California Press, 1966), xvi.

43. J. B. Shank offers a nuanced cultural history of French eighteenth-century reception (really construction) of "Newtonianism." He attributes the foundational importance of the series of debates, or "Newton Wars," to the construction of Enlightenment *philosophe* identity. *The Newton Wars and the Beginning of the French Enlightenment* (Chicago: University of Chicago Press, 2008).

44. Gal and Chen-Morris, *Baroque Science*. Scholarship on the social practice of early modern science has transformed our understanding of the practice and institutions of knowing. See the work of Mary Baine Campbell, Lorraine Daston, Peter Dear, Anthony Grafton, Katharine Park, J. B. Shank, and Steven Shapin, among others.

45. Gal and Chen Morris, *Baroque Science*; Roux, "Two Comets"; Schechner Genuth, *Comets*; van Nouhuys, *The Age*. See also Peter Dear, *Discipline and Experience: The Mathematical Way in the Scientific Revolution* (Chicago: University of Chicago Press, 1995), 86–87.

46. That is not to say that Versailles was ever fully complete. Even after the court moved into the palace in 1682, the construction continued perpetually.

47. The carefully constructed myth of Louis XIV's absolutism has been the subject of voluminous scholarship, starting with chroniclers of his own era such as Philippe de Courcillon, marquis de Dangeau (1638–1720), Sévigné, and Louis de Rouvroy, duc de Saint-Simon (1675–1755) and continuing in the next century with Voltaire's 1751 sociologizing history *Le siècle de Louis XIV*. Twentieth-century classics that study the dynamics of this modern form of mediatized power include Peter Burke's *Fabrication of Louis XIV* (1992), Nicole Ferrier-Cavarivière's *L'Image de Louis XIV dans la littérature française de 1660 à 1715* (1981), Ernst Kantorowicz's *The King's Two Bodies* (1957), Louis Marin's *Le portrait du roi* (1981), and Jean-Marie Apostolidès's *Le roi-machine* (1981) and *Le prince sacrifié* (1985). While scholarship treating absolutism as axiomatic has continued into the current day, for example, in Philip Mansel's *King of the World* (2020), many recent works revise or challenge this understanding of both Louis XIV's sovereignty and absolutism including William Beik's *Louis XIV and Absolutism* (2000), Bjørnstad's *Dream of Absolutism* (2021), my *Vaux and Versailles* (2008), Mitchell Greenberg's *Baroque Bodies* (2001), Chloe Hogg's *Absolutist Attachments* (2019), Ellen McClure's *Sunspots and the Sun King* (2006), and Harriet Stone's *Crowning Glories* (2019).

48. The critical corpus of Michel Foucault has infused our understanding of these issues and this period from "The Eye of Power" to *The Order of Things* (*Les Mots et les Choses*), and beyond. Andrea Frisch traces the emergence in the sixteenth century of modern, epistemic conceptions of witnessing that emphasize firsthand personal experience. Andrea Frisch, *The Invention of the Eyewitness* (Chapel Hill: University of North Carolina Press, 2004).

49. This tendency is exemplified variously by Christophe Schuwey, *Un entrepreneur des lettres au XVIIè siècle: Donneau de Visé, de Molière au Mercure galant* (Paris: Classiques Garnier, 2020) and Alison Stedman, *Rococo Fiction* (Lewisburg, PA: Bucknell University Press, 2012), which both question the unitary construction of the era's literature. Bjørnstad in *Dream of Absolutism* and Hogg in *Absolutist Attachments* reexamine the connections between Louis XIV and his subjects; Juliette Cherbuliez's *In the Wake of Medea* (New York: Fordham, 2020); John Lyons with *Kingdom of Disorder: Theory of Tragedy in Classical France* (West Lafayette, IN: Purdue Studies in Romance Literatures, 1999); and Jennifer Eun-Jung Row in *Queer Velocities: Time Sex, and Biopower on the Early Modern Stage* (Evanston, IL: Northwestern University Press, 2022) unsettle traditional notions of classical theater. Alain Viala attends to the culture and esthetics of *galanterie* in *La France galante* (Paris: Presses universitaires de France, 2008). Numerous other scholars contribute to this reexamination of seventeenth-century French literature and culture.

50. Eileen Reeves, *Evening News. Optics, Astronomy, and Journalism in Early Modern Europe* (Philadelphia: University of Pennsylvania Press, 2014), 16.

51. Speaking of nineteenth-century observatories, David Aubin observes: "From the observatory and its associated constellation of savants and instrument makers radiated various technologies, experimental setups, and ideological constructs that helped to remold not just the public understanding of precision science but the very experience of observing, circulating, and

communicating." "The Fading Star of the Paris Observatory in the Nineteenth Century: Astronomers' Urban Culture of Circulation and Observation," in "Science and the City," special issue, *Osiris*, 2d ser., 18, no. 1 (2003): 81.

Chapter 1

1. *Journal d'Olivier Lefèvre d'Ormesson, et extraits des mémoires d'André Lefèvre d'Ormesson*, ed. Adolphe Chéruel (Paris: Imprimerie Impériale, 1861), 2:285.

2. *Journal d'Olivier Lefèvre d'Ormesson*, 2:285.

3. Sévigné, *Correspondance*, 1:80.

4. Roland Racevskis argues that Sévigné's famous epistolary corpus exemplifies a new kind of temporally inscribed subjectivity, defined by the rhythms of a newly reformed and regularized postal system. "Both the marquise's experiences as she recounted them and her thoughts as she expressed them factored into her daily life, an existence structured in time by the activities of receiving and sending mail." Roland Racevskis, *Time and the Ways of Knowing under Louis XIV* (Lewisburg, PA: Bucknell University Press, 2003), 91.

5. Events like Fouquet's trial and comets that brought people together in public spaces also spawned topical songs. See Nicholas Hammond, *The Powers of Sound and Song in Early Modern Paris* (State College: Penn State University Press, 2019) and www.parissoundscapes.org.

6. Letter 824, January 2, 1681. Sévigné, *Correspondance* 3:39.

7. Throughout this chapter I will be using modern terms in English to refer to seventeenth-century French periodicals as "press," "journalism," and "the news," although these terms were not yet operational, and in retrospect have been contested descriptors for some of the periodicals I will be discussing in this chapter. For Hatin or Tavernier, for example, Loret's dispatches to Longueville "était déjà une forme de 'petite presse' " as would be the *Mercure galant*. Eugène Tavernier, *Du journalisme, son histoire, son rôle politique et religieux* (Paris: H. Oudin, 1902), 11. As the most serious and enduring among the periodicals discussed in this chapter, Denis de Sallo's *Journal des sçavans* is often recognized as an influential step in the development of journalism. For example, Denis-François Camusat opens his 1734 "histoire de tous les journaux depuis leur origine jusqu'à présent" (1:iii) by declaring the *Journal des sçavans* "l'Origine des Journaux": "Les Journaux sont nés en France et comptent à peine soixante-cinq ans d'ancienneté. C'est une des plus heureuses inventions du Siècle de Louis le Grand, et les Historiens de ce Monarque l'ont jugée assez importante pour ne la devoir pas omettre dans les Fastes d'un règne aussi remarquable par les progrès que les Sciences" (1:1–2). See Denis-François Camusat, *Histoire critique des journaux* (Amsterdam: J. F. Bernard, 1734), 1:6–32. Marion Brétéché and Dinah Ribard talk about a certain "formule mercure" publishing current events in a periodical format "pendant le laps de temps dont il s'agit de rendre compte." These *mercures* also function as "les archives du temps present." Marion Brétéché and Dinah Ribard, "Qu'est-ce que les mercures au temps du *Mercure galant*?" *Dix-septième siècle* 270, no. 1 (2016): 9–10, 13. Schuwey takes on the question "Qu'est-ce que le *Mercure galant*?" by looking at the material and commercial specificities of the publication and the way it compared with other publications of its era.

Schuwey, *Un entrepreneur*, 323–75, see esp. 323n1. Rather than focus on classifications, this chapter follows coverage of the comet, something writers and readers of the seventeenth century understood to be a current event of popular interest. How, this chapter asks, do ephemeral topics of interest ramify in the media landscape of the era? I will often use the word "periodical," as the four *journaux* featured in this at least aspired to regular (periodical) publication to cover the constantly evolving possibilities for new content.

8. Ellen Welch locates in Sévigné's accounts of two much-discussed trials—Fouquet's discussed here and that of the marquise de Brinvilliers in the *affaire des poisons*—an emergent depiction of public opinion. "State Truths, Private Letters, and Images of Public Opinion in the *Ancien Régime*: Sévigné on Trials," *French Studies* 67, no. 2 (April 2013): 170–83.

9. In *Evening News*, Reeves traces how in the first decades of the seventeenth century in Italy, the development of optical technology and astronomical observation energized the nascent periodical press. By 1660, the period in which this book starts its inquiry, commercial publishing in France had grown and diversified, fomented by a broadening reading public, infrastructure such as roads and the post, and, especially during the Fronde years, the kind of political turmoil that generates news and opinions. See Christian Jouhad, *Mazarinades: La Fronde des mots* (Paris: Aubier, 2009) and Kathrina LaPorta, *Performative Polemic: Anti-Absolutist Pamphlets in Late Seventeenth-Century France* (Newark: University of Delaware Press, 2021).

10. See Déborah Blocker and Anne Piéjus's introduction to their 2016 special edition of *Dix-septième siècle*, "Auctorialité, voix et publics dans le *Mercure galant*." *Dix-septième siècle* 270, no. 1 (January–March 2016): 3–8. and Brétéché and Ribard's "Qu'est-ce que les mercures au temps du *Mercure galant?*"

11. Chloe Hogg demonstrates that war coverage was another stalwart of the *Mercure galant*, appealing to "a public of individuals eager to see the stories of their friends and families in print." Hogg, *Absolutist Attachments*, 107.

12. Sara Harvey and Geoffrey Turnovsky examine the interactive features that made the *Mercure galant* particularly engaging and dynamic. Sara Harvey, "Les fins de l'obscurité dans les énigmes du *Mercure galant*." In *L'obscurité. Langage et herméneutique sous l'ancien régime*, ed. Delphine Denis (Louvain-la-Neuve: Bruylant-Académia, 2007), 171–82. Geoffrey Turnovsky, "Les lecteurs du *Mercure Galant*: Trois aperçus," *Dix-septième siècle* 270, no. 1 (2016): 65–80.

13. The regular circulation achieved by Sallo's *Journal des sçavans* eluded most periodical projects. Despite attempts to continue publication after Loret's death, *La Muze historique* petered out quickly. Donneau's *Mercure galant* floundered in its initial iteration and reinvented itself before becoming a blockbuster success. After less than a year, Jean-Baptiste Denis was forced to reimagine his popular series of *Mémoires concernant les arts et les sciences* as *Conférences presentées à Monsieur le Dauphin* following an injunction obtained by the editor of the *Journal des sçavans*. Despite the popularity evidenced by reprints and international, ostensibly pirated, editions, publication ceased in 1674. Nicolas de Blégny, whose journal was first published by d'Houry, also encountered opposition—from both the *Journal des sçavans* and

from the medical establishment at the university. On Denis and his transfusion experiments see Holly Tucker, *Blood Work: A Tale of Medicine and Murder in the Scientific Revolution* (New York: Norton, 2012).

14. Ménestrier is explaining how to see whether comets have parallax as a way to determine whether they are part of earth's atmosphere, as Aristotle believed, or whether they are true celestial bodies. Ménestrier, *Lettre.*

15. Jean-Dominique Cassini IV, *Mémoires pour servir à l'histoire des sciences et à celle de l'Observatoire Royal de Paris* (Paris: Bleuet, 1810), 265.

16. Jean-Dominique Cassini, *Observations sur la comète qui a paru au mois de décembre 1680. et en janvier 1681. Presentées au Roy. Par Mr. Cassini, de l'Académie royale des sciences* (Paris: Estienne Michallet, 1681). Cassini observed the 1664–65 comet using a telescope remarkable for the era. His research was sponsored by Queen Christina of Sweden, who accompanied Cassini during his observations and to whom he dedicated his book. Jean-Dominique Cassini, *Theoria motus cometae anni MDCLXIV pars prima . . .* (Rome: Fabio De Falco, 1665).

17. Stanisław Lubieniecki, *Theatrum cometicum* (Amsterdam: Typis Daniel Baccamude apud Francis Cuyper, 1666–68).

18. Söderlund analyzes Lubieniecki's complicated frontispiece. Söderlund, *Taking Possession*, 180–81.

19. Lubieniecki, *Theatrum cometicum* (Leiden: Pieter van der Meersch, 1681). See also Pingré, *Annales célestes*, 275.

20. Boston printer/engraver/editor John Foster's 1681 almanac details the comet's appearance. However, extant almanacs are rare: "Ils sont rares, infiniment plus rares encore que les livrets de la Bibliothèque bleue. Leur périodicité, les prêts dont ils sont l'objet, les nombreuses manipulations dues à leur fréquente utilisation (ils sont imprimés sur papier bon marché de mauvaise qualité, donc fragile), sont autant de raisons qui expliquent cette rareté." Geneviève Bollème, *Les almanachs populaires aux xviie et xviiie siècles: Essai d'histoire sociale* (Paris: Mouton & Co., 1969), 23.

21. Pingré, *Cométographie* 1:162.

22. Pingré, *Cométographie* 1:164.

23. Elizabeth Eisenstein, *The Printing Press as an Agent of Change: Communications and Cultural Transformations in Early-Modern Europe*, 2 vols. (Cambridge: Cambridge University Press, 1979), 631. The proliferation of comet texts differs from the developments traced in Eisenstein's classic chapters "Technical Literature Goes to Press" and "The Copernican Revolution" in that no advancement derives from the circulation and sharing of data and projects. Roux notes this accumulation that builds neither scientific advancement nor consensus in "The Two Comets."

24. Adrian Johns studies the founding of the *Philosophical Transactions.* Adrian Johns, "Miscellaneous Methods: Authors, Societies and Journals in Early Modern England," *British Journal for the History of Science* 33 (2000): 159–86.

25. Like the war coverage Hogg analyzes in *Absolutist Attachments*, coverage of the comets reveals "news useful and entertaining to private subjects gathered together as a public of (self-interested) media consumers. This understanding of the journal's activity nuances the notion of an early modern public

defined solely by the absolutist monarch, with elite subjects obliged to carve out their own spheres of public-forming activity at the margins of or in opposition to the absolutist sphere." Hogg, *Absolutist Attachments*, 109.

26. Schuwey, *Un entrepreneur*, 21–38.

27. Schuwey, *Un entrepreneur*, 53–85.

28. Of course, in the print marketplace, even this novelty could be recycled, as publications relating to past comets were often reissued when new comets appeared.

29. Very fittingly, the *OED* records both meanings of the word "cometic" to the period under study in this book. The figurative usage, "Designating something likened to a comet, esp. in being portentous, brilliant, or transient," is first documented in English in 1661, while the observational usage, "of or relating to a comet or comets; of the nature of or resembling a comet," first appears in 1668 in the *Philosophical Transactions*.

30. Benedict Anderson, *Imagined Communities: Reflections on the Origins and Spread of Nationalism*, rev. ed. (New York: Verso, 1998).

31. Recent scholarship on this question includes Hogg, *Absolutist Attachments* and Ibbett, *Compassion's Edge*.

32. Hatin recognizes Loret's verses as "un recueil dans lequel sont consignés tous les faits remarquables, politiques et littéraires, tous les bruits de la ville et de la cour, pour une période de quinze années, remplie d'événements de toute nature. Une gazette qui a eu le singulier privilège d'intéresser pendant aussi longtemps la société la plus polie et la plus éclairée, ne saurait manquer d'avoir une grande importance historique." Eugène Hatin, *Histoire politique et littéraire de la presse en France* (Paris: Poulet-Malassis et de Broise, 1859), 1:290–91.

33. These original printed volumes of collected verse are quite rare. The fact that the *La Muze historique* was published in folio after an initial quarto volume seems to indicate high prestige and interest at the time of its publication; however, Loret's verse accounts of seventeenth-century day-to-day life fell starkly out of favor in the eighteenth century. A later generation of bibliophiles and historians, among them Hatin and Rothchild, revived interest in his work, and from this we have a nineteenth-century facsimile from which I will be citing. Jean Loret, *La Muze historique, ou Recueil des lettres en vers contentant les nouvelles du temp: Écrites à son altesse Mademoizelle de Longueville, depuis duchesse de Nemours*, 5 vols., ed. Charles-Louis Livet et al. (Paris: P. Jannet, P. Daffis, and H. Champion, 1857–91).

34. Loret, *Muze historique*, 4:295.

35. Loret, *Muze historique*, 4:295.

36. Loret, *Muze historique*, 4:296.

37. Loret, *Muze historique*, 4:296.

38. Loret, *Muze historique*, 4:296.

39. Loret, *Muze historique*, 4:296.

40. Loret, *Muze historique*, 4:296.

41. Loret, *Muze historique*, 4:296.

42. Loret, *Muze historique*, 4:296.

43. From the *Mémoires* of the duc de Saint-Simon to Marie-Christine de Villedieu's 1665 tragicomedy *Le favori*, many accounts and literary works

from the era depict the way seating dramatized favor and influence. Mademoiselle de Montpensier, for example, recounts in her *Mémoires* how her intimates, and fellow exiles, the comtesse de Fiesque and Madame de Frontenac (Anne de La Grange Trianon), are shunned at a ballet: "Elles étaient tout derrière, cachées avec coiffes comme des personnes qui ne s'osent montrer." She ends the anecdote with the queen's withering comment: "Elles étaient derrière, parmi la canaille; le roi ni moi ne nous informons pas des gens qui sont où elles étaient." Anne-Marie-Louise d'Orléans, duchesse de Montpensier, *Mémoires*, in *Oeuvres completes*, 2 vols., ed. Jean Garapon (Paris: Honoré Champion, 2020), 1:744.

44. On the founding of the journal see Jean-Pierre Vittu, "La formation d'une institution scientifique: Le *Journal des savants* de 1665 à 1714," *Journal des savants* 1 (2002): 179–203.

45. "Le dessein de ce Journal étant de faire savoir ce qui se passe de nouveau dans la République des lettres." "L'imprimeur au Lecteur," *Journal des sçavans*, January 5, 1665, n.p. Hélène Merlin follows the development of a notion of a public of individual readers in a *République des Lettres*. Hélène Merlin, *Public et littérature en France au XVIIe siècle* (Paris: Belles Lettres, 1994), 59–112.

46. "L'imprimeur au lecteur," *Journal des sçavans*, January 5, 1665, n.p.

47. "Il sera très-avantageux à ceux qui entreprendront quelque ouvrage considérable, puis qu'ils pourront s'en servir pour publier leur dessein, et inviter tout le monde à leur communiquer les manuscrits, et les pièces fugitives qui pourront contribuer à la perfection des choses qu'ils auront entreprises." "L'imprimeur au lecteur."

48. "L'imprimeur au lecteur."

49. Samuel Sorbière, *Discours de Mr de Sorbière sur la comète* (Paris: s.n., 1665).

50. *Journal des sçavans*, January 5, 1665, 82. Bayle cites this passage in the introduction to his *Pensées diverses sur la comète* as a kind of esthetic manifesto for his work. See chapter 5.

51. *Journal des sçavans*, January 5, 1665, 82.

52. *Journal des sçavans*, January 5, 1665, 82.

53. *Journal des sçavans*, January 5, 1665, 82.

54. See Jean-Pierre Vittu, "Diffusion et reception du *Journal des savants* de 1665 à 1714," in *La Diffusion et la lecture des journaux de langue française sous l'ancien régime*, ed. Hans Bot (Amsterdam and Maarssen: APA-Holland University Press, 1988), 168–69. Several months later, this same collège hosted the performance of *Le Balet [sic] des comètes*. See chapter 2.

55. *Journal des sçavans*, January 26, 1665, 41.

56. Jacques Grandami, *Le cours de la comète qui a paru sur la fin de l'année 1664, avec un traité de sa nature, de son movement et de ses effets and le parallèle des deux comètes qui ont paru les années 1664 & 1665* (Paris: Sébastien Cramoisy and Sébastien Mabre-Cramoisy, 1665). In a second pamphlet, Grandami tries (but fails) to determine whether the comet sightings in winter 1664 and spring 1665 are one comet or two. Jacques Grandami, *Le parallele des deux comètes qui ont paru les années 1664 & 1665* (Paris: Sebastien Cramoisy and Sebastien Mabre-Cramoisy, 1665).

57. Auzout's publication about the comet would lead to the establishment of the Académie royale des sciences and the building of the Observatoire Royal, which is the subject of chapter 3.

58. *Journal des sçavans*, January 26, 1665, 45.

59. *Journal des sçavans*, January 26, 1665, 46.

60. *Journal des sçavans*, January 26, 1665, 46–48.

61. *Journal des sçavans*, February 2, 1665, 58–60.

62. "Chacun parle du phénomène, qui depuis presque deux mois paraît sur notre horizon. Les Astronomes en examinent la nature et en observent toutes les démarches: le peuple en craint les effets, et s'informe des misères qu'il présage. Ceux qui ne sont pas assez éclairés pour en découvrir la nature, ni assez superstitieux pour en appréhender les évènements, se débattent du genre qu'il doit avoir." *Journal des sçavans*, February 2, 1665, 58–60.

63. *Journal des sçavans*, February 2, 1665, 58.

64. *Journal des sçavans*, February 2, 1665, 58.

65. Vittu finds in the articles from the early years of the *Journal des sçavans* "une autre silhouette du lecteur supposé de la revue, amateur de livres et de spéculations scolastiques plus que d'observations et d'expériences: plus curieux de vulgarisation que des derniers travaux." Vittu, "Diffusion," 170.

66. Vittu, "Diffusion," 170–73.

67. *Journal des sçavans*, January 13, 1681, 9.

68. *Journal des sçavans*, January 13, 1681, 11.

69. *Journal des sçavans*, February 24, 1681, 49–52; 59; March 17, 1681, 93–94; May 5, 1681, 145–52. The *Journal des sçavans* review is the only remaining record of Anthelme Voituret's *Explication de la comète qui a paru à la fin de l'année et au commencement de celle-ci* (Dijon, 1681). There is no extant copy. Claude Mallement de Messange, *Dissertation sur les comètes à Monsieur le Procureur Général du Grand Conseil* (Paris: J. Cusson, 1681). Jean-Dominique Cassini, *Abrégé des observations et des réflexions sur la comète qui a paru au mois de décembre 1680 et aux mois de janvier, février, mars de cette année 1681: Présenté au Roy Par Mr. Cassini.* (Paris: Estienne Michallet, 1681).

70. *Journal des sçavans*, May 26, 1681, 178–80. Gallet wrote several reports to the *Journal des sçavans* on various scientific issues.

71. *Journal des sçavans*, June 30, 1681, 299–302.

72. *Journal des sçavans*, July 21, 1681, 325. Jean de Fontaney, *Observations sur la comète de l'année 1680 & 1681, faites au Collège de Clermont* (Paris: Gabriel Martin, 1681).

73. *Journal des sçavans*, May 5, 1681, 152.

74. The revue went through several incarnations in the first half decade of its existence. For the journal's return as a monthly in 1677, Donneau de Visé established a partnership (formalized in 1681 by a contract) with playwright Thomas Corneille. Monique Vincent, *Donneau de Visé et le "Mercure galant,"* 2 vols (Paris: Aux Amateurs de Livres, 1987), 121. Schuwey traces the ingenuity and entrepreneurial energy of Donneau de Visé's quest to refine a successful model of commercial publication in the nearly two decades between his first foray into printing until the *Mercure galant* found footing with its "first stable

formula" in 1678. *Un entrepreneur*, 322. These twenty years between Donneau's first unauthorized yet innovative edition of a Molière play and the eventual editorial success that would make him and the *Mercure galant* key cultural figures of Louis XIV's France were marked by two comets.

75. Deborah Steinberger notes that *Mercure galant* began to feature more science content in 1681 following the comet mania. "*Le Mercure Galant* and Its Student Body: Donneau de Visé's Inclusive Pedagogy," *Cahiers du dix-septième* 17 (2016): 48.

76. Stedman emphasizes the generic diversity, *Rococo Fiction*, 98–99.

77. Sara Harvey and Geoffrey Turnovsky examine the interactive features that made the *Mercure galant* particularly engaging and dynamic, while Elsa Veret-Basty examines the sometimes conflicting conceptions of authority in this collaborative publication space. Sara Harvey, "Les fins de l'obscurité dans les énigmes du *Mercure galant.*" In *L'obscurité: Langage et herméneutique sous l'ancien régime*, ed. Delphine Denis (Louvain-la-Neuve: Bruylant-Académia, 2007), 171–82. Elsa Veret-Basty, "'Être quelque chose comme auteur': le statut paradoxal de l'énigmatiste dans *Mercure galant.*" *Dix-septième siècle* 270, no. 1 (2016): 97–114. Turnovsky, "Les lecteurs."

78. "Avis," *Mercure galant*, January 1681, n.p.

79. "Avis, "*Mercure galant*, January 1681, n.p.

80. "Avis," *Mercure galant*, January 1681, n.p.

81. "Avis," *Mercure galant*, January 1681, n.p.

82. "Avis," *Mercure galant*, January 1681, n.p.

83. *Mercure galant*, January 1681, 93. For this passage and unless noted otherwise, I will cite the Paris edition available in the BNF and digitized on Gallica. Although generally they are identical, the Lyon edition printed by Thomas Amaulry includes only the first 227 pages of the Paris edition's 343 pages for this month and the commercial paratexts are distinct. The Paris edition invites readers to send in letters, warns of the bad quality of pirated editions, gives instructions to readers and provincial booksellers to send orders to Blageart, and advertises the *Extraordinaire* and Nicolas de Blégny's *Journal de médecine*. The Lyon edition advertises old and new editions of the *Mercure galant* for sale by Amaulry, available by the year or by the issue, and also by subscription. *Mercure galant, dédié à Monseigneur le Dauphin* (Paris: C. Blageart, January 1681): 93; *Mercure galant, dédié à Monseigneur le Dauphin* (Lyon: Thomas Amaulry, 1681). See also Schuwey, *Un entrepreneur*, 352–54.

84. *Mercure galant*, January 1681, 93–94. Here the comet might be seen as representing a particular kind of "thing" narrative, the kind of defamilized literary narration popular in *galant* society that would become a staple of Enlightenment philosophical literature. Chloe Hogg studies this phenomenon, which has been underexamined in the seventeenth century French studies in "Some Observations on Seventeenth-Century Object Poems" (manuscript). In *Fiction without Humanity*, Lynn Festa examines how "writers use . . . personified things to investigate humanity's own creaturely and thinglike nature, offering estranging descriptions of the world from perspectives that are not (necessarily) lodged in human beings." Lynn Festa, *Fiction without Humanity: Person, Animal, Thing in Early Enlightenment Literature and Culture* (Philadelphia: University of Pennsylvania Press, 2019), 2.

85. Right after Comiers's *Dissertation*, the *Mercure* announces that La Reynie, having been named a counselor of state, is retaining the position of lieutenant of police. The description of the order La Reynie installs seems the polar opposite of the kind of promiscuous disorder the comet inspires: "Son extrême vigilance, son exactitude à faire observer les Règlements et sa juste sévérité utile au Public, parce qu'elle tient tout le monde dans le devoir, lui ont fait faire des choses dont avant lui aucun Magistrat n'avait pu venir à bout, et qui manquaient seules pour rendre Paris la plus belle Ville, et la mieux policé de toute la Terre." *Mercure galant*, January 1681, 145. Holly Tucker describes La Reynie's multipronged efforts to install systems of surveillance and control in *City of Light, City of Poison: Murder, Magic and the First Police Chief of Paris* (New York: Norton, 2017). Joan De Jean, *How Paris Became Paris: The Invention of the Modern City* (New York: Bloomsbury, 2014) 122–43. See also chapter 2.

86. *Mercure galant*, January 1681, 95.

87. *Mercure galant*, January 1681, 95–96.

88. "Ah, c'est me faire trop d'outrage, / Mon corps est clair et lumineux; / Et, si l'on veut ouvrir les yeux, / L'on verra bien que mon visage / Ne porte rien de furieux / Qui prédise le moindre orage . . . Je ne veux d'aucun le trépas." *Mercure galant*, January 1681, 97.

89. "LOUIS LE GRAND . . . Ce Héros a porté ses pas / Jusques aux plus lointains Climats / Vaincus par son Bras redoutable; / Et dans ses travaux indomptable, / Les chaleurs comme les frimas / L'ont vu toujours infatigable." *Mercure galant*, January 1681, 97.

90. The *Mercure galant* likes to narrate shiny things, for example, the adumbration of the outfits from a Carnival Ball is full of detail and sparkle. *Mercure galant* (Lyon ed.), February 1681, 195–205.

91. *Mercure galant*, January 1681, 99.

92. *Mercure galant*, February 1681, 49.

93. *Mercure galant*, January 1681, 329.

94. *Mercure galant*, January 1681, 98–99.

95. *Mercure galant*, January 1681, 228. See also Mary Elizabeth Storer, "Information Furnished by the *Mercure Galant* on the French Provincial Academies in the Seventeenth Century," *PMLA* 50, no. 2 (1935): 444–68.

96. *Mercure galant*, January 1681, 330.

97. Whereas English distinguishes between news and novellas, in French both are called "nouvelles." *Nouvelles* (novellas) are a staple of Donneau de Visé's early publication projects. See Schuwey, *Un entrepreneur*, 87–103, and Monique Vincent, *Anthologie des nouvelles du "Mercure galant," 1672–1710* (Paris: Société des textes français modernes, 1996).

98. *Mercure galant*, February 1681, 57–58.

99. "Avis," *Mercure galant*, January 1681, n.p.

100. "Laissant prendre la lumière à son Amant, elle monta avec lui jusqu'au Grenier, d'où elle vit fort commodément cette nouvelle Planète." *Mercure galant*, February 1681, 58–59.

101. *Mercure galant*, February 1681, 70.

102. *Mercure galant*, February 1681, 270.

103. *Mercure galant*, January 1681, 271.
104. *Mercure galant*, January 1681, 271.
105. *Mercure galant*, January 1681, 272.
106. *Mercure galant*, January 1681, 272–73.
107. *Mercure galant*, January 1681, 326.
108. *Mercure galant*, January 1681, 327.
109. *Mercure galant*, February 1681, 177–78.
110. Johns asks, "What differences existed between kinds of periodical, and why did some categories endure while others did not?" Johns, "Miscellaneous Methods," 159.
111. A 1936 thesis in medicine traces Blégny's life and publications. Pierre Jean Tellier, *Aventurier médical au XVIIe siècle: Nicolas de Blégny* (Paris: Librairie Louis Arnette, 1932).
112. *Mercure galant*, March 1681, n.p. The Paris edition of the January 1681 *Mercure galant* announces that in the year to come, Blégny's *Journal de médecine* will begin delivery to the provinces "pour satisfaire à la réquisition des Personnes de Province" in its original small format, "seulement le premier jour de chaque mois." *Mercure galant*, January 1681, n.p. The February 1681 issue includes a similar announcement.
113. Natacha Coquery cites Blégny as one of the "pioneers of the publicitary press" along with fellow enterprising apothecary/newsman/entrepreneur Théophraste Renaudot. Her focus is on *Les adresses de Paris* of 1691 and 1692, the ambitiously commercial publication that he launched a decade after his *Journal* and Académie de nouvelles découvertes. Coquery notes that "les libraires et les apothicaires se saisirent les premiers de la presse pour se faire connaitre: les annonces de parutions de livres, de drogues et onguents sont de loin les plus nombreuses. Le discours fantaisiste des apothicaires renforce une réputation de bonimenteurs déjà bien ancrée." Natacha Coquery, *Tenir boutique à Paris au XVIIIe siècle: Luxe et demi-luxe* (Paris: CTHS Histoire, 2011), 45–46, 52.
114. D'Houry seems to have been the sole printer for the journal's first year (1679) before the collaboration with Blageart and the Académie des nouvelles découvertes de médecine, which is Blégny's place of business, a kind of freelance academy for lectures, lessons, and demonstrations, including dissections. D'Houry specialized in medical titles. He published a second edition of Jean-Baptiste Denis's *Mémoires, conférences et observations sur les arts et les sciences: Presentées à Monseigneur le Dauphin* and later founded a successful and long-running almanac, known after 1699 as *L'almanac royal*. Licensed at first solely as a *libraire* (from 1678), he was caught with an unsanctioned printing press in his house in 1708 and forced to sell off his printing materials. He was licensed as a printer in November 1712. Jean-Baptiste Denis, *Memoires, conférences et observations sur les arts et les sciences: Presentées à Monseigneur le Dauphin*, 2d ed. (Paris: Laurent d'Houry, 1682).
115. Isabelle Pantin situates receptions of Galileo and Copernicanism within the context of European political and confessional tensions. Isabelle Pantin, "New Philosophy and Old Prejudices: Aspects of Reception of Copernicanism in a Divided Europe," *Studies in the History and Philosophy of Science* 30, no. 2 (1999): 237–62. Daniel Garber excavates the changing reception of Galileo

in early to mid-seventeenth century France. Garber, "On the Frontlines of the Scientific Revolution: How Mersenne Learned to Love Galileo," *Perspectives on Science* 12, no. 2 (2004): 157.

116. The journal first appeared as *Nouvelles découvertes sur toutes les parties de la médecine: Recueillies en l'année 1679; Par N.D.B. chirugien du Roy, maistre et juré à Paris* with approbation from the king's first physician, Antoine d'Aquin, to whom the volume is dedicated, and a *privilège* dated February 2, 1679, and permission to publish for six years.

117. "Avertissement," *Nouvelles descouvertes sur toutes les parties de la Médecine: Recueillies en l'année 1679* (Paris: Laurent d'Hourry, 1679), n.p.

118. "Avertissement," *Nouvelles descouvertes*, January 1679, n.p.

119. "Avertissement," *Nouvelles descouvertes*, January 1679, n.p.

120. "Avertissement," *Nouvelles descouvertes*, January 1679, n.p. Needless to say, punctual publishing is easier said than done. The March letter begins with excuses for the issue's tardiness. *Nouvelles découvertes*, March 1679, 79.

121. Blégny, "Avertissement," *Nouvelles descouvertes*, January 1679, n.p.

122. *Nouvelles découvertes*, May 1679, 175–76; *Nouvelles découvertes*, July 1679, 272.

123. *Monatliche neueröffnete Anmerckungen, über alle Theile der Artzney-Kunst: Zusammen gebracht im Jahr 1679. Allen der Leib und Wundartzney Zugethanen und Liebhabern zu sonderbahren Gefallen aus dem Französischen ins Teutsche übersetzet durch J.L.M.C.* (Hamburg: Gottfried Schultzens Kosten, 1680). *Zodiacus Medico-Gallicus, sive Miscellaneorum Medico Physicorum Gallicorum . . . annus primus scilicet M.DC.LXXIX* (Geneva: Leonard Chouët, 1680).

124. "Comme les Impressions contre-faites se font toujours trop précipitamment pour être sans fautes, et que dans les essais des remèdes qu'on publiera, la moindre chose de plus ou de moins en pourrait rendre les suites funestes, ou du moins les effets inutiles, on avertit ceux qui achèteront ces Nouvelles Découvertes, de ne prendre que les cahiers qui seront recouverts d'un papier blanc, sur lequel l'Auteur aura écrit les mots suivants de sa propre main. VEU PAR L'AUTHEUR, avec sa paraphe, qu'on reconnaitra aisément dans la suite aussi bien que son écriture, et on promet vingt Louis d'or a celui qui indiquera les Libraires, les Imprimeurs, ou les autres gens qui en vendront de contrefaits." "Avertissement," *Nouvelles descouvertes*, January 1789, n.p. Blégny's brother, Étienne de Blégny,was an expert scribe and calligrapher, whose published work includes a guide to verifying. Étienne de Blégny, *Traité concernant la manière de procéder à toutes les verifications d'écritures contestées en justice, par le Sr. de Blégny, juré expert écrivain* (Paris: Guillaume Cavelier, 1698).

125. Nicolas de Blégny, *Le Temple d'Esculape, ou Le dépositaire des nouvelles découvertes qui se font journellement dans toutes les parties de la médecine*, January 1680. Riffaud minutely traces Blageart's career as Molière's printer in his history of publisher-bookseller Jean Ribou. Alain Riffaud, *Le libraire de Molière* (Arles: Les Sources Portaparole, 2022).

126. Under the title *Nouveautez journalières concernant les sciences et les arts, qui font parties de la médecine*, Blégny, Blageart, and d'Houry experimented with releasing, in parallel with the small monthly volumes, a weekly

edition in quarto format in the summer of 1680, before returning to monthly publication in the fall and then returning to only duodecimo format for the following year. *Nouveautez journalières concernant les sciences et les arts, qui font parties de la médecine: Recueillies par Nicolas de Blegny, chirurgien ordinaire du corps de Monsieur; Pendant les années 1680 & 1681* (Paris: Nicolas Blégny, Claude Blageart, et Laurent d'Hourry, 1680).

127. A notice printed at the end of the December 1679 issue, just before the alphabetical table of the year's articles, alerts readers to the typeface set to debut in the new year: "Ceux qui craignent d'être trompés ne seront pas pour cela moins en sureté, parce qu'il a fait faire à ses dépenses des nouveaux caractères, qui serviront aux titres de tout ce qu'il fera imprimer d'orénavant, et qui feront aisément distinguer les véritables des faux exemplaires." "Avis," *Nouvelles découvertes*, December 1679, n.p.

128. We can see the *Mercure galant*'s zigzag type on the title page of the play, *La comète*, mentioned in Blageart's promotional text, and which I discuss in chapter 4.

129. Alain Riffaud, *L'aventure éditoriale du théâtre français au XVIIe siècle* (Paris: Presses de l'Université Paris-Sorbonne, 2018), 153.

130. Nicolas de Blégny, *Journal des nouvelles découvertes conçernant les sciences et les arts qui font parties de la médecine. Tome III. Contenant toutes celles qui ont esté recueillies durant le cours de l'année 1681. Par NICOLAS DE BLEGNY, à present chirugien ordinaire du corps de Monsieur, et juré à Paris.* (Paris: Nicolas de Blégny, Claude Blageart, and Laurent D'Houry, 1681).

131. In the "Avertissement" to the 1680 journal year, Blégny offers instructions for readers who need to fill in missing issues from the previous year to find them directly from him, in the rue de Guénégaud. Blégny, *Le Temple d'Esculape*, January 1680, n.p.

132. "Pièces," the term Blégny uses to refer to the articles in his introductory epistle ("vous êtes de trop bon goût pour ne pas connaître la valeur des pièces que je vous envoi"), constituted for his contemporaries a discernable textual unity. Schuwey challenges readers to attend to this "littérature en pièces" in ways that fundamentally alter our understanding of literary works and the classical esthetic. "Les ouvrages n'étaient donc ni conçus dans un geste créateur unique ni abordés comme un tout parfait et définitif. Ils résultaient d'un assemblage d'unités de genres et de longueur variables, allant d'une brève remarque ou d'une pointe, à une scène, une pièce, un opuscule entier." Schuwey, *Un entrepreneur*, 165. As Dinah Ribard discusses, Bayle (the subject of this book's final chapter) also traded in *pièces* in his reading and writing practices. Dinah Ribard, "La philosophie mise en recueils: les 'pièces fugitives,' in *GRIHL, De la publication: Entre Renaissance et Lumières* (Paris: Fayard, 2002), 61–75

133. "L'Ordre que vous m'avez donné, Monsieur, touchant la recherche des Nouveautés de l'Astronomie a été ponctuellement exécuté." *Extraordinaire du Journal de médecine*, October 1681, n.p.

134. On early modern conceptions of translation see Roger Zuber, *Les "belles infidèles" et la formation du goût classique*, rev. ed. (Paris: Albin Michel, 1995).

135. "M. de Blegny, par qui elle est dirigée, a l'avantage d'avoir pratiqué et enseigné successivement toutes les parties de la philosophie et de la médecine. Il a composé dix-huit volumes très-curieux." [Nicolas de Blégny], *Les adresses de la ville de Paris avec le trésor des almanachs, livre commode en tous lieux, tous temps et toutes conditions, par Abraham Du Pradel, astronome Lyonnais* (Paris: La veuve de Denis Nion, 1691), 12. Note the proximity of the printer-bookseller Nion, "sur le quai de Nesle, au coin de la rue Guénégaud," to Blégny's rue Guénégaud address.

136. "La Société Royale de Médecine est encore une espèce d'Académie en laquelle on passe des règles à la pratique, ce qui fait qu'elle est composée de Philosophes, de Médecins, de Chirurgiens et d'Apothicaires artistes. Elle tient des Conférences publiques tous les Dimanches après Vêpres, rue de Fauxbourg Saint Antoine chez Monsieur de Blegny qui en est Directeur, et qui a commencé cet établissement par ordre du Roy sous la protection de M. Daquin premier Médecin de S.M. Il a déjà publié plusieurs volumes d'Observations et d'Expériences, et il travaille sans relâche à faire de nouvelles découvertes. Cette Société a des membres en plusieurs Villes de Provinces, qui travaillent utilement à la fin commune, qui est la perfection du plus important de tous les Arts." [Nicolas de Blégny], *Le livre commode contenant les addresses de la ville de Paris et le trésor des almanachs pour l'année . . . 1692 . . . par Abraham Pradel* (Paris: La veuve de Denis Nion, 1692), 40.

137. He published two polemical letters: a fourteen-page quarto addressed to Bourdelot in 1677 refuting Descartes's theory of the tides and an undated eight-page octovo (after 1681) to Claude Mallemant de Messange on "les deux nouveaux systèmes qu'ils ont inventés."

138. Joseph Jérôme de La Lande, *Bibliographie astronomique: Avec l'histoire de l'astronomie depuis 1781 jusqu'à 1803* (Paris: Imprimerie de la République, 1803), 296–97. Louis Moréri et al., *Le grand dictionnaire historique . . . nouvelle edition*, 7 vols. (Paris: Les libraires associés, 1759), 509.

139. Tinelis's Galileo does not resemble Mario Biagioli's *Galileo, Courtier* (Chicago: University of Chicago Press, 2018) or the careful manager of credit, innovation, and secrecy—also Biagioli (*Galileo's Instruments of Credit* [Chicago: University of Chicago Press, 2007])—although we can see shades of the maverick trailblazer Galileo figure Napoleon idolized and that John Lewis traces in *Galileo in France* (New York: Peter Lang, 2006). It is certainly not the Galileo that "Mersenne came to love" described by Garber in "Frontlines of the Scientific Revolution."

140. Roux, "An Empire Divided: French Natural Philosophy (1670–1690)," in *The Mechanization of Natural Philosophy*, ed. Daniel Garber and Sophie Roux (Dordrecht: Springer, 2013), 55–95, 95.

141. Roux, "An Empire Divided," 68.

142. While not wildly insightful as a philosophical intervention, it is nonetheless revealing as an example of a middlebrow reading from the period. Tinelis, *Messager*, n.p.

143. Tinelis, *Messager*, n.p.

144. Tinelis, *Messager*, 62.

145. Tinelis, *Messager*, 78.

146. Tinelis, *Messager*, 88.

147. Tinelis, *Messager*, 89–90.

148. For Tinelis the continued development of the telescope reconfigures the optics of imagination and thus transforms the relationship of moderns to classical authority. Tinelis proposes that had Epicurius or Eudoxus lived today, they would have looked at the cosmos through a telescope: "Si Épicure vivait on pourrait lui faire voir les petites étoiles dont les lueurs ramassées composent la voie de lait, et Eudoxe qui désirait de voir le Soleil d'un lieu fort proche au hasard d'être brulé de ses rayons, pourrait dans notre siècle sans s'exposer à ce hasard, remarquer dans cet Astre des particularités qu'il désirait passionnément de connaître, et on pourrait lui faire voir le mouvement circulaire de ce brillant Astre, ses macules, ses facules, &c." Tinelis, *Messager*, 93–94.

149. Tinelis, *Messager*, 95.

150. Tinelis, *Messager*, 124–26.

151. Tinelis, *Messager*, 149.

152. Tinelis, *Messager*, 150. If a comet observed at the same moment from two different locations (or the same location at two different moments) appears to be in different positions, the comet is farther from the earth than the moon. Using trigonometry, the difference between the two apparent positions can be used to calculate the comet's distance from the earth. This question had already been the source of major controversy in 1618. Gal and Chan-Morris, *Baroque Science*, 81–113.

153. "Je leur ai ici donné une idée claire et satisfaisante d'une chose dont leurs oreilles avaient été rebattues durant longtemps, et qui peut-être leur aurait toujours été inconnue." Tinelis, *Messager*, 154–55.

154. Tinelis, *Messager*, 159.

155. This seems to be a reference to Anthelme Voituret's *Explication de la comète*, which was discussed in the *Journal des sçavans*. See chapter 1, note 69.

156. The synthesis comes in the last piece of the *Messager céleste* of 1681. Tinelis de Castelet's "Dissertation Contenant des Réflexions Curieuses et Nouvelles sur la question, si l'Astrologie judiciaire doit être d'usage dans la pratique de la Médecine" (163–258) is the culmination of the volume and the most bizarre and delightful read. Tinelis asserts that only someone versed in mathematics and medicine can root out the pervasive remains of astrology from medicine. He is not trained in medicine; rather, he is an autodidact. With the knowledge gained by reading medical books, he has transformed his health from feeble to robust, a feat which "quelques-uns de mes amis ont pris pour une espèce d'enchantement," Tinelis, *Messager*, 166.

157. Gautier may also have been a pseudonym for Blégny. Blégny/Gautier's *Mercure sçavant* was part of a generation of "mercures" published by Huguenots exiles. Marion Brétéché has brought scholarly attention to a dozen of these understudied publications, and in particular the ways they pursue political journalism. Marion Brétéché, *Les compagnons de Mercure: Journalisme et politique dans l'Europe de Louis XIV* (Ceyzérieux: Champ Vallon, 2015).

158. As Coquery demonstrates, city guides exist at least a century earlier, as Montaigne himself comments, but the intersection of print and these *bureaux d'adresses* is a seventeenth-century phenomenon. Coquéry, *Tenir boutique*, 51.

159. *Au Roy, et à nosseigneurs les commissaires députez, par sa majesté, pour l'execution de l'edit du mois de mars 1693: Concernant l'ordre hospitalier du Saint-Esprit.* Factum, Ordre hospitalier du Saint-Esprit, 1693.

160. The few historians of medicine who unearth Blégny's periodical display the same reticence about him as evidenced, for example, by Achille Chéreau writing under the Seconde Empire: "Moi, médecin, j'éprouve un grand embarras pour juger ce fameux personnage. S'il était de notre époque, si nous le voyions à l'oeuvre, il ne parait pas douteux que nous fussions amenés à attacher sur son épaule la pancarte du charlatan émérite. Il a, en effet, tous les attributs du charlatan: le verbiage, l'assurance, l'intempérance de paroles, l'invention de remèdes nombreux, bizarres, à étiquettes ruolzées [*sic*], l'encolure du batailleur, de l'histrion, la témérité, l'ardeur intempestive de l'ambitieux, le clinquant de l'homme de la foire." Achille Chéreau, "Essai sur les origines du journalisme médical français" (Paris: Aux Bureaux de l'Union Médicale, 1867), 5.

161. Jean Berryat et al., eds., *Collection académique, composée des mémoires, actes ou journaux des plus célèbres académies et sociétés littéraires, des extraits des meilleurs ouvrages périodiques, des traités particuliers et des pieces fugitives les plus rares, . . . Tome Septième de la partie etrangère. Et le premier de la Médecine séparée* (Dijon: François Desventes; Paris: Michel Lambert, 1766), ii.

162. "Notre illustre auteur fut indigné de ce qu'une invention aussi utile que le sont les journaux, devenue dans des mains viles un instrument de calomnie et de méchanceté, tournait ainsi à la honte des lettres." Berryat et al., *Collection académique*, ii. In the preface to his *Nouvelles de la République des Lettres*, Bayle called Blégny and Gautier's *Mercure sçavant* a "Bureau d'adresse de Médisance." Pierre Bayle, préface to *Nouvelles de la République des Lettres* (March 1684): 1:n.p.

Chapter 2

1. Rose Pruiksma offers a provocative reevaluation of the factors contributing to the establishment of the Académie. "Generational Conflict and the Foundation of the *Académie Royale de la Danse*: A Reexamination," *Dance Chronicle* 26, no. 2 (2003): 169–87.

2. Julia Prest, "The Politics of Ballet in the Court of Louis XIV," in *Dance, Spectacle, and the Body Politick, 1250–1750*, ed. Jennifer Nevile (Bloomington: Indiana University Press, 2008), 229.

3. Mark Franko, *Dance as Text: Ideologies of the Baroque Body* (Cambridge: Cambridge University Press, 1993), 1.

4. Ellen Welch demonstrates that dance was fundamental to the practice of early modern diplomacy (107–30) and for articulating an increasingly absolutist vision of French statecraft. *A Theater of Diplomacy. International Relations and the Performing Arts in Early Modern France* (Philadelphia: University of Pennsylvania Press, 2017), 131–55. Skiles Howard examines the way dance functioned as a site of contestation, social differentiation, and political expression in Tudor and Stuart England. *The Politics of Courtly Dancing in Early Modern England* (Amherst: University of Massachusetts Press, 1998).

5. See Jean-Vincent Blanchard and Hélène Visentin, eds., *L'invraisemblance du pouvoir: Mises en scène de la souveraineté au XVIIe siècle* (Paris: Schena Editore/Presses de l'Université de Paris-Sorbonne, 2005); Benoît Bolduc, *La fête imprimée: Cérémonies et spectacles politiques (1549–1662)* (Paris: Garnier, 2016); Nicolas Russell and Hélène Visentin eds., *French Ceremonial Entries in the Sixteenth Century: Event, Image, Text* (Toronto: Center for Reformation and Renaissance Studies, 2007); Abby Zanger, *Scenes from the Marriage of Louis XIV: Nuptial Fictions and the Making of Absolutist Power* (Stanford, CA: Stanford University Press, 1997), and my *Vaux and Versailles*, chapter 1. Prest examines Louis XIV's appearance in *ballets de cour* from his appearance at age twelve in the *Ballet de Cassandre* (1651) to the final *ballet de cour* of his reign in 1669. Prest, "The Politics of Ballet at the Court of Louis XIV," 229–40. Cohen emphasizes "the intense corporality of the politics enacted on stage." *Art, Dance, and the Body*, 52. Michael Burden remarks "the major achievement of the *Ballet de la nuit* lay in the use to which the character of Apollo, god of light and sun, was put; danced by the young Louis XIV, his image as the Sun King, the *roi soleil*, was assured." "A Spectacle for the King," in *Ballet de la nuit*, ed. Michael Burden and Jennifer Thorp (Hillsdale, NY: Pendragon Press, 2009), 6. In popular culture the *Ballet de la nuit* scene in Gérard Corbiau's *Le roi danse* (2000), which was based on Philippe Beaussant's biography of Lully, offers a cinematic staging of Louis XIV's political use of dance.

6. Marie-Françoise Christout, *Le ballet de cour au XVIIe siècle: The Ballet de Cour in the 17th Century* (Geneva: Minkoff, 1987), 7. As Renaudot's description of *the Ballet de la nuit* rather awkwardly theorizes, ballet was one of the most powerful ways the young king defined himself and made himself the sun, that is, the center of attention, the organizing principle, and the motor of social action: "Les spectateurs auraient difficilement discerné la plus charmante si celles où notre jeune monarque ne se faisait pas moins connaitre sous ses vêtements que le Soleil se fait voir au travers des nuages qui voilent quelquefois sa lumière, n'en eussent reçu un caractère particulier d'éclatante Majesté qui en marquait la différence ; et faisait dire par un sentiment très-juste que si toutes les parties de ce rare divertissement ne se rapportaient qu'à la gloire de ce Prince, ce n'était que par une nécessaire réflexion de celle qu'il leur communiquait." Théophraste Renaudot, *La gazette de France* 1653, 222–23.

7. See Plato's *Timaeus* 40a-d, and Graham Pont, "Plato's Philosophy of Dance" in *Dance, Spectacle, and the Body Politick, 1250–1750*, ed. Jennifer Nevile (Bloomington: Indiana University Press, 2008), 267–81. Sir Thomas Elyot's theorization of dance articulates some of these views. See also Howard, *Politics of Courtly Dancing*.

8. The year following his appearance as the sun in the *Ballet de la nuit*, the king danced as Apollo in *Les nopces de Pélée et de Thétis* (1654). He played the sun in *Hercule amoureux* (1662), the Carousel of 1662, and the *Ballet de flore* (1669). The role of Apollo in the comédie-ballet *Les amants magnifiques* (1670) was also written for the king, although he retired from the public stage the year prior and did not dance this role. See Cohen, *Art, Dance, and the Body*, 79–88, and Georgia Cowart, *Triumph of Pleasure: Louis XIV and the Politics of Spectacle* (Chicago: University of Chicago Press, 2008), 1–83.

9. See chapters 1 and 3.

10. I am grateful to Ellen Welch, who generously shared the *Ballet des comètes* with me, and to Juliette Cherbuliez, who found the *Ballet de la comète*.

11. Soissons is approximately one hundred kilometers northeast of Paris, and over 120 kilometers from Versailles.

12. Franko published a transcription of the *Lettres patentes* along with an English translation in appendix 3 of *Dance as Text*, 166–85.

13. Franko, *Dance as Text*, 109.

14. Franko, *Dance as Text*, appendix 3, 168.

15. According to Franko, the king "effectively repossessed the noble body . . . thus laying claim to all bodies to come by founding dance as a discipline. This form of aesthetic autonomy reasserts the privileged status of dance while effectively removing it from public access." *Dance as Text*, 111.

16. Statute VII of the *Lettres patentes* stipulates that "tous ceux qui voudront faire profession de Danse en ladite ville et Faux-bourgs [de Paris], seront tenus de faire enregistrer leurs noms et demeures sur un registre." Franko, *Dance as Text*, appendix 3, 178 and 169.

17. Rock distinguishes between printed programs and libretti: "The libretti tell us more about basic structure, action, and characters. Programs, especially the later ones, often contain the ballet's spoken and sung poetry, and names of the dancers in relation to the roles they played, descriptions of the dances as solos, duets, trios, and the like, and the name of the ballet master." *Terpsichore at Louis-Le-Grand: Baroque Dance on the Jesuit Stage in Paris* (Boston: Institute of Jesuit Sources, 1996), 53.

18. Calendrier électronique des spectacles sous l'ancien régime et sous la revolution (CESAR), https://cesar.huma-num.fr/cesar2/titles/titles.php?fct=edit&script_UOID=168620. This collège, founded in 1214 as a hostel for poor students, from around 1524 functioned as the local cathedral school ("remplace véritablement l'école capitulaire"). Louis Carolus-Barré, "Le collège Saint-Nicolas-aux-Clercs de Soissons au moyen age: Histoire et archéologie," *Bulletin de la Société Nationale des Antiquaires de France* (1973, 1975): 29. The collège was confided to the direction of the Oratorians in 1676. See Paul Lallemand, *Histoire de l'éducation dans l'ancien oratoire de France* (Paris: Ernest Thorin, 1888), 102.

19. How to reconstruct or reimagine historical dance has been the subject of inventive scholarship some of which is highlighted in Mark Franko, ed., *The Oxford Handbook of Dance and Reenactment* (Oxford: Oxford University Press, 2018). Seeta Chaganti juxtaposes medieval poems and images about dance with contemporary dance performances to imaginatively reenact "the experience of participating in and watching" medieval dance. *Strange Footing: Poetic Form and Dance in the Late Middle Ages* (Chicago: University of Chicago Press, 2018), 6. On historical dance and experience, see Chaganti, *Strange Footing*, 36–45.

20. *Ballet de la comète*, n.p.

21. On chance in seventeenth-century France, see chapter 4, note 60; and John Lyons, *The Phantom of Chance: From Fortune to Randomness in Seventeenth-Century French Literature* (Edinburgh: Edinburgh University Press, 2012).

22. *Ballet de la comète*, 7.
23. *Ballet de la comète*, 5.
24. *Ballet de la comète*, 5.
25. *Ballet de la comète*, 5–6.
26. *Ballet de la comète*, 6.
27. *Ballet de la comète*, 2.
28. *Ballet de la comète*, 2.
29. On the sun in the royal ballet, see Renaudot, *La gazette de France* 1653, 222–23.
30. *Ballet de la comète*, 8.
31. *Ballet de la comète*, 9.
32. *Ballet de la comète*, 8.
33. *Ballet de la comète*, 10.
34. *Ballet de la comète*, 10.
35. Jennifer Eun-Jung Row examines what she identifies as the queer temporal strategies of haste and delay in seventeenth-century French theater. Row, *Queer Velocities.*
36. *Ballet de la comète*, 10–11.
37. *Ballet de la comète*, 11. It is also worth noting that other figures who appear at various points in this book, the Fouquet family and Olivier Lefèvre d'Ormesson, also had strong ties to Soissons.
38. *Ballet de la comète*, 11.
39. *Ballet de la comète*, 12.
40. *Ballet de la comète*, 12.
41. Noémie Ndiaye, *Scripts of Blackness: Early Modern Performance Culture and the Making of Race (Philadelphia: University of Pennsylvania Press, 2022)*, 83-102.
42. *Ballet de la comète*, 13.
43. *Ballet de la comète*, 14.
44. *Ballet de la comète*, 16.
45. *Ballet de la comète*, 16.
46. *Ballet de la comète*, 4.
47. See chapter 4, note 7. A July 1700 obituary in the *Mercure galant* celebrates "Mr Bertherand de l'Académie Royale de Soissons": "Quoi que Mr Bertherand eût un talent fort heureux pour la Poésie, et qu'il ait fait beaucoup de Vers très agréables, on n'a rien d'imprimé de lui qu'une petite Pièce fort ingénieuse, intitule le Ballet de la Commette; et le Portrait d'Armande de Lorraine d'Harcourt, Abbesse de Notre Dame de Soissons." *Mercure galant*, July 1700, 177.
48. In Soissons, Asseline published various local ephemera, a reprint of *Discours sur le renfermenent des pauvres*, and a history of Soissons. E. Stewart Saunders offers a detailed account of Fouquet's library—the investment he lavished on it, the political and cultural implications of his ambitious vision, and the fate of the precious volumes and erudite staff after Fouquet's fall from favor in 1661. "Politics and Scholarship in Seventeenth-Century France: The Library of Nicolas Fouquet and the College Royal," *Journal of Library History* 20, no. 1 (1985): 1–24. Asseline's work as printer would have been part of the cultural project Sauders describes.

49. Tucker, *City of Light*. De Jean, *How Paris Became Paris*, 122–43.

50. Anne Piéjus remarks that "le théâtre, la musique et la danse assumèrent dans les collèges, sous Louis XIV, un rôle fondamental à la fois par l'ambition qu'ils reflétaient dans la formation des jeunes gens, et souvent même des enfants, et par leur exceptionnel perméabilité au monde social, artistique et politique." Introduction to *Plaire et instruire le spectacle dans les collèges de l'ancien régime* (Rennes: Presses universitaires de Rennes, 2007), 3. Jennifer Neville likewise comments, "Aristocratic men and women, courtiers, ambassadors, princes of the church, all were both participants and spectators on an almost daily basis." Introduction to *Dance, Spectacle, and the Body Politick*, 33. Judith Rock observes, "In the adult lives of upper-class students, physical self-presentation and mastery in dancing would be crucial for achieving a place in society, maintaining reputation and forming social relationships. The test for taking one's place at Louis XIV's court was to dance with a partner before the king, with all the courtiers watching." "The Jesuit College Ballets: What We Know and What's Next," *Journal of Jesuit Studies* 4 (2017): 436–37. Piéjus also notes that unlike the Jesuit tragedies written by Jesuit professors, college ballets often drew on outside talents "on perçoit aisément non seulement une certaine rivalité entre établissements scolaires, mais un souci d'égaler, voire de surpasser la magnificence des spectacles de la ville et de la cour." "Un spectacle collectif," *Dix-septième siècle* 238, no. 1 (2008): 9.

51. Joseph de Jouvency, *De la manière d'apprendre et d'enseigner* [*De ratione discendi et docendi*], trans. Henri Ferté (Paris: Hachette, 1892), 57.

52. Ernest Boysse, *Le théâtre des jésuites* (Paris: Henri Vaton, 1880), 31.

53. In the years directly following *the Balet des Comètes*, some of France's most accomplished dancers took part in the prize-day ballets at le Collège de Clermont, renamed in 1682 le Collège Louis-le-Grand. By identifying the dancers listed in livrets from slightly later ballets, Nathalie Lecomte is able to identify the professional dancers in the *ballets de collège* as well as the ages, sometimes quite young, of some of the student dancers. "Les interprètes des ballets dansés au collège Louis-le-Grand de 1684–1699," in *Plaire et instruire*, ed. Anne Piéjus (Rennes: Presses universitaires de Rennes, 2007): 114–37.

54. "The plays and ballets were understood as a part of the Jesuit theological and moral mission to king, court, nobility, and the upper classes in general, the people making and influencing what we would call policy decisions at every level. Just as the Society of Jesus routinely used elements of foreign cultures to help them communicate in their missions, they used the production of plays and ballets to communicate with, influence and convert." Rock, "Jesuit College Ballets," 438–39. Laura Naudeix emphasizes that the Jesuits were atypical in using dance. "Tragédie et ballet: Enjeux d'un spectacle mixte," in *Plaire et instruire*, ed. Anne Piéjus (Rennes: Presses universitaires de Rennes, 2007): 55–69.

55. Gofflot, "Le théâtre au college," 91. Rock, "Jesuit College Ballets," 440. La Gravette de Mayolas observes "Quantité de gens remarquables, / Témoins de ces plaisirs aimables, / Avec les autres spectateurs/ En furent les admirateurs." *Continuateurs de Loret. Lettres en vers de La Gravette de Mayolas, et al.*, ed. James Armand de Rothschild (Paris, D. Morgand et Ch. Fatout, 1881), 1:171. Jean Loret depicts the annual prize-day ceremony in his August

21, 1660, dispatch: https://www2.unil.ch/ncd17/index.php?extractCode=726. Marie-Claude Canova-Green follows the Jesuit collèges' theatrical productions for and about the king during his long wedding voyage in "Scènes jésuites du marriage de Louis XIV et de l'infante Marie-Thérèse d'Autriche" in *Plaire et instruire*, 70–94.

56. Jeannine Dorvane asserts that "as many as four thousand people gathered in the main courtyard of Louis-le-Grand [the later name of the Collège de Clermont]." "Ballet de Collège" in *International Encyclopedia of Dance*, ed. Selma Jeanne Cohen (Oxford: Oxford University Press, 1998), 1:349.

57. In a *lettre en vers* (August 16, 1665) La Gravette de Mayolas attributes the play to François Diez (1629–87), who wrote a succession of prize-day plays for the collège, including in 1664 (*Hermenigilde*) and 1666 (*Gusman*). *Continuateurs de Loret*, 1:171. The BNF record lists a coattribution on all of these works to François Ridelle (1627–94). According to the *Spectacles du XVIIème* database, Diez collaborated with fellow Jesuits Louis Jobert and Vincent Pauteraux on the 1667 tragedy and with Jobert alone on the tragedies from 1668 to 1670.

58. "Selon la coutume ordonnée / Dans cette Maison chaque année, / Le Révérend Père DIEZ, / Un Esprit des plus déliés, / Dont la veine docte et fertile / Égale Sénèque et Virgile. / Ce Poète et grand Orateur, /De cet Ouvrage fut l'Auteur, /Dont le nom fameux est IRLANDE, / Histoire belle, vraie et grande, / Ayant fait choix de bons Acteurs / Et fait instruire les Danseurs, / Qui dans leurs postures discrètes, Dansaient le Ballet des Comètes." *Lettres en vers à Madame la Duchesse de Nemours, 16 août 1665. Continuateurs de Loret* 1:171.

59. Most sincere thanks to Seeta Chaganti, Noah Guynn, Anna Klosowska, and Claire Waters, who consulted with me on the medieval plotlines and references in *Irlande*, and to Fran Dolan and Tiffany Worth, who generously elucidated the events of 1665 in Ireland before I found the play summary, which clarified that the subtitle of the ballet was a reference to the tragedy (about an eponymous queen), not the place. As is fitting for the arguments I am making about comets, researching this livret included forays in many chaotic directions, and I am grateful to the wonderful colleagues who have accompanied my errant research strands and ideas.

60. Claude François Ménestrier, *Des ballets anciens et modernes selon les règles du théâtre* (Paris: René Guignard, 1682), 279. Critics often cite this passage, as well as Ménestrier's commentary on what he calls "ballets d'attache" (279). The exact nature of this "attache" between the ballet and the drama into which the ballet is intercalated can be unclear, as Naudeix explores in "Tragédie et ballet." Piéjus traces the theory and practice of Jesuit collège performances through the career of Joseph de Jouvancy, who authored the tragedies at Louis-le-Grand (formerly the Collège de Clermont) from 1674 to 1698 and whose *Ratio docendi et discendi* (1685) offered a practical guide to Jesuit pedagogy. Jouvancy recommends that ballets should have a unified composition—"en toute composition, il faut que les parties concordent entre elles, et forment un ensemble où tout se tient." Piéjus, "Un spectacle collectif," 58.

61. Ménestrier, *Des ballets anciens et modernes*, 282.

62. Boysse enumerates a list of "ballets de circonstance" inspired by current events, from Louis XIV's marriage to military victories. *Théâtre des jésuites*, 53–54.

63. Ménestrier transcribes the summary from the livret, with a mention only of the Collège de Clermont performance. *Des ballets anciens et modernes*, 282–83.

64. Naudeix delves into how key Jesuit pedagogues negotiated the fascinating (and sometimes vexing) issue of ballet's expressive and communicative potentialities. "La danse constitue une forme particulièrement élaborée de représentation qui s'apparente à l'énigme: parce qu'elle se donne à déchiffrer, alors que les jésuites revendiquent une danse mimétique ; parce qu'elle n'explicite pas le sens qu'il faut lui donner, mais au contraire le brouille, elle place le spectateur dans une position active, guidée par le programme, et le renvoie véritablement à l'école." Naudeix, "Tragédie et ballet," 65–66.

65. Anne Surgers's analysis of Father Andrea Pozzo's treatises on theatrical perspective imputes theological importance to the symmetry of Jesuit theatrical design. "Une image unifiée, un regard captif: Le décor du théâtre jésuite dans les traités du F. Andrea Pozzo S.J. (1642–1709)," in *Plaire et instruire*, 85–98.

66. *Balet des comètes*, n.p.

67. Welch, *Theater of Diplomacy*; Cowart, *Triumph of Pleasure*.

68. *Balet des comètes*, n.p.

69. Rock asserts that the ballets were constructed as a kind of rhetoric, with the notation for each dance detailing the "figures" and "case" through which the dancer attempts to move or persuade the audience. Rock, "Jesuit College Ballets," 444–45.

70. Meredith Martin and Gillian Weiss identify this visual theme of the four corners of the earth as a frequent feature in what they discover to be the widespread iconography of galley slavery and the court's use, to symbolic ends, of the pervasive presence of galley slaves, in Louis XIV's France. *The Sun King at Sea. Maritime Art and Galley Slavery in Louis XIV's France* (Los Angeles: Getty Research Institute, 2022).

71. *Balet des comètes*, n.p.

72. *Balet des comètes*, n.p.

73. Rock, "Jesuit College Ballets," 434–35. Ndiaye, *Scripts*. McDowell E. Kenley examines the comic *matassin* characters in courtly ballets in this period, the music that accompanied their dances, and their association with commedia dell'arte characters and with the burlesque. While clearly the Clermont students would not have engaged in comic choreography as tinged with bawdy miming as the *matassins* described by Kenley, who brandish enemas at Monsieur de Pourceaugac, the livret clearly indicates a different style of movement associated with a transgressive presence. "Mad Fools and the Praise of Folly: Matassins and the Ballets of Lully, Destouches and Campra (1660–1718)," *Early Music* 45, no. 2 (August 2017): 449.

74. According to Rock, "In the Jesuit college ballets characters dancing in that style were often those acting outside the moral and Christian virtues the Jesuits hoped to instill in their students." "Jesuit College Ballets" 435. *Ratio docendi et discendi*, 56–57, quoted in Piéjus, n24.

75. Franko, *Dance as Text*, 5.
76. Franko, *Dance as Text*, 5.
77. Juliette Cherbuliez uses Jacques Callot's series known as "Bohémiens on the Move" or "The Life of Egyptians" to examine early modern notions of exile. "Yesterday's Exiles," *Self and State in the Early Modern World* (Cambridge: University of Cambridge Press, 2021), 24–25.
78. Christout, *Ballet*, 22–23. Gofflot notes, "Les sujets de ces pièces sont presque toujours religieux. Pourtant, on trouve quelquefois des thèmes profanes, voire païens, car le culte que les Jésuites professaient pour l'antiquité les faisait mettre en scène les héros de Rome et d'Athènes. Parfois même, ils les habillaient à l'antique, les faisaient penser et agir comme des païens et parler comme Sénèque." *Le théâtre au college*, 99. Norman revises critical understanding of the Quarrel of the Ancients and the Moderns and thereby recuperates the ancient as a kind of esthetic resistance. See Larry Norman, *The Shock of the Ancient: Literature and History in Early Modern France* (Chicago: University of Chicago Press, 2011).
79. Ellen McClure traces the underappreciated importance in seventeenth-century literature of "the concept of idolatry, with its slippery ambiguity, its profound implications simultaneously for authorship and artistic representation, and its terrifying implications for the fate of the human soul" in order to "interrogate authorship and artistry in a rapidly changing world." McClure, *The Logic of Idolatry in Seventeenth-Century French Literature* (Rochester, NY: Boydell and Brewer, 2020), 23.
80. *Balet des Comètes*, n.p.
81. *Balet des Comètes*, n.p.
82. *Balet des Comètes*, n.p.
83. *Irlande tragédie sera representee*, n.p.; *Balet des Comètes*, n.p.
84. *Balet des Comètes*, n.p.
85. The king visited the collège in 1674 (donating a tableau by Jean Jouvenet of *Alexander and the Family of Darius*), and in 1682 the collège was renamed after him: Louis-le-Grand. Hogg reads Charles Le Brun's rendering of the same Alexandre scene, *Les reines de Perses aux pieds d'Alexandre* (1660–61), as a demonstration of "a new kind of subject-sovereign relationship, and a new vision of the monarch, at the beginning of the young king's personal rule." Hogg, *Absolutist Attachments*, 17.
86. Sara Melzer explores how "culture and colonization were always conjoined, so interdependent that each enabled and shaped the other." See, Melzer, *Colonizer or Colonized: The Hidden Stories of Early Modern French Culture* (Philadelphia: University of Pennsylvania Press, 2012), 3. On the interconnected categories of gender and colonial others in the French Caribbean, see Ashley Williard, *Engendering Islands: Sexuality, Reproduction, and Violence in the Early French Caribbean* (Lincoln: University of Nebraska Press, 2021).
87. Jacques de Nambuc also figured in the 1663 *Ballet de la verité*. Boysse, *Théâtre des jésuites*, 147. According to Piéjus, "Ces jeunes interprètes étaient nécessairement issus des milieux aristocratiques, la précocité de la formation à la Belle Dance étant souvent l'un des signes de distinction que l'on qualifierait aujourd'hui de 'marqueur social.'" Piéjus, introduction, 8.

88. "Deux nouvelles statues déboulonnées en Martinique," *Le Monde*, July 27, 2020, https://www.lemonde.fr/societe/article/2020/07/27/deux-nouvelles-statues-deboulonnees-en-martinique_6047405_3224.html.

Chapter 3

1. Auzout was not alone in considering "comète" to be a masculine noun. The debate about the gender of the word "comet" was a popular topic of discussion. This question appears in the *Journal des sçavans* as discussed in chapter 1 and in the play *La comète*, which is the focus of chapter 4.

2. Pascal Descampes, "A l'origine de l'Observatoire Royal," https://www.obspm.fr/a-l-origine-de-l-observatoire.html?lang=fr (accessed June 2, 2017). The page is illustrated by the Le Clerc image of Louis XIV with the Académie royale des sciences that I discuss later in this chapter. I quote the English version of the website rather than my own translation of the French website.

3. Louis XIV's political use of the spectacular has been an important thread in scholarship from Apostolidès and Marin to more current work. For example, Hogg's analysis of *Les reines de Perses* foregrounds the importance of visual engagement in creating the affective bonds that she argues linked subjects and sovereign. *Absolutist Attachments*, 21–66. See also, Blanchard, *L'optique du discours au XVIIe siècle: De la rhétorique des jésuites au style de la raison modern* (Québec: Presses de l'Université Laval, 2005); Harriet Stone, *Crowning Glories: Netherlandish Realism and the French Imagination during the Reign of Louis XIV* (Toronto: University of Toronto Press, 2019); and Zanger, *Scenes from the Marriage*.

4. There is already a long scholarly tradition of challenging the epistemological primacy of the visual. In *Vanities of the Eye*, for example, Stuart Clark traces a counter history (often Protestant but also French Pyrrhonian) of "how ambiguously sight was regarded even in an already ocularocentric age," contending that "during the early modern period, and more especially between the Reformation and the Scientific Revolution, vision was anything but objectively established or secure in its supposed relationship to external fact. Many intellectuals, at least . . . seem to have been preoccupied precisely with questions to do with whether human vision did give reliable access to the real world after all—and whether vision was indeed veridical." Stuart Clark, *Vanities of the Eye: Vision in Early Modern European Culture* (Oxford: Oxford University Press, 2007), 1. Harriet Stone reconsiders epistemology through the lens of Dutch still life painting. See *Tables of Knowledge: Descartes in Vermeer's Studio* (Ithaca, NY: Cornell University Press, 1996). Likewise, early modern writers, from Descartes to La Fontaine (in *Les amours de Psyché et de Cupidon*, 1669), challenged readers to question what their eyes seemed to tell them.

5. A draft "Projet de la compagnie des sciences et des arts" is included in the Huygens papers. Brown, who analyzes the "Projet" at length, judges "that Huygens received it early in 1665, about the time of the publication of the *Éphéméride*, seems probable." Christiaan Huygens, *Correspondance 1662–1663*, vol. 4 of *Oeuvres complètes de Christiaan Huygens* (La Haye: Martinus Nijhoff, 1891), 325–29. Harcourt Brown, *Scientific Organizations*

in Seventeenth Century France (1620–1680) (New York: Russell and Russell, 1967), 145. Charles Perrault also proposed an academy around this time. David S. Lux, "Colbert's Plan for the Grand Académie," *Seventeenth-Century French Studies*, 12 (1990): 177–88.

6. See the special issue "Les academies avant l'Académie: L'essor des sociétés savants en France avant la foundation de l'Academie des sciences," ed. Domenico Callacciani and Alberto Frigo, *Dix-septième siècle* 292 (July 2021); Brown, *Scientific Organizations*, 75; Roger Hahn, *The Anatomy of a Scientific Institution: The Paris Academy of Sciences, 1666–1803* (Berkeley and Los Angeles: University of California Press, 1971), 4–6; Roger Hahn, "La fondation de l'Académie royale des science," in *Sciences and Curiosités à la Cour de Versailles*, ed. Béatrix Saule and Catherine Arminjon (Paris and Versailles: Réunion des musées nationaux/Château de Versailles, 2011), 31–35; David Sturdy, *Science and Social Status: The Members of the Academie des Sciences 1666–1750* (Woodbridge, UK: Boydell, 1995), 63–79.

7. In recent decades a vibrant critical literature on salons includes important works by Faith Beasley, Dena Goodman, Erica Harth, Ian Maclean, and Linda Timmermans. Faith Beasley's *Versailles Meets the Taj Mahal: François Bernier, Marguerite de La Sablière, and Enlightening Conversations in Seventeenth-Century France* (Toronto: University of Toronto Press, 2018) foregrounds the intellectual connections fostered in Marguerite de La Sablière's salon.

8. In the short self-portrait published posthumously with the 1694 catalog of Melchisédec Thevenot's library (which was absorbed by the king's when Thevenot was named royal librarian in 1684), Thevenot recalls setting up his own informal center of research upon returning to Paris from a series of "very dangerous" and "important" missions for the crown. He struggles to afford to house his colleagues and support their research program. Melchisédec Thevenot, *Bibliotheca Thevenotiana, sive catalogus impressorum et manuscriptorum librorum bibliothecae viri clarissimi D. Melchisedicis Thevenot* (Paris: Delaulne, 1694), n.p. See also Hahn, *Anatomy*, 7.

9. Soll, *Information Master*, 7.

10. On Colbert and the Academies, see Soll, *Information Master*, 99–101.

11. Daston and Park focus primarily on the Royal Society in chapter 6 of *Wonders and the Order of Nature*, but their argument about the social bases for turning from theorizing systems to examining individual occurrences is likewise enlightening for the Académie royale des sciences. *Wonders*, 215–53. See also Roux, "The Two Comets," 119–26.

12. "Epistre," Adrien Auzout, *Éphéméride du comète de 1664*. s.l., n.p. Sturdy emphasizes the degree to which the *savants* waged a charm campaign on Colbert, appealing to his desire for glory. Sturdy, *Science and Social Status*, 63–75.

13. At the end of the "Epistre," Auzout follows a similar line of persuasion in relation to his work toward building longer telescopes. He is planning to publish a discourse on telescopy, and in particular his invention of a tubeless aerial telescope. "I will describe the uses that make them worthy to be used by Kings as well as Astronomers. I thought it was my duty to work on it after having had the honor of showing Your Majesty the Moon with the beautiful Telescope that the King of England presented to Monsieur, only Brother of

Your Majesty, and so that it be known that the biggest Telescopes that have been made up to the present, were primarily made in your Kingdom." "Epistre," Auzout, *Éphéméride*, n.p.

14. "Epistre," Auzout, *Éphéméride*, n.p.

15. In fact, the pamphlet was published later, although Auzout insists that it was sent to the printer in advance. Furetière's *Dictionnaire universel* defines "éphémérides" in the following way: "Ce sont des tables calculées par des Astronomes, qui marquent l'état du ciel tous les jours à midi, c'est à dire le lieu où à midi se trouvent toutes les Planètes: et ce sont ces tables qui servent à dresser les horoscopes ou thèmes célestes."

16. "Epistre," Auzout, *Éphéméride*, n.p.

17. "Epistre," Auzout, *Éphéméride*, n.p.

18. Auzout, *Éphéméride*, 1.

19. Auzout was working on a "tuyau," or optical tube, to hold bigger telescopes to hone this predictive model. He imagines the increasingly exact observation that royal investment will make possible: "My plan not being to give Observations of the past, but Predictions for the future, I will not speak of Observation." Auzout, *Éphéméride*, 2.

20. Auzout, *Éphéméride*, 2.

21. On the political aspirations of seventeenth-century cartography, see Tom Conley, *The Self-Made Map: Cartographic Writing in Early Modern France* (Minneapolis: University of Minnesota Press, 2010); Thierry Mariage, *The World of André Le Nôtre*, trans. Graham Larkin (Philadelphia: University of Pennsylvania Press, 2010); and Chandra Mukerji, *Territorial Ambitions and the Gardens of Versailles* (Cambridge: Cambridge University Press, 1997).

22. "Epistre," Auzout, *Éphéméride*, n.p.

23. Brown, *Scientific*, 91–116. On the Accademia del Cimento see Luciano Boschiero, *Experiment and Natural Philosophy in Seventeenth-Century Tuscany: The History of the Accademia del Cimento*, Australasian Studies in History and Philosophy of Science 21 (Dordrecht: Springer Verlag, 2007); and W. E. Knowles Middleton, *The Experimenters: A Study of the Accademia del Cimento* (Baltimore: Johns Hopkins University Press, 1971).

24. "Epistre," Auzout, *Éphéméride*, n.p.

25. *Journal des sçavans*, January 26, 1665, 44.

26. In the introduction to a volume of essays about nineteenth-century observatories, David Aubin, Charlotte Bigg, and H. Otto Sibum note that "while they insisted on the importance of observation in the pursuit of scientific truths, scientists and philosophers widely differed on what they meant by it." David Aubin, Charlotte Bigg, and H. Otto Sibum, introduction to *The Heavens on Earth: Observatories and Astronomy in Nineteenth-Century Science and Culture* (Durham, NC: Duke University Press, 2010), 4–6.

27. Aubin, Biggs, and Sibum contend that observatories not only served as "key nodes" in the construction of scientific and political networks but furthermore served as the model for such networks. Aubin, Biggs, and Sibum, introduction, 15–16.

28. Brown, *Scientific*, 150–54; Hahn, *Anatomy*, 11–14; Sturdy, *Science and Social Status*, 77; and Lux, "Colbert's Plan."

29. Sturdy, *Science and Social Status*, 77–78.

30. Voltaire highlights the establishment of the Académie des sciences as a reaction to important discoveries by the Royal Society. Voltaire, *Le siècle de Louis XIV*, 161–62.

31. Anita Guerrini, *The Courtiers' Anatomists: Animals and Humans in Louis XIV's Paris* (Chicago: University of Chicago Press, 2015), 93. The academy moved to the Louvre in 1699.

32. Charles and Claude Perrault explain, "On pensa d'abord à placer l'Observatoire sur le tertre de Montmartre, comme le plus proche de Paris qui fut assez élevé pour bien découvrir tout l'horizon. Mais on trouva que les fumées qui s'élèvent continuellement de Paris, situé au midi de Montmartre, étaient un obstacle perpétuel à toutes sortes d'observatoire. De sorte qu'il fut conclu qu'il fallait choisir une situation tout opposée à celle-là et que fut Paris à son nord, pour avoir le levant, le midi et le couchant entièrement libres et dégagés de toutes les fumées et de toutes les vapeurs qui s'élèvent continuellement de la ville." Charles and Claude Perrault, "Pourquoi et comment l'observatoire a esté basty," in *Fortifications, sciences, lettres, beaux-arts, bâtiments,* vol. 5 of *Lettres, instructions, et memoires de Colbert*, ed. Pierre Clément (Paris: Imprimerie impériale, 1873), 515.

33. Olaf Pedersen, "Some Early European Observatories," *Vistas in Astronomy* 20, no. 1 (1976): 18–19.

34. Aubin, Bigg, and Sibum, introduction, 22–23.

35. The inclusion of the observatory in de Nicolas de Fer's 1704 *Atlas curieux* is an interesting mise en abyme. The maps collected in the atlas were based on observations by the Académie royale des sciences, some taken from the observatory as Auzout had predicted. De Fer, *L'atlas curieux*, 2:n.p.

36. René Taton, "Les origines et les débuts de l'Observatoire de Paris," *Vistas in Astronomy* 20, no. 1 (1976): 67.

37. Hahn, *Anatomy*, 19. On specifics of the building, see Guerrini, *Courtiers' Anatomists*, 93nn2–3.

38. Aubin, Bigg, and Sibum, introduction, 24.

39. Auzout was one of many *savants* to work on telescope design with the hopes of increasing magnification without jeopardizing accuracy. Examples of previous observatories in Europe might include Hevelius's rooftop observatory, the Round Tower of Copenhagen, and Tycho Brahe's Uraniborg and Stjärneborg. See Florian Cajori, "Four Old Astronomical Observatory Buildings," *Scientific Monthly* 26, no. 4 (April 1928): 372–76. Pedersen credits Tycho's published works (he had a printing press at Uraniborg) for inventing the modern observatory. Pedersen, "Some Early," 24.

40. Oded Rabinovitch discusses how "Versailles served as link between the glory of Louis XIV and the personal status of men of letters" and was thus important, for example, in association with Claude Perrault's dissection of animals from the royal menagerie." Oded Rabinovitch, *The Perraults: A Family of Letters in Early Modern Science* (Ithaca, NY: Cornell University Press, 2018), esp. chapter 5, 104, 114–17.

41. Perrault, "Pourquoi," 515.

42. See Rybczynsky's discussion in *Home*, picked up in my *Vaux and Versailles* and by DeJean in *The Age of Comfort*. Witold Rybczynski, *Home: A Short History of an Idea* (New York: Viking, 1986); Joan De Jean, *The Age of*

Comfort: When Paris Discovered Casual and the Modern Home Began (New York: Bloomsbury, 2009).

43. Charles Perrault's posthumously published memoires are a primary source, although doubtless a biased one, regarding his brother Claude Perrault's life and works. These largely serve as the basis for André Hallays's *Les Perrault* (Paris: Perrin et Cie,1926). More modern works on Perrault include Wolfgang Hermann's *The Theory of Claude Perrault* (London: Frances Schram, 1973); a chapter on Claude Perrault's architectural theory in Alberto Pérez-Gómez's *Architecture and the Crisis of Modern Science* (Cambridge, MA: MIT Press, 1983), 17–47; Antoine Picon's *Claude Perrault, 1613–1688, ou la curiosité d'un classique* (Paris: Picard, 1988), which focuses on various aspects of Perrault's career, from medicine and machines to architecture and theory; Michael Petzet's *Claude Perrault und die Architektur des Sonnenkönigs: Der Louvre König Ludwigs XIV und das Werk Claude Perraults* (Munich: Deutscher Kustverlag, 2000); and Oded Rabinovitch's fine history of the Perrault family: *The Perraults*.

44. Paul Bonnefon, introduction to *Mémoires de ma vie*, by Charles Perrault (Paris: Renouard, 1909), 5.

45. Rabinovitch, *The Perraults*, 1–36.

46. As early as 1662 Colbert, explains Charles Perrault, recognized the central role architectural and festive constructions—and images of these—played in the establishment of Louis XIV's adult reign, so in early 1663 he convokes a "kind of privy council" to help him shape a coherent and powerful message. Charles Perrault, *Mémoires de ma vie*, 34–35. Perrault also recounts (or conjures) the king's own recognition of the central importance of the Petite Académie's work: "'Je vous confie la chose du monde qui m'est la plus précieuse, qui est ma gloire. Je suis sûr que vous ferez des merveilles ; je tâcherai de ma part de vous fournir de la matière qui mérite d'être mise en oeuvre par des gens aussi habiles que vous êtes." Charles Perrault, *Mémoires de ma vie*, 41.

47. The Louvre colonnade design process was contentious and intrigue filled, so the resulting structure remains difficult to attribute. Petzet offers the most comprehensive history. Claude Perrault had submitted a Louvre project in 1664, but the crown pursued a proposal from noted Italian sculptor and architect Gian Lorenzo Bernini (1598–1680), who made a famous visit to Paris in 1665 to work on the plans. Robert Berger notes: "What then ensued between late 1665 and the spring of 1667 is largely hidden from the historian's purview but must have resembled diplomatic intrigue. A cabal, already formed against Bernini in 1665, went to work to persuade Colbert against the construction of Bernini's Louvre, probably on the grounds of both aesthetics and French national honor." Robert Berger, *A Royal Passion: Louis XIV as Patron of Architecture* (Cambridge: Cambridge University Press, 1994), 33. In April 1667, after Bernini's plans were abandoned, Claude Perrault was named to the Petit conseil with Louis XIV's architect Louis Le Vau and artist Charles Le Brun. Berger, *Royal Passion*, 34.

48. Admittedly this is a slightly ironic distinction to inscribe under an engraved portrait that offers Claude Perrault's likeness to our view, but such inconsistencies aside, the lozenge celebrates Perrault as an avatar of Modern

sensibility, which would be politicized when the Quarrel of the Ancients and the Moderns consumed France about a decade later, with Charles Perrault at the head of the Moderns. Joan De Jean, *Ancients and Moderns* (Chicago: University of Chicago Press, 1997), and Larry Norman, *The Shock of the Ancient.*

49. Picon, *Claude Perrault*, 19–21. Domenico Bertoloni Meli, *Thinking with Objects: The Transformation of Mechanics in the Seventeenth Century* (Baltimore: Johns Hopkins University Press, 2006), 161–65. Although certain practitioners and theoreticians presented medicine in this light, other scholars reexamine the supposed primacy of visuality in medicine and medical education. See Cynthia Klestinec, "Practical Experience in Anatomy," in *The Body as Object and Instrument of Knowledge: Embodied Empiricism in Early Modern Science*, ed. Ofer Gal and Charles T. Wolfe (Sydney: Springer, 2010), 33–58.

50. Claude Perrault, *Les dix livres d'architecture de Vitruve, corrigés et traduits nouvellement en François avec des notes et des figures* (Paris: Jean-Baptiste Coignard, 1673), n.p.

51. Hilary Ballon argues that Perrault's theory of the arbitrary nature of architectural beauty, adumbrated in his introduction, prepares a unique role for French sovereign power vis à vis architecture: "Claude Perrault provided theoretical support for the extension of royal authority into the regulation of classical architecture. . . . In the domain of architectural theory, Perrault mapped out a space into which the Bourbon crown moved, a space constituted for the societal function of defining aesthetic conventions whose authority derived neither from nature nor God but from the king of France." Hilary Ballon, "Constructions of the Bourbon State: Classical Architecture in Seventeenth Century France," *Studies in the History of Art* 27 (1989): 141–42.

52. Picon, *Claude Perrault*, 20 et passim.

53. The Du Hamel Latin history of the academy was published in 1698. Fontenelle's history did not appear in print until 1733. Guerrini traces the contradictions between these two early histories of the academy.

54. Bernard Le Bovier de Fontenelle, *Depuis son établissement en 1661 jusqu'à 1686*, vol. 1 of *Histoire de l'Académie royale des sciences* (Paris: Gabriel Martin, Jean-Baptiste Coignard, fils, Hippolyte-Louis Guerin, 1733).

55. The secretary of the Accademia del Cimento, Lorenzo Magalotti, expresses a similar experimentalist project in the preface to the Accademia's published *Saggi di naturali esperienze fatte nell'Accademia del Cimento sotto la protezione del serenissimo Principe Leopoldo di Toscano e descritte dal segretario dessa accademia* (1666): "If sometimes in passing from one experiment to another, or for any other reason whatever, some slight hint of speculation is given, this is always to be taken as the opinion or private sentiment of the academicians, never that of the Academy, whose only task is to make experiments and to tell about them" (trans W. E. K. Middleton, quoted in Boschiero, *Experiment*, 2). Boscheiro calls this "an example of the experimentalist rhetoric that appears to have been sweeping across Europe during the latter half of the seventeenth century." Boschiero, *Experiment*, 3.

56. I follow a significant critical field in noting that the *claim* of scientific observation is often more an ideal than a reality, one that becomes increasingly complex the closer one scrutinizes it. For example, Oded Rabinovitch

observes that "the literary opportunities and constraints created by the Academy's relation to the monarchy and to literary circles in Paris shaped the nature of the cutting-edge scientific work in natural history done in mid-seventeenth century Paris." "Chameleons between Science and Literature: Observation, Writing, and the Early Parisian Academy of Sciences in the Literary Field," *History of Science* 51, no. 1 (2013), 33. See also Gal and Chen-Morris, *Baroque Science*, and Lorraine Daston and Peter Gallison, *Objectivity* (New York: Zone Books, 2010).

57. Fontenelle, *Histoire de l'Académie royale des sciences*, 1:43; see also Picon, *Claude Perrault*, 202nn12–13; 273. A seventeenth-century guidebook observes, for example, "The four facades exactly face the four Corners of the Earth." Charles Le Maire, *Paris ancient et nouveau, ouvrage très curieux où l'on voit la fondation, les accroissemens, le nombre des habitans et des maisons de cette grande ville, avec une description de ce qu'il y a de plus remarquable* (Paris: Chez Michel Vaugon, 1685), 3:296.

58. On the difficulty of attributing the ubiquitous but understudied topographical prints of the Pérelle family, see William Roberts, "Perelle's Topographical Albums: Problems and Solutions," in "Relations and Relationships in Seventeenth-Century French Literature," Special issue, *Biblio* 17, no. 166 (2006): 163–76.

59. Berger writes, "There is the long-recognized demand that royal edifices 'speak,' in some fashion, about the political power, material wealth, and artistic taste of the ruler. No potentate understood this requirement better than Louis XIV." Berger, *A Royal Passion*, 2.

60. Furetière's second example of "observatoire" in his *Dictionnaire universel* slyly undermines Louis XIV's "singulier et magnifique" star-castle, however. He states that the Beijing Observatory, which predates the Observatoire Royal by more than two centuries, is the new institution's equal: "Lieu propre à observer les astres. Le Roy a fait bâtir auprès de Paris un *Observatoire* Royal, un château pour l'observation, qui est un édifice singulier et magnifique. On dit qu'à la Chine il y a un *Observatoire* admirable, et des instruments aussi excellents que ceux de l'Europe, comme témoigne le Père Greslon Jésuite en son *Histoire de la Chine*."

61. Charles Wolf calculates the cost of the observatory construction at 713,954 livres. *Histoire de l'Observatoire de Paris de sa foundation à 1793* (Paris: Gauthier-Villars, 1902).

62. Charles Sevin de Quincy, *Histoire militaire du règne de Louis-le-Grand, Roy de France enrichie des plans necessaires: On ya joint un traité particulier de pratiques et de maxims de l'art militaire* (Paris: Chez Denis Mariette, Jean-Baptiste Delespine, Jean-Baptiste Coignard, 1726), 1:285.

63. "By these two means, the Petite Académie endeavored to provide a comprehensive and chronological numismatic history of the reign." Robert Wellington, *Antiquarianism and the Visual Histories of Louis XIV* (Farnham: Ashgate, 2015), 52. On the Petite Académie and the *Histoire métallique* see chapter 2. In chapter 5 Wellington traces the elaboration of Sébastien Leclerc, Antoine Coypel, and Charles-Louis Simonneau, *Médailles sur les principaux événements du règne de Louis Le Grand, avec des explications historiques: Par l'Académie royale des médailles et des inscriptions* (Paris: Imprimerie royale, 1702).

64. The entry on the Observatoire Royal celebrates the building's design and siting, magnificence, and usefulness in almost identical terms to the Pérelle prints and likewise in terms that could extend to the book's printed page, a simultaneous view of both sides of the medal (the king's head and the observatory) and the short text and explication of the medal's Latin motto. Leclerc, Coypel, and Simonneau, *Médailles sur les principaux événements du règne de Louis Le Grand*, 103. A catalog to a 1966 exhibition celebrating the tricentenary of the academy notes, "Les devises et les médailles furent composés par l'Académie royale des Inscriptions très postérieurement, comme ce fut le cas pour la majorité des médailles, à la date de fondation de cette Académie." Pierre Gauja and Marie-Thérèse Delom, *Troisième Centenaire, 1666–1966* (Paris: Imprimerie Tournon et Cie, 1966), 3.

65. "Arc de triomphe pour les conquêtes de terre—observatoire pour les cieux." Quoted in Michael Petzet, Claude Perrault als Architekt des Pariser Observatoriums," *Zeitschrift für Kunstgeschichte* 30, no. 1 (1967): 1–54.

66. Jean Picard's *Voyage* documents his trip to Uraniborg to measure the exact longitude and latitude of Tycho Brahe's observatory at Hven, so that the French Academy could use his observations. Jean Picard, *Voyage d'Uranibourg, ou Observations faites en Danemarck* (Paris: Imprimerie royale, 1680), 1.

67. Jean-Dominique Cassini, "De L'origine et du progress de l'astronomie dans la géographie et dans la Navigation," in *Recueil d'observations faites en plusieurs voyages par ordre de Sa Majesté pour perfectionner l'astronomie et la géographie: Avec divers traitez astronomiques par messieurs de l'Académie royale des sciences* (Paris: Imprimerie Royale, 1693), 1–42. The collected texts can be found in the Smithsonian Institution. Cassini's text appears bound separately in collections as well. An allegorical vignette used in a few other titles in the same collection features a robed figure measuring a distance on a globe with calipers. http://www.e-rara.ch/zut/content/pageview/1951484.

68. Berger, *A Royal Passion*, 46.

69. *Mercure galant*, November 1688, 1:226–28.

70. Una Roman D'Elia, "Classicism and the Problem of National Style: The Illustrations of Claude Perrault's Vitruvius," in *Coming About . . . : A Festschrift for John Shearman*, ed. Lars Jones and Louisa Matthews (Cambridge, MA: Harvard University Art Museums, 2001), 335.

71. Pérez-Gómez, introduction to *Ordonnance for the Five Kinds of Columns after the Method of the Ancients*, by Claude Perrault. Translated by Indra Kagis McEwen. Los Angeles: Getty Publications, 1993, 7–8.

72. Petzet comments on Sébastien Leclerc's frontispiece to Claude Perrault's *Les dix livres* in "Claude Perrault als Architekt des Pariser Observatoriums." Diana Rowell discusses the image, although she incorrectly attributes the frontispiece to the *Abrégé*. Diana Rowell, *Paris: The "New Rome" of Napoleon I* (New York: Bloomsbury, 2012), n77 et passim.

73. Ballon, "Constructions," 144.

74. Claude Perrault, *Les dix livres*, n.p.

75. Roman D'Elia attributes the prints in the Vitruvius translation to Perrault. "While LeClerc signed some of the prints, specifically the allegorical scenes, 'S. Le Clerc Inventor,' and while he and others sign some others as the engravers, 'sculp.,' many other prints are unsigned. Perrault claims

responsibility for the illustrations in the preface. Indeed, the uniform quality of the prints, despite the different engravers, suggests one controlling designer." Roman D'Elia, "Classicism," 337–38n6.

76. Contemporary descriptions such as the one in the *Mercure galant* attest that Cassini and his family and assistants lived in much of the space originally designed to house the laboratories and conferences of the Academy.

77. Léopold Durangel, *Portrait of Jean-Dominique Cassini*, 1879. Bibliothèque de l'Observatoire de Paris.

78. In a 1903 catalog of prints representing the Jardin des Plantes, Louis Denise attests to the reprinting of this frontispiece, still in superfolio five years later, as the frontispiece for the *Mémoires pour servir à l'histoire des plantes, dressez par M. Dodart* (Paris: Imprimerie Royale, 1676) and the *Recueil des voyages des académiciens*. Louis Denise, *Bibliographie historique et iconographique du Jardin des plantes, Jardin royal des plantes médicinales, et Muséum d'histoire naturelle* (Paris: H. Daragon, 1903), 50–52. I have not been able to locate this last work in the BNF catalog or elsewhere. In a 1911 article on engravings of the observatory, Félix Boquet cites the same titles. Both follow the Jombert catalog of Le Clerc's engravings. According to Jombert, in 1730 this image was reduced to quarto and engraved by Cl Duflos, père, to serve as frontispiece to *Memoires de l'Académie des sciences*, although it was ultimately not used for this purpose. Boquet locates copies of this rare print in museum collections. Félix Boquet, "L'Observatoire de Paris: A propos de deux gravures relatives à sa foundation," *L'astronomie* 25 (1911): 51. Charles-Antoine Jombert, *Catalogue raisonné de l'oeuvre de Sébastien Le Clerc, chevalier Romain, dessinateur et graveur du cabinet du Roi*, 2 vols. (Paris: Charles-Antoine Jombert, 1774), 1:150–51.

79. My thanks to Juliette Cherbuliez and Christopher Semk for help identifying and interpreting these flowers, many of which are corms or bulbs. Dodart's *Mémoires pour servir à l'histoire des plantes* was not based on direct observation of plants but was a compilation from books.

80. According to the *Mercure galant* account, Louis XIV acquired at least two burning mirrors.

81. Measuring nearly a meter in diameter, the mirror could magnify the light of a single candle to illuminate the Petite Gallerie de Versailles. It is still held in the collections of the Observatoire de Paris.

82. Jean Donneau de Visé, *Suite du voyage des ambassadeurs de Siam en France . . . Novembre 1686* (Lyon: Thomas Amaulry, 1687).

83. Donneau de Visé, *Suite du voyage*, 30.

84. Donneau de Visé, *Suite du voyage*, 36–41.

85. Donneau de Visé, *Suite du voyage*, 42.

86. Donneau de Visé, *Suite du voyage*, 42.

87. Donneau de Visé, *Suite du voyage*, 43.

88. "The Ambassador marveled at the clarity of a Telescope of twenty-five feet, with which he observed the most distant objects." Donneau de Visé shares a promotional aside about an aerial telescope system described by Comiers in his 1665 book on comets (and republished in the 1683 and 1684 *Extraordinaires* du *Mercure galant*) that would allow the construction of accurate tubeless telescopes even eight times the size of the one the ambassador admires. Donneau de Visé, *Suite du voyage*, 43–46.

89. Donneau de Visé, *Suite du voyage*, 50–54.

90. Donneau de Visé, *Suite du voyage*, 52.

91. Donneau de Visé, *Suite du voyage*, 52.

92. Donneau de Visé, *Suite du voyage*, 54.

93. Donneau de Visé, *Suite du voyage*, 53–54. Donneau references Father Jean de Fontaney, whose published comet observations from the Collège de Clermont were reviewed in the *Journal des sçavans*. See chapter 1, n72.

94. Donneau de Visé, *Suite du voyage*, 55–56.

95. Fontenelle began his literary career under the aegis of Donneau de Visé. See chapter 4.

96. On imagination in Fontenelle's *Entretiens sur la pluralité des mondes*, see Juliette Cherbuliez, "On Letting Sleeping Blonds Lie: Gender, Leisure Literature, and the Imagination in Fontenelle," *Romanic Review* 102, no. 1/2 (January–March 2011): 145–68; and Natasha Lee, "Planetary Perspectives in Enlightenment Fiction and Science," in *French Global: A New Approach to Literary History*, ed. Christie McDonald and Susan Rubin Suleiman (New York: Columbia University Press, 2010), 94–107.

97. "Next he saw a Planetary Machine following the Copernican System. It can be called talking Ephemerides, and can find the state of the Sky at whatever time one chooses, either past, present or future, the longitude and latitude of each Planet, and by consequence its real place in the Sky on whatever day one wishes, simply by turning a crank. . . . The Ambassador spent a long time considering how much slower Saturn travelled than the other Planets." Donneau de Visé, *Suite du voyage*, 58–60.

98. Donneau de Visé, *Suite du voyage*, 60–62.

99. Donneau de Visé, *Suite du voyage*, 61.

100. The episode ends on an assertion of witness that is entirely auditory: "I do not report anything here that I did not witness; everything I cite from the Ambassador, I heard myself" ("Je ne vous rapporte rien ici dont je n'ai été témoin; tout ce que je vous cite de l'Ambassadeur, je l'ai entendu moi-même.") Donneau de Visé, *Suite du voyage*, 65–66.

101. Quoted and translated in Johann-Christian Klamt, "Astronomical Observatories in the Seventeenth and Eighteenth Centuries," in *Public Buildings in Early Modern Europe*, ed. Koen Ottenhym, Krista De Jonge, and Monique Chatenet, Architectectura Moderna 9 (Turnhout: Brepolis, 2010), 373.

102. Huygens, *Oeuvres*, 8:390

103. Jean-Dominique Cassini IV, *Mémoires pour servir à l'histoire des Sciences et à celle de l'Observatoire Royal de Paris* (Paris: Bleuet, 1810), 185. This theme is echoed by Wolf in his 1902 history and in more recent work by David Aubin. Wolf, *Histoire de l'Observatoire*.

104. Cassini IV, *Mémoires*, 185.

105. Cassini IV, *Mémoires*, 185.

106. "Il paraît qui l'architecte, qui avait présidé à l'ordonnance et à la distribution de l'Observatoire, n'avait qu'une faible notion de la pratique des observations, et qu'il avait fort peu consulté les astronomes sur les commodités qu'il devait leur procurer. Il crut sans doute avoir tout fait pour l'astronomie en lui bâtissant un vaste édifice fort élevé, d'une belle masse et d'un style d'architecture sage, sévère et convenable au genre de la science ; mais ce n'était

rien de tout cela qu'il fallait. La hauteur et l'étendue de l'édifice n'étaient qu'un inconvénient d'autant plus considérable que, dans quelqu'endroit que l'on pût se placer, la masse du bâtiment dérobait à la vue la plus grande partie du ciel. A moins de monter et de rester en plein air sur la plate-forme, on ne pouvait suivre le cours d'un astre élevé depuis son lever jusqu'à son coucher ; et pour observer au levant ou au couchant, il fallait transporter un instrument d'un bout à l'autre du bâtiment ; de plus, les voûtes massives, qui couvraient toutes les salles, ne permettaient de découvrir le méridien depuis le zénith jusqu'à l'horizon, dans aucun endroit de l'Observatoire ; enfin, ce qui paraitra fort singulier, on n'avait pas ménagé une seule place d'où l'on pût prendre des hauteurs correspondantes sans déranger considérablement l'instrument." Cassini IV, *Mémoires*, 185.

107. Josephine Touma very helpfully pointed out this difference.

108. "It is altogether remarkable that the creation of the Paris Observatory happens at such a marked turning point in this discipline, a circumstance that allowed its astronomers to play a decisive role in the final refinement of these innovations and in the collection of the first results that used them. Of course, the precision of the instruments, the quality of research programs and the competence of the observers were in reality much more determinant than the building itself, but its creation nevertheless occupied symbolic value." Taton, "Les origines," 67.

109. Huygens, *Oeuvres Complètes*, 8:312–13.

110. Jean-Dominique Cassini, *Abrégé des observations et des réflexions sur la comète qui a paru au mois de décembre 1680 et aux mois de janvier, février, mars de cette année 1681: Présenté au Roy Par Mr. Cassini* (Paris: Estienne Michallet, 1681), xx.

111. Jean-Dominique Cassini, *Premières observations de la comète de ce mois d'aoust 1682, présentées au Roy par M. Cassini de l'Académie royale des sciences* (Paris: Sébastien Mabre-Cramoisy, 1682), 3.

112. "Ce premier crayon servira pour montrer à VOSTRE MAJESTÉ mon obéissance, en lui présentant, selon le commandement dont il lui a plu de m'honorer, ce qu'on découvre de plus rare et de plus remarquable à son Observatoire Royal." Cassini, *Premières observations*, 7.

113. I add just for the ironic twist to this story that by the end of the ancien régime due to strife between the Cassini heir, Jean-Dominique IV, and the rest of the astronomers in the academy, the Observatoire Royal was not even the site of most of the academicians' observations. Roger Hahn, "Les Observatoires en France," in *Enseignement et diffusion des sciences en France au XVIIIe siècle*, ed. René Taton (Paris: École pratique des hautes études, 1964), 653–58.

114. The account of the meeting is in the *Mercure galant*, May 1699, 1–14, here 2–3.

115. *Mercure galant*, May 1699, 3–4.

116. *Mercure galant*, May 1699, 4–5.

117. *Mercure galant*, May 1699, 7.

118. *Mercure galant*, May 1699, 9.

119. "Tout ce qui se passa dans cette séance fut fort applaudi. Parmi ceux qui composaient la nombreuse assemblée qui s'y trouva, il y avait des Résidents,

des Envoyés, quantité d'Étrangers de distinction, et même des Princes." *Mercure galant*, May 1699, 13.

120. Boschiero, for example, argues that a strict focus on direct observation in the Accademia del Cimento downplayed philosophizing as a way of depoliticizing the academy's activities. He also shows that there were big stakes for the projection of the comet. Bosciero, *Experiment*, 222–28.

Chapter 4

1. See chapter 1.

2. *Mercure galant*, January 1681, 328.

3. Julie Boch reads the play as a precursor to Fontenelle's later preoccupations: "À la fois synthèse des arguments du temps contre la superstition astrologique et plaidoyer pour la physique contre la rhétorique, *La comète*, malgré la modestie affichée de son propos et de ses dimensions, non seulement apparaît comme le reflet d'une époque, mais annonce une oeuvre à venir." Julie Boch, "Une rhétorique de la superstition: *La comète* de Fontenelle (1681)," in *Approches de la pensee des Lumières* (Reims: Éditions et presses universitaires de Reims, 2012), 108. The play falls outside the time frame covered in Schuwey's masterful synthesis of Donneau de Visé's early career. He mentions it in a list of Donneau de Visé's plays as a "possible collaboration" with Fontenelle (124) and draws attention to the zigzag typography it shares with the *Mercure galant*. Schuwey, *Un entrepreneur*, 124 and 370–71. Riffaud likewise discusses Claude Blageart's use of the *Mercure galant* type on the title page of *La comète*, which he attributes to Donneau de Visé and Thomas Corneille. Riffaud, *L'aventure*, 153, 155, 177.

4. Vincent documents the collaboration agreement in *Donneau de Visé et "Le Mercure galant,"* 2:13–21.

5. It might be helpful to think of the play alongside the novellas published in the *Mercure galant* (see my reading of the novella "La comète" in chap. 1) or the novellas from Donneau de Visé's earlier collections, for example, the *Nouvelles galantes, comiques et tragiques* or the third volume of *L'amour échappé*, which Schuwey characterizes as "des histoires thématisant des sujets d'actualité tels que les contrôles de noblesse ou les débats astrologiques." Schuwey, *Un entrepreneur*, 121.

6. Louis XIV's royal declaration of August 1680 commanded the consolidation of the two troupes and gave them a monopoly on theatrical performances in Paris. In early January 1681 the troupes, which had merged their repertoires, came to an official agreement about pensions and royalties. *Registre de La Grange (1658–1685)*, ed. Edouard Thierry (Paris: J. Claye, 1876). See Jan Clarke, *The Guénégaud Theatre in Paris (1673–1680)*, 3 vols. (Lampeter, Wales: Edwin Mellen Press, 1998), 2:397–400.

7. Mélèse counts six performances of *La comète*, but seven are listed in La Grange. *Registre de La Grange*, 253. Mélèse lists *La comète* as a comedy-ballet and attributes the music to Bertherand. Pierre Mélèse, *Un homme de lettres au temps du grand roi: Donneau de Visé, fondateur du "Mercure galant"* (Paris: Droz, 1936), 158 and Mélèse, *Répertoire analytique des documents contemporains d'information et de critique concernant le théâtre à Paris sous Louis XIV 1659–1715* (Paris: Droz, 1934), 172. The reference to Bertherand

is repeated on the CESAR database; however, judging from the July 1681 obituary for Charles Bertherand (1631–1700) in the *Mercure galant*, I think it most likely that Mélèse was understandably confused by a similarity in titles and that Bertherand, who was a member of the cultured elite in Soissons, was the author of the *Ballet de la comète*, which was the subject of chapter 2. *Mercure galant*, July 1700, 174–82.

8. *Mercure galant*, February 1681, 265–66.

9. In chapter 1 I discussed how Blageart and d'Houry developed another distinctive typeface for Nicolas de Blégny's publications. About the zigzag typeface, Schuwey asserts, "Ces caractères spéciaux contribuèrent ainsi à créer une gamme 'Mercure galant' immédiatement identifiable à sa typographie. . . . Les ouvrages héritaient ainsi de la réputation du *Mercure galant*, devenu une sorte de marque commerciale aisément identifiable." Schuwey, *Un entrepreneur*, 370.

10. Riffaud, *L'aventure*, 155.

11. "Permis d'imprimer. Fait à Paris ce 9. Fevrier 1681. DE LA REYNIE."

12. Schuwey discusses Donneau de Visé's "véritables spéculations dans le domaine de la librairie" as a signal aspect of his professional engagements. Schuwey, *Un entrepreneur*, 459.

13. *Histoire de l'imprimerie et de la librairie: Où l'on voit son origine et son progrés jusqu'en 1689, Divisée en deux livres* (Paris: chez Jean de la Caille, 1689), 2:275–76. Riffaud, *L'aventure*, 151. Schuwey identifies the booksellers of the Galerie du Palais as "les principaux marchands de nouveautés littéraires" who specialize in publishing successful plays and "des formules éditoriales innovantes." Schuwey, *Un entrepreneur*, 35.

14. Blageart printed, published, and/or sold plays by authors including Claude Boyer, Thomas Corneille, Cyrano de Bergerac, Claude Deschamps Villiers, Molière, Nicolas Pradon, Philippe Quinault, and Paul Scarron. See Riffaud, *L'aventure*, 151–55. Claude Blageart's participation in theatrical publication as printer and bookseller is documented in Riffaud's and Schuwey's online *Répertoire du théâtre français imprimé au XVIIe siècle*, http://repertoiretheatreimprime.yale.edu/.

15. Riffaud emphasizes the novelty of Molière's decision to hold his own *privilège* and negotiate directly with the printer, rather than cede it to an editor. Riffaud, *L'aventure*, 152. See also Riffaud's meticulous reconstruction of Ribou's career in *Le libraire de Molière*.

16. Vincent documents the collaboration agreement in *Donneau de Visé et "Le Mercure galant,"* 2:13–21.

17. As we will see again with the example of Bayle in chapter 5, anonymous publication and the subsequent curiosity of readers to discover authors'—and characters'—identities were standard features of the era. Matilde Bombart, *Le savoir des clefs*, forthcoming.

18. *Mercure galant*, August 1689, n.p.

19. BNF Richelieu theater volume 8-RF-6032 and 8-BL-12939 at the Arsenal.

20. 8-YTH-42306. The copy of the play in the Bibliothèque Carré d'art in Nîmes also includes an ink inscription "Vizé" on the title page.

21. This copy of the play, bound separately in a plain leather binding, includes no other marginalia, save what appears to be a laundry list. *La*

comète: Comédie (Paris: C. Blageart, 681). ETH-Bibliothek Zürich, Rar 6553, http://dx.doi.org/10.3931/e-rara-25883.

22. *Registre de La Grange*, 253.

23. Mélèse asserts that this constitutes "la première critique de l'auteur," while for Schuwey, Donneau de Visé's theater criticism begins with Molière and Corneille. Mélèse, *Donneau de Visé*, 7; Schuwey, *Un entrepreneur*, 73–83.

24. See Jan Clark, "*La Devineresse* and the Affaire des poisons," *Seventeenth Century French Studies* 28 (2006): 221–34.

25. Critics from Mélèse to Clarke assert that the authors wrote the play with direction from chief of police La Reynie (1625–1709).

26. *Mercure galant*, January 1710, 270–99.

27. *Mercure galant*, January 1710, 281–82.

28. Schuwey emphasizes that "Indépendamment de son importance ou du succès qu'elle rencontrait, chaque pièce de Donneau de Visé bénéficiait en effet d'un important effort promotionnel." Schuwey, *Un entrepreneur*, 136. See also 143–49 on publicizing his plays in the *Mercure galant*. Clarke notes Donneau de Visé and T. Corneille's self-promotion, "the extent to which they were prepared to exploit the possibilities provided by *Le Mercure galant* when it came to publicising their own productions." *The Demise of the Machine Play*, vol. 3 of *The Guénégaud Theatre in Paris (1673–1680)*, 48–56, here 49. Philippe Cornuaille traces Donneau de Visé's publicity for the play via the *Mercure galant*, an illustrated almanac, and printed handheld fire screens. Philippe Cornuaille, "L'écran de *La devineresse*, ou Le miroir déformant de l'affaire des poisons," *Nouvelles de l'estampe* 258 (2017): 258.

29. A complete example of the almanac is conserved in the Louvre collections (Collection Edmond de Rothschild 26919 LR/ Recto).

30. The almanac was followed by printed handheld fire screens resembling the promotional paper fans sometimes distributed by political candidates to American church congregations. Cornuaille hypothesizes that these might have been sold as souvenirs to spectators following the performance. Cornuaille, "L'écran," 258.

31. Cornuaille, "L'écran," 258.

32. Mélèse, *Donneau de Visé*, 160–62; Clarke, *Guénégaud*, 3:371–83.

33. *La Pierre Philosophale: Comédie mêlée de spectacles* (Paris: C. Blageart, 1681).

34. Pierre-François Godart de Beauchamps, *Recherches sur les théâtres de France: Depuis l'année onze cent soixante et un; jusques à present*, 3 vols. (Paris: Prault, 1735), 2:372. François Parfaict and Claude Parfaict, *Dictionnaire des théâtres de Paris contenant toutes les pièces qui on été représentées jusqu'à present sur les différens théâtres François* (Paris: Lambert, 1756), 2:133.

35. Antoine de Léris, *Dictionnaire portatif des théâtres: Contenant l'origine des differens théâtres de Paris* (Paris: C.A. Jombert, 1754), 88

36. J. M. B. Clément and J. de Laporte, *Anecdotes dramatiques*, 3 vols. (Paris: Veuve Duchensne, 1775), 1:220–21.

37. The play does not appear, for example, in the 1707 London edition by P. and I. Vaillant of *Oeuvres diverses de M. de Fontenelle* or any of its frequent reeditions; the Paris 1715 eight-volume edition by M. Brunet; the 1716

Amsterdam *Oeuvres diverses de M. de Fontenelle*; or the 1728–29 *Oeuvres diverses de M. de Fontenelle* published by Gosse and Neaulme in The Hague.

38. *Oeuvres de Monsieur de Fontenelle*, 10 vols. (Paris: B. Brunet, 1758).

39. For example. *La comète* is included in the eleven-volume *Oeuvres de Monsieur de Fontenelle, nouvelle edition* (Paris: Saillant, 1761–67). No mention is made of collaboration or Donneau de Visé.

40. Mélèse published a critical biography of Donneau de Visé in 1936. Recent scholarly interest in the *Mercure galant* was pioneered by Monique Vincent, who wrote a 1987 thesis and published a 1998 index and a 2005 monograph on the *Mercure galant*, which she calls in the title of her book, "*La première revue feminine d'information et de culture*." Innovative critical work on the *Mercure galant* continues in the scholarship of Sara Harvey, Chloé Hogg, Anne Piéjus, Christophe Schuwey, Deborah Steinberger, Geoffrey Turnovsky, and others. Schuwey includes a bibliography of critical work on Donneau de Visé and a list of Donneau de Visé's plays. *Un entrepreneur*, 494–500 and 124.

41. Schuwey, *Un entrepreneur*, 43–46, 124.

42. Schuwey notes: "Il compose des pièces en fonction des besoins des troupes et de l'État, divise le travail d'écriture en collaborant avec Thomas Corneille et surtout, déploie un appareil promotionnel massif et innovant. . . . Comme la structure de ses nouvelles, la dramaturgie de Donneau sert avant tout à lier des scènes inspirées de tendances littéraires et de sujets d'actualité." Schuwey, *Un entrepreneur*, 123.

43. Cornuaille is here referring to Donneau de Visé's publicity for *La devineresse*. Cornuaille, "L'écran," 258. Schuwey remarks on the recourse Donneau de Visé made of fictional characters to praise his own plays without seeming to infringe on the norms of *bienséance*. For example, when Donneau de Visé reports on the audience impression of *La devineresse* in October 1679, he promotes the way the play dispels belief in fortune telling. Schuwey, *Un entrepreneur*, 143–47.

44. This is the era in which modern conceptions of both authors and reading publics developed. See Alain Viala, *Naissance de l'écrivain: Sociologie de la littérature à l'âge classique (*Paris: Minuit, 1985), and Merlin, *Public et littérature*. On the development of modern notions of time, see Racevskis, *Time*.

45. Mélèse, *Donneau de Visé*, 4.

46. Mélèse, *Donneau de Visé*, 4; Schuwey, *Un entrepreneur*, 34–38.

47. An online edition, created under the supervision of Bourqui and Schuwey at the University of Fribourg (Switzerland), represents the first critical edition of *Les nouvelles nouvelles*, http://www.unifr.ch/nouvellesnouvelles/index.html.

48. For Schuwey, *Les nouvelles nouvelles* unsettle the very notion of what a book is: "Aucun autre de ses ouvrages ne remet aussi directement en question nos présupposés sur ce qu'est, fondamentalement, un livre." Schuwey, *Un entrepreneur*, 54. "La structure des *Nouvelles nouvelles* ne relève pas d'enjeux esthétiques. Elle vise à multiplier l'intérêt et les actions de l'ouvrage; individuellement, chaque pièce agissait en effet sur la sphère publique et sociale ainsi que sur le marché du livre et du divertissement." Schuwey, *Un entrepreneur*, 69.

49. Mélèse, *Donneau de Visé*, 159.

50. The *Entretiens* were an incredible publishing phenomenon, appearing in thirty-three editions in the author's long lifetime. See Frédérique Aït-Touati, *Fictions of the Cosmos: Science and Literature in the Seventeenth Century*, trans. Susan Emanuel (Chicago: University of Chicago Press, 2011); and Cherbuliez, "Sleeping Blonds."

51. Nicolas-Charles-Joseph Trublet, *Mémoires pour servir à l'histoire de la vie et des ouvrages de Mr de Fontenelle*, 2d rev. ed. (Amsterdam: Marc Michel Rey, 1759), 18.

52. Claude Nicolas Le Cat, *Éloge de Monsieur de Fontenelle des trois académies de Paris, secrétaire de celles des sciences, &c. Par M. Le Cat, secrétaire perpétuel de l'Académie de Rouen, &c.* (Rouen: J. N. Besogne, 1759), 11.

53. The first edition of the *Grand dictionnaire historique* was published in 1674, and Moréri died in 1680, so the commentary on *La comète* was the addition of a subsequent editor.

54. This edition and subsequent editions of Moréri also attribute Thomas Corneille's operas *Psyché* and *Bellerophon* to Fontenelle. Moréri et al., *Grand dictionnaire*, 230–31.

55. Boch, "Une rhétorique," 103. Following from her project of reading the play as a precursor to Fontenelle's better-known works, Boch attempts to show how *La comète* does live up to its serious billing in the *Mercure galant*. "Comme le système copernicien tout juste accepté avait chassé la terre du centre de l'univers, Fontenelle prétend bouter l'homme hors des finalités premières de la nature, dont il n'est qu'une parcelle et qui s'organise indépendamment de lui." Boch, "Une rhétorique," 103.

56. On a literature of "pièces" see chapter 1.

57. *La comète*, i.

58. *La comète*, i.

59. *La comète*, i.

60. Games of chance, extremely popular in the late seventeenth century, inspired reflection (notably by Girolamo Cardano and Galileo). Many consider the correspondence between Blaise Pascal and Pierre de Fermat discussing a problem about gambling posed by the chevalier de Méré as the birth of statistics and probability. Cards and dice are common symbols in still-life paintings and genre scenes. Michael Call has studied theatrical depictions of games of chance from this period and Molière's meditation on risk, probability, faith, and doubt in *L'avare*. Michael Call, "Fortuna Goes to the Theater: Lottery Comedies in Seventeenth-Century France," *French Forum* 40, no. 1 (2015): 1–15; and "Money for Nothing: Molière's Miser and the Risky World of Seventeenth-Century France," *Quidditas* 32 (2011): 8–29. See also Lyons, *Phantom of Chance*. Shakespearean and digital humanities specialist Gina Bloom uncovers epistemological connections between games and stages in early seventeenth-century England and the present day. Gina Bloom, *Gaming the Stage: Playable Media and the Rise of English Commercial Theater* (Ann Arbor: University of Michigan Press, 2018).

61. *La comète*, ii.

62. *La comète*, iii.

63. *La comète*, iii.

64. *La comète*, iii.

65. *La comète*, iii.
66. *La comète*, iv.
67. *La comète*, iv.
68. On the practice of using *tablettes*, see Chloe Hogg, "Pour une esthétique des tablettes: Clélie et les tablettes à écrire au XVIIe siècle," *Seventeenth-Century French Studies* 28 (2006): 117–33.
69. *La comète*, iv.
70. *La comète*, iv.
71. *La comète*, v.
72. *La comète*, vi.
73. Savien de Cyrano de Bergerac, *Histoire comique par Monsieur Cyrano de Bergerac, contenant les Estats & Empires de la Lune* (Paris: Charles de Sercy, 1657).
74. *La comète*, vi.
75. *La comète*, vi.
76. *La comète*, vi.
77. This kind of humorous indictment of a gravely misguided paterfamilias would not be unfamiliar to comic theatergoers of seventeenth-century France. One need only think of Orgon in Molière's *Tartuffe* (1664), especially in the disturbing but comically genius act 1, scene 4, where the master of the house, told repeatedly of his poor wife's illness, can only ask, like a lovelorn sot, "et Tartuffe?" after the charlatan who has entranced him.
78. *La comète*, x.
79. *La comète*, x.
80. *La comète*, ix.
81. *La comète*, xii.
82. See chapter 1.
83. *La comète*, xii.
84. *La comète*, xvii.
85. *La comète*, xvii.
86. *La comète*, xvii.

Chapter 5

1. As I explain below, Bayle's text underwent several editions and transformations even during his lifetime. To simplify I will follow the common practice of referring to it most often with the abbreviated title *Pensées diverses* or *Pensées diverses sur la comète* unless I am making a point specific to one of the particular editions.
2. The journal was published by fellow Huguenot refugee Henry Desbordes. In a preface to the first issue, Bayle emphasizes that he is adapting to the robust and open Dutch publishing context the periodical format pioneered in Denis de Sallo's *Journal des sçavans* (founded 1665) and picked up in other European intellectual centers. Following these models his journal profiled recently published books, scientific news, and obituaries of famous people. Bayle particularly emphasizes the journal's inclusive stance: "On doit donc mettre bas tous les termes qui divisent les hommes en différentes factions, et considérer seulement le point dans lequel ils se réunissent, qui est la qualité d'Homme Illustre dans la République des Lettres. En ce sens-là tous les

Savants se doivent regarder comme frères." Pierre Bayle, preface to *Nouvelles de la République des Lettres* (Amsterdam: Henry Desbordes, 1684), n.p.

3. The text of *Pensées diverses* is also digressive and complex and could interestingly be studied in terms of literary form. The scope necessary for such an inquiry exceeds a single book chapter.

4. "*Aujourd'hui* . . . ce que l'on appelle proprement Romans sont des fictions d'aventures amoureuses, écrites en prose avec art, pour le plaisir et l'instruction des lecteurs. Je dis des fictions, pour les distinguer des histoires véritables." Pierre-Daniel Huet's treatise (*Traité*) was published in 1670 in the first volume of Lafayette's *Zayde* (or, *Zaïde*). *Zayde histoire espagnole par Monsieur de Segrais: Avec un traité de l'origine des romans par Monsieur Huet* (Paris: Claude Barbin, 1670), 3.

5. Catherine Gallagher asserts that the novelistic fictionality established in eighteenth-century literature became in the nineteenth century and beyond the signal mode of the novel. The historical narrative of Gallagher's analysis has been variously challenged, for example, in the winter 2020 forum on medieval fictionalities in *New Literary History*. Nicholas Paige's *Before Fiction* also rejects the origin story narrative of the French novel traced by authors such as Beasley or DeJean. Within Paige's schema, Bayle's *Pensées diverses* dates from the very beginning of the second of three "regimes" in which "novelists cease posing as Aristotelian poets and instead pretend to offer their readers real documents ripped straight from history" and "the pseudofactual pact demanded that readers pretend to regard novels as true" (x). For the purposes of this study, the differences in the "history of the novel" narrative per se are less significant than the ways that Bayle harnesses the techniques and vocabularies he found in the literary works of his contemporaries and in the critical debates of the time. Catherine Gallagher, "The Rise of Fictionality," in *The Novel*, ed. Franco Moretti (Princeton, NJ: Princeton University Press, 2006), 337; Julie Orlemanski, "Who Has fictions? Modernity, Fictionality, and the Middle Ages," *New Literary History* 51, no 1 (2020): 145–70; Nicholas Paige, *Before Fiction: The Ancien Régime of the Novel* (Philadelphia: University of Pennsylvania Press, 2011), x.

6. In chapter 5 of *Revising Memory*, "Lafayette H/historienne: Rescripting Plausibility," Beasley examines the ways Lafayette puts pressure on generic expectations regarding *vraisemblance* to circumvent the masculinist constraints of history and the historical novel. Faith Beasley, *Revising Memory: Women's Fiction and Memoirs in Seventeenth-Century France* (New Brunswick, NJ: Rutgers University Press, 1990), 190–243.

7. Huet, *Traité*, 10.

8. Huet, *Traité*, 11. For Gallagher, the novel "tried, for at least two centuries, to hide its fictionality behind verisimilitude or realism, insisting on certain kinds of referentiality and even making extensive truth claims. . . . The novel has seemed ambivalent towards its fictionality—at once inventing it as an ontological ground and placing severe constraints upon it. . . . Candidly and explicitly differentiating their works from the kinds of referentiality proffered by neighboring genres, these writers coaxed their readers to accept the imaginative status of their characters." Gallagher, "Rise," 337.

9. In *Before Imagination*, Lyons recovers pre-Romantic conceptions of imagination, which was conceived in Aristotelian terms, or often in regard to

Stoic practice, as the bringing to mind of sensorial perceptions. John Lyons, *Before Imagination: Embodied Thought from Montaigne to Rousseau* (Stanford, CA: Stanford University Press, 2005).

10. Richard Popkin, who introduced Bayle to anglophone readers, characterized Bayle as "a superSceptic who attacked a vast variety of theories and opinions in his many, many writings." Richard Popkin, *The History of Scepticism from Savonarola to Bayle* (Oxford: Oxford University Press, 2003), 283. My focus on Bayle's relatively understudied first published work, Bayle's engagement with literature, and the literary context and stylings of the *Pensées diverses* offers a different way of situating Bayle in the intellectual context of his era. In the last section of this chapter, I explore the question of Bayle's scholarly reception.

11. Beasley, *Salons* and *Versailles Meets the Taj Mahal.*

12. Popkin includes Huet in *The History of Scepticism* as the last step in a progression toward Bayle, who he positions as the end point of this history. "Huet's scepticism involved doubt in that genuine knowledge about the real world can be gained by human beings. He saw the hypothetical experimental science of the Royal Society of England as the proper attitude for a sceptic and, along with the appeal to faith, the only means of making discoveries about God, Nature, and man, instead of the arrogant, illiterate, plagiarized, dogmatic view of the late René Descartes." Popkin, *Scepticism*, 280–81.

13. Aït-Touati, *Fictions*, 11. Aït-Touati draws on scholarship by Peter Dear, who established the "literary techniques and strategies" of natural science (Peter Dear, "Narratives, Anecdotes, and Experiments: Turning Experience into Science in the Seventeenth Century," *The Literary Structure of Scientific Argument*, ed. Peter Dear, 135–63 [Philadelphia: University of Pennsylvania Press, 1991], 163) and Steven Shapin's attention to the "literary technology" underpinning natural science. Dear's paper examines "scientific literary practice, regarded as a crucial feature of scientific practice as a whole, and it contends that an account of an experiment is an essential part of its performance." Dear, "Narratives, Anecdotes, and Experiments," 135.

14. Joyce Bost and Hubert Bost, introduction to *Pensées diverses sur la comète*, by Pierre Bayle (Paris: GF Flammarion, 2007), 18–19.

15. McKenna emphasizes that while Bayle's "écriture codée" and meandering, indirect style frustrated his philosophical heirs, "Chez Bayle, cependant, ce 'fatras' est essential: on ne saurait réduire la promenade à un 'but' restreint, car le but est précisément la promenade elle-même, à laquelle le point d'arrivée donne son unité et sa cohérence." Antony McKenna, *Études sur Pierre Bayle* (Paris: Honoré Champion, 2015), 48.

16. Festa, *Fiction*, 15.

17. Élisabeth Labrousse inaugurated the scholarly rediscovery of Bayle's thought in the last fifty years with a highly erudite series of studies: *Inventaire Critique de la Correspondance de Pierre Bayle* (Paris: Vrin, 1961); *Pierre Bayle: Du pays de foix à la cité D'Erasme* (La Haye: Nijhoff, 1963); *Pierre Bayle: Hétérodoxie et rigorisme* (La Haye: Nijhoff, 1964); and *Notes sur Bayle* (Paris: Vrin, 1987).

18. According to Pierre Des Maizeaux (or Desmaizeaux), a fellow Huguenot exile who translated Bayle into English and wrote his biography, "Il donna

sa Lettre sur les Comètes à Mr. Leers, Libraire de Rotterdam, homme d'esprit et de mérite, afin qu'il la fit imprimer. Et comme il prit toute sorte de précautions pour n'être pas reconnu Auteur de cet ouvrage, il ne changea rien dans le style de Catholique Romain, ni dans le langage et les éloges imités du Mercure Galant. Il crut que rien ne serait plus propre qu'un tel langage à faire juger que cette Lettre n'était point l'Écrit d'un homme sorti de France pour la Religion." Pierre Des Maizeaux, *La vie de Mr Bayle*, rev. ed., 2 vols. (La Haye: P. Gosse et J. Neaulme, 1732), 1:71–72.

19. The *avertissement* to the 1732 La Haye edition of Des Maizeaux's *Vie de Mr Bayle* notes Bayle's publishing success: "Le goût du Public pour les Oeuvres de Mr Bayle, est bien marqué par les Éditions fréquentes qu'on en a faites coup sur coup. Les deux premières, qui ont paru du vivant de l'Auteur, ont été élevées en assez peu de temps avec beaucoup d'empressement." Des Maizeaux, *Vie*, n.p.

20. According to Lucien Febvre and Henri-Jean Martin, "The great Dutch booksellers like the Leers of Rotterdam were able, through their extensive business contacts and the favoured geographical position of their country, to promote throughout Europe, from London to Berlin, the works of Bayle and pirated versions of Paris editions of the best French writers." *The Coming of the Book: The Impact of Printing 1450–1800*, trans. David Gerard (New York: Verso, 1997), 197.

21. "It is this great printer-bookseller who will be Bayle's ally throughout the vicissitudes of his career at the École Illustre, until the end of his life." McKenna, *Études*, 48.

22. There is some scholarly disagreement about the birth year of Des Maizeaux, a Huguenot writer and intellectual originally from central France. Des Maizeaux left France with his family for Switzerland to avoid persecution and from there emigrated to London, passing through Holland and meeting Bayle in Rotterdam. From London, Des Maizeaux maintained an active role in the epistolary network of the Huguenot diaspora. He worked as a kind of cultural journalist for French-language periodicals and published edited works, translations, biographies, and editions including works on Nicolas Boileau-Despréaux, Leibniz, John Locke, and Charles de Saint-Évremond in addition to his works on Bayle, which included the translation, collected letters, and biography.

23. André Prat's critical edition was first published in 1911–12 and, revised by Pierre Rétat, was republished in 1984 and 1994. *Pensées diverses* also figures in Alain Niderst's 1971 edition of Bayle's *Oeuvres diverses*. Robert Bartlett's 2000 State University of New York Press edition is the first modern English translation of the work. In 2006 Michiel Wielema's abridged Dutch translation, with a preface by Bart Leeuwenburgh, was published by Ten Have. Pierre Bayle, *Pensées diverses sur la comète*, ed. A. Prat, Société des textes français modernes, 2 vols. (Paris: Edouard Cornely et Cie, 1911–12); Pierre Bayle, *Oeuvres diverses*, ed. Alain Niderst (Paris: Éditions sociales, 1971); Pierre Bayle, *Various Thoughts on the Occasion of a Comet*, trans. Robert Bartlett (Albany: State University of New York Press, 2000). Pierre Bayle, *De inslag van een komeet*, trans. Michiel Wielema (Kampen: Ten Have, 2006).

24. Jonathan Israel contextualizes the mammoth endeavor of the fifteen-volume edition of Bayle's letters and the discoveries that it can provide regarding the famously cryptic thinker: "The many intriguing details it provides concerning the intellectual and ideological controversies he obsessively preoccupied himself with for more than a quarter of a century—what problems he focused on and when, what he read, with whom he debated, his opinions of opponents and viewpoints, and finally, what he sought to accomplish in the intellectual sphere." Jonathan Israel, "Pierre Bayle's Correspondence and Its Significance for the History of Ideas," *Journal of the History of Ideas* 80, no. 3 (2019): 480. *Correspondance de Pierre Bayle*, ed. Élisabeth Labrousse, Antony McKenna, et al., 15 vols. (Oxford: Voltaire Foundation, 1999–2017). An online version is being developed under the direction of McKenna and Fabienne Vial-Bonacci by Pierre Mounier and Cindy Tessier. *Correspondance de Pierre Bayle*, http://bayle-correspondance.univ-st-etienne.fr; *Correspondence de Pierre Bayle (1791 letters)*, Early Modern Letters Online, http://emlo-portal.bodleian.ox.ac.uk/collections/?catalogue=pierre-bayle; and *Lumières electroniques* (Electronic enlightenment) of the Voltaire Foundation in Oxford, as well as ePistolarium, Circulation of Knowledge and Learned Practices in the 17th-Century Dutch Republic, University of Utrecht, http://ckcc.huygens.knaw.nl. Letters quoted below follow the numbering of this edition.

25. Israel notes, "Bayle the individual and thinker will assuredly become better known; yet it has long been obvious that, even with his most longstanding and trusted correspondents, he remained always exceedingly cautious, even cagey concerning his core philosophical views and convictions." While the correspondence is also guarded, Israel also concludes, "The correspondence can only strengthen the impression that Bayle passed through several stages in his intellectual development, keeping an open and searching mind while increasingly keeping his core views largely to himself." Israel, "Pierre Bayle's Correspondence," 482 and 483.

26. DeJean cites the definition of the novel in Bayle's *Dictionnaire historique et critique* as evidence for the primacy of a French women's novel tradition that was written out of literary histories through the efforts of critics such as the militant proponent of the Ancients, Nicolas Boileau-Despréaux (1636–1711), and subsequent generations of male intellectuals including the *philosophes*. DeJean, *Tender*, 162–63.

27. Des Maizeaux lists the book among the manuscripts Bayle left behind after death: "'Jugements, ou Journal de Littérature.' Ce recueil contient des réflexions critiques sur les Livres qu'il avait lus, et celles qu'on lui avait communiquées par lettres ou de vive voix." Des Maizeaux, *Vie*, 2:327.

28. "Il avait beaucoup travaillé dans sa jeunesse à faire des extraits des Livres qu'il lisait, et à faire des observations sur ces Livres. . . . Ses recueils lui furent d'un grand secours lorsqu'il travailla pour le public." Des Maizeaux, *Vie*, 2:327.

29. "Les Lettres qu'il leur écrivait roulaient sur tout ce qui lui venait dans l'esprit, Philosophie, Littérature, Nouvelles politiques qu'il aimait passionnément: il avouait lui-même qu'il écrivait sans s'attacher à une suite régulière de pensées." Des Maizeaux *Vie*, 1:34. McKenna, one of the editors of Bayle's letters, notes that for Bayle, "the true *raison d'être* of correspondence is to supply

literary and political news . . . to allow his family to access the world of information, power, science, philosophy, and literature." McKenna, *Études*, 36.

30. Bayle's letters frequently document the cost, time and incertitude of sending letters and books across muddy roads to provincial destinations. Bayle, letter 163 to Joseph Bayle, Sedan, December 17, 1678. In 1678, Bayle offers detailed advice to his brother about the rates by *poste* versus *messagerie*. Letter 151 to Joseph Bayle, April 1678.

31. Bayle, letter 13 to Jacob Bayle, September 21, 1671.

32. Bayle and his brothers exchanged lists of books they purchased and read. Bayle asks about an author he sees mentioned: "Dites moi, je vous prie, dans votre p[remi]ere lettre qui e[st] ce Mr. Le Bret dont il fait mention dans sa preface, et dont je vois des pieces dans la liste des livres achetez par n[ot]re f[rere.] [C]ette liste et celle des livres q[ue] vo[us] avez leus m'a pleu extremem[en]t, et quant aux 2 lettres que vous me p[ro]mettés par mois, je vous assure qu'elles me seront tres agreables, et que je souhaitte ardemm[en]t q[ue] vous me teniez parole." Bayle, letter 151 to Joseph Bayle, April 25, 1678.

33. McKenna, *Études*, 42.

34. According to McKenna, "This group of friends is enthusiastic for recent literature—[Geuz de] Balzac, Voiture, Ménage, Cyrano—and sees itself as a part of the larger Republic of Letters." McKenna, *Études*, 37.

35. From his post in Coppet as a private tutor, for example, he wrote to his friend Vincent Minutoli about the books mentioned in Jean-Baptiste Dennis's *Mémoires concernant les arts et les sciences*. Bayle, letter 23 to Vincent Minutoli, October 9, 1672.

36. McKenna, *Études*, 36.

37. Bayle, letter 13 to Jacob Bayle, September 21, 1671.

38. Although Gallagher is primarily interested in later narratives, Villedieu's is related in some way to the category she calls the "alternative-history novel." Her 2018 book *Telling It Like It Wasn't* "explores why and how we conduct these counterfactual thought experiments. . . . When . . . did this mode of speculation start and what forms has it taken in previous centuries and in our own? What uses does it have, and what contexts stimulate its growth?" Catherine Gallagher, *Telling It Like It Wasn't: The Counterfactual Imagination* (Chicago: University of Chicago Press, 2018), 2–3.

39. Pierre Bayle, *Dictionnaire historique et critique* (Rotterdam: Reinier Leers, 1697) 2:833nA.

40. Micheline Cuénin and Nathalie Grande and Edvige Keller-Rahbé brought critical attention to Villedieu's literary career, including her novels. Beasley's 1990 *Revising Memory*, DeJean's 1991 *Tender Geographies*, and Myriam Dufour-Maître's 1999 *Les précieuses* cast new light on the contributions of fiction written by ancien régime women, many of whom were recognized in their own era but later written out of literary history. Beasley has subsequently explored the close relationship between this fiction tradition and the world of the salons, where the authors conversed and collaborated in *Salons, History, and the Creation of Seventeenth-Century France* and in *Versailles Meets the Taj Mahal*. Micheline Cuénin, *Roman et société sous Louis XIV: Madame de Villedieu*, 2 vols. (Paris: Champion, 1979); Myriam

Dufour-Maître, *Les précieuses: Naissance des femmes de lettres en France au XVIIe siècle*, rev. ed. (Paris: Honoré Champion, 2008); Nathalie Grande and Edvige Keller-Rahbé, eds., "Madame de Villedieu, ou Les audaces du roman," special issue, *Littératures classiques*, no. 61 (Spring 2007).

41. According to DeJean, "Huet and his followers, both male and female, are partisans of women's right to literary authority; in addition, these bourgeois are more or less vocal proponents of the destabilization of social stratification." DeJean, *Tender*, 160. Viala turned a sociological lens on the *galant* esthetic. See *La France galante* and *L'esthetique galante*. Fumaroli's *Poète et le roi* and my *Vaux and Versailles* also examine the late seventeenth century through the lens of competing artistic values systems. Marc Fumaroli, *Le poète et le roi: Jean de La Fontaine et son siècle* (Paris: Fallois, 1997).

42. Beasley, *Salons*, 134.

43. Like Édmond Lacoste, whose *Bayle nouvelliste et critique littéraire* precedes her work by ten years, Ruth Cowdrick mines Bayle's correspondence to portray Bayle as a passionate and astonishingly wide-ranging reader. "His comments were directed towards writings both old and new, presenting many different viewpoints and a wide variety of subjects. Not only works of erudition, but books on art, music, human behavior, military science and other miscellaneous topics came to his attention, and contributed to the wide range of knowledge which formed an important part of the mental equipment for his later activities." Ruth Elizabeth Cowdrick, "The Early Reading of Pierre Bayle," PhD diss., Columbia University, 1939, 47.

44. Cowdrick, "Early Reading," 40.

45. Beasley examines this literary style in *Salons* chapter 2, "Defining a Literary Culture: The *Ruelles* and Literary Innovation" as does Viala in *La France galante*.

46. Cowdrick notes this engagement lasted throughout Bayle's career: "For years before the publication of the *Nouvelles de la République des Lettres* Bayle was interested in the technique of literary production and in the principles and methods of literary criticism. This was a constant interest which developed early and varied little." Cowdrick, "Early Reading," 47. On the literary debates, see Maurice Laugaa, *Lectures de Madame de Lafayette* (Paris: Armand Colin, 1971).

47. "Unlike the tumultuous years of the Fronde and the early years of the Sun King's reign, the next three decades were a high-risk environment for those who sought a literary career. Literary form and content were then the subject of intense critical debate, a debate with strong political underpinnings. . . . These decades initiated the most influential moment of canon formation in French history; only literature useful to the absolute monarchy was promoted by the critics who worked in collaboration with the State." De Jean, *Tender*, 96.

48. Beasley, *Salons*, 134–47.

49. Bayle, letter 23 to Vincent Minutoli, October 9, 1672.

50. Bayle, letter 36 to Jean Bayle, July 31, 1673.

51. Desmaizeaux, *Life*, 23–24.

52. Des Maizeaux penned a first biographical sketch of Bayle at the request of the Earl of Shaftesbury shortly after Bayle's death in 1706 and published it

with his 1708 English translation, *Miscellaneous Reflections, Occasion'd by the Comet Which Appear'd in December 1680: Chiefly to Explode Popular Superstitions. Written to a Doctor of the Sorbon, by Mr. Bayle. Translated from the French. To Which I Added the Author's Life (*London: J. Morphew, 1708). He dates the completion of the more complete *La vie de Mr Bayle* December 13, 1729, in a letter to François de la Mothe-Fénelon published as a foreword to the French biography. Des Maizeaux, *La vie*, 64.

53. Bost, *Pierre Bayle*, 182.

54. Bost and Bost, introduction, 11. Alexandra Willaume Albertini, who situates Bayle's work in a religious and scientific context, notes "la position dissidente de Bayle à l'égard de la posture calviniste en ce qui concerne la superstition astrologique." "Rhétorique du discours contre la superstition dans les 'Pensées diverses sur la comète' de Pierre Bayle: astronomie et littérature au service de la raison," *Studi Francesi*, 174 no. 58.3 (2014): 456, http://journals.openedition.org/studifrancesi/1031.

55. Bost and Bost, introduction, 14.

56. Desmaizeaux, *Life*, 24. The passage from the French version reads: "Il crut donc qu'il serait très inutile de raisonner davantage, a moins qu'il n'employât un argument qui fit voir que les attributs de Dieu ne permettent pas qu'il destine les Comètes à un tel effet. Il médita là-dessus et il s'avisa bientôt de cette raison théologique que *si les Comètes étaient un présage de malheurs, Dieu aurait fait des miracles pour confirmer l'idolâtrie dans le monde*." Des Maizeaux, *La vie*, 64, original emphasis.

57. Huet, *Traité*, 3–4.

58. Cherbuliez argues that "the way women read novels and how they imagined and processed innovative thinking are seen in Fontenelle to be intimately linked." Cherbuliez, "Sleeping Blonds," 149 and 146.

59. "L'imagination, qui agit avec facilité, a moins de part que l'entendement, dont les opérations sont plus laborieuses. Et parce que naturellement le travail nous rebute, l'âme ne se porte à ces connaissances épineuses que dans la vue du fruit, ou dans l'espérance d'un Plaisir éloigné, ou par nécessité." Huet, *Traité*, 112.

60. "Mais les connaissances qui l'attirent et la flattent davantage, sont celles qu'elle acquiert sans peine, et ou l'imagination agit Presque seule, et sur des matières semblables a celles qui tombent d'ordinaire sous nos sens, et particulièrement si ces connaissances excitent nos passions, qui sont les grands mobiles de toutes les actions de notre vie. C'est ce que font les romans." Huet, *Traité*, 112.

61. Lyons emphasizes that "the close association of imagination and the senses" is an important element of the seventeenth-century conception of the imagination. Lyons, *Before Imagination*, xii.

62. Huet, *Traité*, 115.

63. Huet, *Traité*, 115. Given the ambivalence of Augustine's various reflections on the faculty of imagination, Huet is clearly cherry-picking here. Marianne Djuth, "Veiled and Unveiled Beauty: The Role of the Imagination in Augustine's Esthetics," *Theological Studies* 68 (2007), offers a helpful understanding of Augustine's complicated attitudes toward the faculty of imagination by accounting for Augustine's understanding of "how the mind

functions" (78) and the distinction between *phantasia* and *phantasma*, as well as his attitudes toward divine and created beauty. "The empirical strain in Augustine's thought enables him to maintain the vital link between the imagination and the sensible order. When used responsibly, both the imagination and the senses are instrumental in the acquisition of a true understanding of the Christian faith, as is the mind itself in which these activities occur" (91).

64. Bayle, letter 244 to Jacob Bayle, January 1684.

65. According to Beasley, "*La Princesse de Clèves* is the only prose work to have elicited so much critical furor during the classical period." *Salons*, 127.

66. Lyons contrasts the capacious role Madeleine de Scudéry gives to imagination in *Clélie* (1654–61) compared with the very circumscribed role of imagination in Lafayette's *Princesse de Clèves*: "Instead of recognizing and training their imaginative faculty in a selective way, Lafayette's characters devote exceptional energy and ingenuity to the quest for *significance*." Lyons, *Before Imagination*, 166.

67. Bayle, letter 151 to Joseph Bayle, April 1678.

68. Laugaa presents the critical furor over *La Princesse de Clèves*. Laugaa, *Lectures*, 20–111.

69. Bayle, letter 162 to Vincent Minutoli, December 15, 1678.

70. Bayle, letter 164 to Jacob Bayle, January 1679.

71. Bayle, like many of his contemporaries, attributed the critique to Bouhours (letter to Vincent Minutoli, December 15, 1678). Only after the author's death was the text attributed to Jean Baptiste Henri Du Trousset de Valincour (1653–1730).

72. Charnes's *Conversations sur la critique* was published in Paris by Claude Barbin and bears the *achevé d'imprimé* date February 1679. The *privilège du roi* was dated May 1 and recorded on May 3, 1679. Introduction to Jean-Antoine de Charnes, *Conversations sur la critique de La Princesse de Clèves*, ed. François Weil et al. (Tours: Groupe d'Étude du XVIIe Siècle de l'Université François Rabelais, 1973), 1n2 and x–xi). The editors of the 1973 Tours facsimile edition of Charnes cannot speculate, based on Bayle's advanced knowledge of the *Conversations*, on how widespread news of the manuscript was before its publication, "Doit-on penser qu'on leur a donné quelque publicité? Bayle est si bien informé de l'activité des imprimeurs par des sources particulières qu'on n'oserait l'affirmer." Charnes, introduction, x.

Bayle could not have known about the publication through its author because he did not become acquainted (in letters) with Charnes until decades later, when the two corresponded with great warmth and mutual admiration. Bayle and Charnes corresponded in 1697 regarding Bayle's note in the *Dictionnaire historique et critique* questioning Charnes's claim in his 1690 *Vie de Tasse* about the birthplace of Italian writer (and Tasso contemporary) Paul Beni. The surviving correspondence and the notes in both editions of Bayle's dictionary—before and after the epistolary exchange—testify to a great sense of mutual admiration and intellectual warmth between Charnes and Bayle. See introduction and appendices 1–3 in Charnes, *Conversations*, x–xii.

73. "The debate over *La Princesse de Clèves* is thus much more than a discussion over a novel. It is a forum to address more wide-ranging concerns: What is the role of a literary critic and from what realm should the legitimate

critic emanate? Is the role of a critic such as Valincour or Charnes to reflect public opinion, or should the critic's role be more pedagogical? Should the critic actually guide and form the reader and consequently control the pleasure of the reader by defining it according to certain criteria?" Beasley, *Salons*, 125–26.

74. Charnes quoted in Beasley, *Salons*, 129.

75. "Ceci répond à plusieurs endroits de la lettre, et vous délivre de la peine d'imaginer cent choses, et de chercher des voies d'application. Il n'y a point d'autre mystère dans tout cela que le dessein de varier les matières, et de faire un plus beau spectacle aux yeux du lecteur." Bayle, letter 244 to Jacob Bayle, January 1684.

76. "Le cartésianisme ne fera pas une affaire, je le regarde simplement comme une hypothèse ingénieuse qui peut servir à expliquer certains effets naturels, mais au reste j'en suis si peu entêté, que je ne risquerai pas la moindre chose pour soutenir que la nature se règle et se gouverne selon ces principes-là. Plus j'étudie la philosophie, plus j'y trouve d'incertitude: la différence entre les sectes ne va qu'à quelque probabilité de plus ou de moins: il n'y en a point qui ait frappé au but et jamais on n'y frappera apparemment, tant sont grandes les profondeurs de Dieu dans les oeuvres de la nature, aussi bien que dans celles de la grâce. Ainsi, vous pouvez dire à Mr Gaillard que je suis un philosophe sans entêtement et qui regarde Aristote, Épicure, Des-Cartes, comme des inventeurs de conjectures que l'on suit ou que l'on quitte selon que l'on veut chercher plutôt un tel qu'un tel amusement à l'esprit." Bayle, letter 190 to Jacob Bayle, May 1681. McKenna, however, cautions that "we should not draw too rigorously skeptical conclusions from these words written to reassure his brother and Jacques Gaillard, the friend likely to find him a professorship in the Netherlands." Preface to volume 3 of the *Correspondence*, electronic edition.

77. Bayle, *Lettre à M.L.A.D.C. Docteur de Sorbonne. Où il est prouvé par plusieurs raisons tirées de la Philosophie, et de la Théologie, que les Comètes ne sont point le présage d'aucun malheur: Avec plusieurs Réflexions Morales et Politiques, et plusieurs Observations Historiques; Et la Réfutation de quelques erreurs populaires* (Cologne [Rotterdam]: P. Marteau [Reinier Leers], 1682), n.p.; Bost and Bost, introduction, 45.

78. Katherine Ann Jensen traces the emergence of this tradition in *Writing Love: Letters, Women, and the Novel in France, 1605–1776* (Carbondale and Edwardsville: Southern Illinois University Press, 1995).

79. Bost traces Bayle's development into a landmark cultural journalist and linchpin of a more widely disseminated, printed portal into the Republic of Letters. Hubert Bost, *Un intellectual avant la lettre: Le journaliste Pierre Bayle, 1647–1706; L'actualité religieuse dans "Les nouvelles de la République des Lettres," 1684–1687* (Amsterdam: APA-Holland University Press, 1994).

80. McKenna, *Études*, 69.

81. "Bayle suit l'épanouissement de 'l'esthétique galante,' la littérature de l'élite parisienne, dont le ton était donné par Pellisson et par La Fontaine, adopté aussi par Molière lui-même, et dont le modèle épistolaire était fourni par Voiture. Pour le style—et pour tout ce qui fonde et exige un style—la correspondance a été un apprentissage précieux pour l'écrivain." McKenna, *Études*, 41. Jean-Louis Guez de Balzac (1597–1654) and Vincent Voiture

(1597–1648) were the great prose stylists of the previous generation. Voiture, a poet and active participant in the famous Rambouillet salon, was noted for his urbane and witty letters, which were widely circulated in manuscript during his lifetime and then published posthumously. Guez de Balzac, an original member of the Academie française, published two collections of highly influential epistolary essays, *Lettres* (1636) and *Recueil de nouvelles lettres* (1637), in a more ponderous style than Voiture's. Poet Paul Pellisson (1624–93) was an intimate of Madeleine de Scudéry and her *précieuse* circle. Secretary to Nicolas Fouquet before his spectacular arrest by Louis XIV, Pellisson helped his patron assemble an impressive stable of artists and writers associated with the *style galant*.

82. McKenna, *Études*, 40.

83. Bayle, *Lettre à M.L.A.D.C.*, n.p.

84. The *Journal des sçavans* remained an important reference for Bayle. He cites it as the model upon which he based his own editorial practices in *Nouvelles de la République des Lettres*.

85. Bayle, *Lettre à M.L.A.D.C.*, n.p.

86. Bayle, *Lettre à M.L.A.D.C.*, n.p.

87. The last genre Bayle mentions is comedy. The passage very much seems a reaction to the *Mercure galant*'s claim that the play by Donneau de Visé that is the subject of chapter 4 positively cures viewers of superstition. "La comédie, qui se vante d'être le souverain remède des maladies de l'esprit, s'est enfin mise sur les rangs et a joué les comètes avec la même liberté qu'elle joue les autres choses. Qui croirait après cela qu'on ne se fût pas accommodé à toute sorte de goûts et qu'on ne fût pas entré dans tous les expédients capables de mettre le monde à la raison sur ce sujet ?" Bayle, *Lettre à M.L.A.D.C.*, n.p.

88. "Je dis la même chose pour l'autre grand retranchement de l'opinion populaire, c'est à dire l'expérience, dont on se glorifie beaucoup." Bayle, *Lettre à M.L.A.D.C.*, n.p.

89. Bayle, *Lettre à M.L.A.D.C.*, n.p.

90. Bayle, *Lettre à M.L.A.D.C.*, n.p.

91. Bayle, *Lettre à M.L.A.D.C.*, n.p.

92. Bayle, *Lettre à M.L.A.D.C.*, n.p.

93. Bayle, *Lettre à M.L.A.D.C.*, n.p.

94. When comparing novels to poems, for example, Huet observes "Les romans ont plus du vraisemblable, quoiqu'ils aient quelquefois du merveilleux. Les poèmes sont plus réglés, et plus châtiés dans l'ordonnance, et reçoivent moins de matière, d'évènements, et d'Épisodes: les Romans en reçoivent davantage . . . ils ne tendent pas tant l'esprit, et le laissent en état de se charger d'un plus grand nombre de différentes idées." Huet, *Traité*, 6–7.

95. William Carlson emphasizes this passage's references to Pascal's comments on language in *Les Pensées*: "The problems of digression and textual fragmentation, the question of the appropriateness of a certain kind of discourse for a certain material or kinds of material, and the matter of anonymity and disguise of the author are the most useful for the elucidation of the deep affinities . . . which point beyond these two individual authors to larger questions . . . of the perception and study of language in France in the seventeenth

century." William Carlson, "'Pensées' and 'Pensées diverses': The art of persuasion in Pascal and Bayle," *French Forum* 4, no. 2 (1979): 140.

96. Bost, *Pierre Bayle*, 184.

97. Bayle, *Lettre à M.L.A.D.C.*, n.p.

98. Charnes, *Conversations*, 32.

99. Aït-Touati, *Fictions*, 95 and 102.

100. Bayle, *Lettre à M.L.A.D.C.*, n.p.

101. Bayle, *Lettre à M.L.A.D.C.*, n.p.

102. Bayle, *Lettre à M.L.A.D.C.*, n.p.

103. Bayle, *Lettre à M.L.A.D.C.*, n.p.

104. See Robert Berger, *In the Garden of the Sun King: Studies on the Park of Versailles under Louis XIV* (Washington, DC: Dumbarton Oaks, 1985), 17–19.

105. Bayle, *Lettre à M.L.A.D.C.*, n.p.

106. Bayle, *Lettre à M.L.A.D.C.*, n.p.

107. Bayle, "Le Libraire au Lecteur," *Pensées diverses*, n.p.

108. Bayle, "Libraire," n.p.

109. Letter 680, Hullin to Bayle, 1686.

110. Letter 680, Hullin to Bayle, 1686.

111. Letter 680, Hullin to Bayle, 1686.

112. Bayle, "Avertissement," *Pensées diverses sur la comète* (1699), n.p.

113. "Comme j'étais Professeur en Philosophie à Sedan lors qu'il parut une Comète aux mois de Décembre mil six cens et quatre-vingt je me trouvais incessamment exposé aux questions de plusieurs personnes curieuses, ou alarmées. Je rassurais autant qu'il m'était possible ceux qui s'inquiétaient de ce prétendu mauvais présage, mais je ne gagnais peu de choses par les raisonnements philosophiques ; on me répondait toujours que Dieu montre ces grands Phénomènes, afin de donner le temps aux pêcheurs de prévenir par leur pénitence les maux qui leur pendent sur la tête." Bayle, "Avertissement," n.p.

114. Bayle, "Avertissement," n.p.

115. Bayle, "Avertissement," n.p.

116. Schuwey, *Un entrepreneur*, 321.

117. Turnovsky "Les lecteurs."

118. Bayle, letter 160, to Jacob Bayle, Sedan, November 26, 1678. He comments to his brother Jacob in February 1678 that he has discovered the name of the *Mercure galant*'s author/editor and the following month, writing to his brother Joseph, describes how the counterfeit editions available in the far-flung provinces are sadly missing the engravings (Bayle, letter 149, February 23, 1678; letter 150, March 1678). Eight months later, he writes to Jacob again about his quest to know more about Donneau de Visé, who he is, whether he is writing to a real woman or whether she is a fictional construction: "J'ai demandé à plus de cinq ou six Parisiens qui était Mr. de Vizé, de quelle pays, de quelle âge, de quelle profession, si la dame à qui il écrit n'est pas une supposition pour avoir lieu de donner l'air de lettre à son Mercure, personne ne sait cela. . . . Heureux ceux qui peuvent puiser dans la source, et malheur aux provinciaux qui n'ont pas le moyen d'avoir un bon et savant facteur pour le commerce de leur curiosité." Bayle, letter 160 to Jacob Bayle, Sedan, November 26, 1678.

119. Hazard, *Crisis*, trans. Mays, 284.

120. Anthony Gottleib, "The Tolerant Philosopher: Why Pierre Bayle Is the Forgotten Figure of the Enlightenment," *New Statesman*, August 13, 2016.

121. The *Dictionnaire* was first published by Leers in 1697 in two folio volumes, then in 1702, again by Leers, in three volumes. It was republished widely throughout the eighteenth and into the nineteenth centuries.

122. In May 1684, Bayle announces to Jacques Lenfant: "Je travaille tous les mois à donner un *Journal des sçavans*." Letter 270, Rotterdam, May 9, 1684. The following month, Bayle expands: "Plusieurs personnes, et surtout de Paris, m'avaient puissamment exhorté à ne point faire un journal uniquement pour les savants; elles m'avaient dit qu'il fallait tenir un milieu entre les nouvelles des gazettes et les nouvelles de pure science, afin que les cavaliers et les dames, et en général milles personnes qui lisent et qui ont de l'esprit sans être savantes, se divertissent à la lecture des *Nouvelles*." Bayle to Jacques Lenfant, letter June 16, 1684, quoted in Eugène Hatin, *Les gazettes de Hollande et la presse clandestine au 17e et 18e siècles* (Paris: René Pincebourde, 1865, 130).

123. "L'esprit de sagesse et de critique qui se communiquait de proche en proche, détruisit insensiblement beaucoup de superstitions." Voltaire, *Siècle*, 2:276.

124. "Les idées superstitieuses étaient tellement enracinées chez les hommes que les comètes les effrayaient encore en 1680. On osait à peine combattre cette crainte populaire. . . . Il fallut que Bayle écrivît contre le préjugé vulgaire un livre fameux, que les progrès de la raison ont rendu aujourd'hui moins piquant qu'il ne l'était alors." Voltaire, *Siècle*, 2:276.

125. The superstition narrative has been repeated for example Christian Biet's document-based popular history asserts "C'est un certain XVIIe siècle, critique et raisonnable, qui balbutie ses premiers mots sceptiques" (103). Schechner traces how in the seventeenth century, "new patterns of idea consumption and amalgamation promoted both the decline of the interpretation of comets as portents and the rise of their interpretation as cosmological agents." Christian Biet, *Les miroirs du Soleil: Le Roi Louis XIV et ses artistes* (Paris: Gallimard, 2000), 103. Schechner Genuth, *Comets*, 12–13.

126. Desmaizeaux, *Life*, iv.

127. "On en estime l'auteur, et l'ouvrage est bien reçu; il a beaucoup de sel, ce qui ne contribue pas peu à son succès." *Mercure galant*, November 1683.

128. Ludwig Feuerbach, *Pierre Bayle, nach seinen für die Geschichte der Philosophie und Menschheit interessantesten Momenten* (Ansbach: C. Brügel, 1838), iv.

129. Louis-Paul Betz, *Pierre Bayle und die "Nouvelles de la République des Lettres" (Este populärwissenschaftliche Zeitschrift) (1684–1687).* Geneva: Slatkine Reprints, 1970.

130. Anthony Gottlieb, *The Dream of Enlightenment: The Rise of Modern Philosophy* (New York: Liveright, 2016), 156.

131. Edmond Lacoste, *Bayle nouvelliste et critique littéraire* (Brussels: Maurice Lamertin, 1929), 4.

132. Popkin, introduction, xi.

133. Popkin, introduction, xi.

134. Anthony Gottlieb, "The Tolerant Philosopher: Why Pierre Bayle Is the Forgotten Figure of the Enlightenment," *New Statesman*, August 13, 2016.

135. Eric Jorink, "Comets in Context: Some Thoughts on Bayle's *Pensées diverses*." In *Le philosophe de Rotterdam: Philosophy, Religion and Reception; Selected Papers of the Tercentenary Conference Held at Rotterdam, 7–8 December 2006*, ed. Wiep van Bunge and Hans Bots, 51–68 (Leiden: Brill, 2008), 52.

136. In addition to studies focusing on his oeuvre, Bayle's work has been granted new attention in the context of recent scholarship on the Enlightenment, for example in the work of Jonathan Israel.

137. Thomas M. Lennon, *Reading Bayle* (Toronto: University of Toronto Press, 1999), 15. In the *Avant-propos* to his 2015 collection of essays, *Écrits sur Bayle*, McKenna provides a lucid exposition to the myriad and often contradictory interpretations of Bayle's philosophy and theology. McKenna, *Écrits*, 7–15.

Conclusion

1. I do not use the term *esquive* as a reference to Abdellatif Kechiche's 2005 movie by that name, but this book does share the movie's desire to examine canonical French culture from viewpoints outside traditional centers of power. Neither is the term "swerve" meant as a reference to Stephen Greenblatt's meditation on Lucretian modernity, although there are clear affinities between Greenblatt's insights and the aesthetic features and modes of association that emerge from an exploration of late seventeenth-century France from the perspective of a comet. Stephen Greenblatt, *The Swerve: How the World Became Modern* (New York: Norton, 2012).

2. Beasley, *Versailles Meets the Taj Mahal.*

3. Pelletier, "Les Globes de Louis XIV," 76–77.

4. Pelletier, "Scientific Connections, Ambitions and Financial Difficulties of the d'Estrées Family," *Globe Studies* 55/56 (2009): 157–58.

5. Streaking across the V of "Voilà," a comet invites readers to *look* (voir-là): "VOILÀ, Messieurs, voilà." The motto "Foetu[s] Clara Suo" means "Shining with Its Own Offspring." Antoine Gallouis, *Oraison Funèbre de Tres-Auguste Princess Marie-Therese d'Austriche, Reine de France et de Navarre, Epouse du Roy Louis Le Grand XIV du Nom* (Paris: Guillaume De Luynes, 1683), n.p.

6. *Mercure galant*, June 1687, 1:17–18.

7. *Mercure galant*, June 1687, 1:17.

8. "Tous les deux seront plus beaux, plus exacts, plus corrects, et plus curieux que tout ce qui a paru jusqu'à présent. Il est ainsi marqué dans le Privilège que le Roy a accordé au Père Coronelli, et il y a même dans ce Privilège, *que ces Globes ne seront pas un des moindres ornements du Château de Versailles*." *Mercure galant*, June 1687, 1:49.

9. Pelletier, "Les Globes de Louis XIV," 74–75; fig. 3, 76. Catherine Hofmann, "'Incomparable Globes' or 'Old Devices as Cumbersome as They Are Useless'? The Vicissitudes of the Great Globes of Coronelli 1683–1915," *Globe Studies* 53/54 (2005/2006): 25. See also Edmond Pognon, "Les vicissitudes des globes de Marly," *Histoire des instruments scientifiques* (1971): 85–91.

10. Hofmann, "'Incomparable Globes,'" 28.

11. http://expositions.bnf.fr/globes/index.htm.

BIBLIOGRAPHY

Premodern Sources

Auzout, Adrien. *Éphéméride du comète de 1664*. s.l.

[Bayle, Pierre]. *Addition aux Pensées diverses sur les comètes, ou Réponse à un libelle intitulé, Courte revüe des maximes de morale & des principes de religion de l'auteur des Pensées diverses sur les comètes, &c.* Rotterdam: Reinier Leers, 1694.

[Bayle, Pierre]. *Continuation des Pensées sur la comète.* Rotterdam: Reinier Leers, 1704.

Bayle, Pierre. *Correspondance de Pierre Bayle.* Edited by Élisabeth Labrousse, Antony McKenna, et al. Oxford: Voltaire Foundation, 1999–2017.

Bayle, Pierre. *Dictionnaire historique et critique.* Rotterdam: Reinier Leers, 1697.

Bayle, Pierre. *De inslag van een komeet: beschowingen over geloof en bijgeloof.* Translated by Michiel Wielema. Preface by Bart Leeuwenburgh. Kampen: Ten Have, 2006.

[Bayle, Pierre]. *Lettre à M.L.A.D.C. Docteur de Sorbonne. Où il est prouvé par plusieurs raisons tirées de la Philosophie, et de la Théologie, que les Comètes ne sont point le présage d'aucun malheur: Avec plusieurs Réflexions Morales et Politiques, et plusieurs Observations Historiques; Et la Réfutation de quelques erreurs populaires.* Cologne [Rotterdam]: P. Marteau [Reinier Leers], 1682.

Bayle, Pierre. *Miscellaneous Reflections, Occasion'd by the Comet Which Appear'd in December 1680. Chiefly to Explode Popular Superstitions. Written to a Doctor of the Sorbon, by Mr. Bayle. Translated from the French. To Which I Added the Author's Life.* Translated by Pierre Des Maizeaux. London: J. Morphew, 1708.

[Bayle, Pierre]. *Nouvelles de la République des Lettres.* Amsterdam: Henry Desbordes, 1684.

Bayle, Pierre. *Oeuvres diverses.* Edited by Alain Niderst. Paris: Éditions sociales, 1971.

[Bayle, Pierre]. *Pensées diverses, écrites à un docteur de Sorbonne, a l'occasion de la cométe qui parut au mois de décembre 1680.* Rotterdam: Reinier Leers, 1683.

[Bayle, Pierre]. *Pensées diverses, écrites à un docteur de Sorbonne, a l'occasion de la cométe qui parut au mois de décembre 1680.* 3d ed. Rotterdam: Reinier Leers, 1699.

Bayle, Pierre. *Pensées diverses sur la comète.* Edited by A. Prat. Société des textes français modernes. 2 vols. Paris: Edouard Cornely et Cie, 1911–12.

Bayle, Pierre. *Various Thoughts on the Occasion of a Comet.* Translated by Robert Bartlett. Albany: State University of New York Press, 2000.

[Bertherand, Charles]. *Ballet de la comete, divisé en deux parties: Et dansé à Soissons le 9 Février 1665*. Soissons: Nicolas Asseline, Imprimeur et Marchand Libraire, Rue des Cordeliers, 1665.

Berryat, Jean et al., eds. *Collection académique, composée des mémoires, actes ou journaux des plus célèbres académies et sociétés littéraires, des extraits des meilleurs ouvrages périodiques, des traités particuliers et des pièces fugitives les plus rares*. Vol. 7. Dijon: François Desventes; Paris: Michel Lambert, 1766.

Blégny, Étienne. *Traité concernant la manière de procéder à toutes les verifications d'écritures contestées en justice, par le Sr. de Blégny, juré expert écrivain*. Paris: Guillaume Cavelier, 1698.

[Blégny, Nicolas]. *Les adresses de la ville de Paris avec le trésor des almanachs, livre commode en tous lieux, tous temps et toutes conditions, par Abraham Du Pradel, astronome Lyonnais*. Paris: La veuve de Denis Nion, 1691.

Blégny, Nicolas. *Journal des nouvelles découvertes conçernant les sciences et les arts qui font parties de la médecine. Tome III. Contenant toutes celles qui ont esté recueillies durant le cours de l'année 1681. Par NICOLAS DE BLEGNY, à present chirugien ordinaire du corps de Monsieur, et juré à Paris.*

[Blégny, Nicolas]. *Le livre commode contenant les addresses de la ville de Paris et le trésor des almanachs pour l'année . . . 1692 . . . par Abraham Pradel.* Paris: La veuve de Denis Nion, 1692.

[Blégny, Nicolas]. *Monatliche neueröffnete Anmerckungen, über alle Theile der Artzney-Kunst. Zusammen gebracht im Jahr 1679: Allen der Leib und Wundartzney Zugethanen und Liebhabern zu sonderbahren Gefallen aus dem Französischen ins Teutsche übersetzet durch J.L.M.C.* Hamburg: Gottfried Schultzens Kosten, Merseburg, drukts Caspar Forberger, F.S. Hoff buchdrucker, 1680.

Blégny, Nicolas. *Nouveautez journalières concernant les sciences et les arts, qui font parties de la médecine: Recueillies par Nicolas de Blegny, chirurgien ordinaire du corps de Monsieur; Pendant les années 1680 & 1681.* Paris: chez l'autheur, Claude Blageart, et Laurent d'Hourry, 1680.

Blégny, Nicolas. *Nouvelles découvertes sur toutes les parties de la médecine: Recueillies en l'année 1679; Par N.D.B. chirugien du Roy, maistre et juré à Paris.* Paris: Laurent d'Hourry, 1679.

[Blégny, Nicolas]. *Au Roy, et à nosseigneurs les commissaires députez, par Sa Majesté, pour l'execution de l'edit du mois de mars 1693: Concernant l'ordre hospitalier du Saint-Esprit*, 1683. Requête de Nicolas de Blegny, administrateur général de l'ordre, incarcéré au For-l'Evêque, le 4 juin 1693, sur plainte des sieurs Du Boullay, de Lessac et autres. Factum, Ordre hospitalier du Saint-Esprit, 1693.

Blégny, Nicolas. *Le Temple d'Esculape ou le dépositaire des nouvelles découvertes qui se font journellement dans toutes les parties de la médecine contenant celles qui ont été recueillies durant le cours de l'année 1680 par Nicolas Blegny, chirurgien du Roi.* Paris: l'autheur, Claude Blageart, et Laurent d'Hourry, 1680.

Blégny, Nicolas. *Zodiacus Medico-Gallicus, sive Miscellaneorum Medico Physicorum Gallicorum . . . annus primus scilicet M.DC.LXXIX.* Geneva: Leonard Chouët, 1680.

Bouhours, Dominique. *Entretiens d'Ariste et d'Eugène*. Paris: Sébastien Mabre-Cramoisy, 1671.

Boileau, Nicolas, trans. *Oeuvres diverses du Sieur D*** avec le traité du sublime ou du merveilleux dans le discours: Traduit du Grec de Longin*. Paris: Denis Thierry, 1674.

Camusat. *Histoire critique des journaux*. Amsterdam: J. F. Bernard, 1734.

Cassini, Jean-Dominique IV. *Mémoires pour servir à l'histoire des sciences et à celle de l'Observatoire Royal de Paris*. Paris: Bleuet, 1810.

Cassini, Jean-Dominique. *Abrégé des observations et des réflexions sur la comète qui a paru au mois de décembre 1680 et aux mois de janvier, février, mars de cette année 1681: Présenté au Roy Par Mr. Cassini*. Paris: Estienne Michallet, 1681.

Cassini, Jean-Dominique. "De l'origine et du progress de l'astronomie dans la géographie et dans la Navigation." In *Recueil d'observations faites en plusieurs voyages par ordre de sa Majesté pour perfectionner l'astronomie et la géographie: Avec divers traitez astronomiques par messieurs de l'Académie royale des sciences*, 1–42. Paris: Imprimerie Royale, 1693.

Cassini, Jean-Dominique. *Observations sur la comète qui a paru au mois de décembre 1680. et en janvier 1681. Presentées au Roy. Par Mr. Cassini, de l'Académie royale des sciences* Paris: Estienne Michallet, 1681.

Cassini, Jean-Dominique. *Premières observations de la comète de ce mois d'aoust 1682, présentées au Roy par M. Cassini de l'Académie royale des sciences*. Paris: Sébastien Mabre-Cramoisy, Imprimeur du Roy, 1682.

Cassini, Jean-Dominique. *Theoria motus cometae anni MDCLXIV pars prima . . .* Rome: Typographia Fabii De Falco, 1665.

Charnes, Jean-Antoine de. *Conversations sur la critique de la Princesse de Clèves (1679)*. Edited by François Weil et al. Tours: Groupe d'Étude du XVIIe Siècle de l'Université François Rabelais, 1973.

Clément, J. M. B., and J. de Laporte. *Anecdotes dramatiques*. 3 vols. Paris: Veuve Duchensne, 1775.

La comète, comédie. Paris: C. Blageart, 1681.

[Corneille, Thomas and Jean Donneau de Visé?]. *La pierre philosophale: Comédie mêlée de spectacles*. Paris: C. Blageart, 1681.

Colbert, Jean-Baptiste. *Fortifications, sciences, lettres, beaux-arts, bâtiments*. Vol. 5 of *Lettres, instructions, et memoires de Colbert*. Edited by Pierre Clément. Paris: Imprimerie impériale, 1873.

Cyrano de Bergerac, Savien de. *Histoire comique par Monsieur de Cyrano Bergerac, contenant les Estats et Empires de la Lune*. Paris: Charles de Sercy, 1657.

De La Lande, Joseph Jérôme. *Bibliographie astronomique: Avec l'histoire de l'astronomie depuis 1781 jusqu'à 1803*. Paris: Imprimerie de la République, 1803.

Denis, Jean-Baptiste. *Mémoire concernant les arts et les sciences, présenté au Monsieur le Dauphin: 1 Février 1672*. Paris: Frédérique Léonard, 1672.

Denis, Jean-Baptiste. *Memoires, conférences et observations sur les arts et les sciences: Presentées à Monseigneur le Dauphin*. 2d ed. Paris: Laurent d'Houry, 1682.

Des Maizeaux, Pierre. *La vie de Mr Bayle.* Rev. ed. 2 vols. La Haye: P. Gosse et J. Neaulme, 1732.

Desmaizeaux, Pierre. *The Life of Mr Bayle: In a Letter to a Peer of Great Britain.* London: [s.n.], 1708.

[Diez, François, and François Ridelle]. *Balet des Comètes. Pour la tragédie d'Irlande.* Paris, 1665.

[Diez, François, and François Ridelle]. *Irlande tragédie sera représentée au collège de Clermont de la Compagnie de Jésus: Pour la distribution des prix fondez par Sa Majesté.* Paris, le VI. Aoust, à une heure après midi, 1665.

Dodart, Denis. *Mémoires pour servir à l'histoire des plantes.* Paris: Imprimerie Royale, 1676.

[Donneau de Visé, Jean]. *Mercure galant, dédié à Monseigneur le Dauphin.* Paris: Claude Blageart and Lyon: Thomas Amaulry, 1681.

[Donneau de Visé, Jean]. *Mercure galant, dédié à Monseigneur le Dauphin.* Lyon: Thomas Amaulry, May 1682.

[Donneau de Visé, Jean]. *Mercure galant, dédié à Monseigneur le Dauphin.* Paris: Claude Blageart, June 1687.

Donneau de Visé, Jean. *Suite du voyage des ambassadeurs de Siam en France . . . Novembre 1686.* Lyon: Thomas Amaulry, 1687.

Drake, Stillman, and C. D. O'Malley, trans. *The Controversy on the Comets of 1618: Galileo Galilei, Horatio Grassi, Mario Guiducci, Johan Kepler.* Philadelphia: University of Pennsylvania Press, 1960.

Félibien, André. *Description sommaire du chasteau de Versailles.* Paris: Guillaume Desprez, 1674

Fer, Nicolas de. *L'atlas curieux, ou le monde représenté dans les cartes générales et particuliéres du ciel et de la terre divisé tant en ses quatre principales parties que par états et provinces et orné par des plans et descriptions des villes capitales et principales et des plus superbes edifices qui les embelissent. Par N. de Fer géografe de Monseigneur le Dauphin.* 2 vols. Paris: Nicolas de Fer, 1705.

Fer, Nicolas de. "Description de l'Observatoire Royal." Bibliothèque Nationale de France Estampes Va 304 t.1/ H79463 bis, 1704.

Fontaney, Jean de. *Observations sur la comète de l'année 1680 & 1681, faites au Collège de Clermont.* Paris: Gabriel Martin, 1681.

Fontenelle, Bernard le Bovier de. *Entretiens sur la pluralité des mondes.* Paris: Veuve C. Blageart, 1686.

Fontenelle, Bernard le Bovier de. *Histoire de l'Académie royale des sciences.* Vol 1: *Depuis son établissement en 1661 jusqu'à 1686.* Paris: Gabriel Martin, Jean-Baptiste Coignard, fils, Hippolyte-Louis Guerin, 1733.

Furetière, Antoine. *Dictionnaire universel: contenant généralement tous les mots françois tant vieux que modernes, et les termes de toutes les sciences et des arts.* 3 vols. Geneva: Slatkine Reprints, 1971.

Gallois, Antoine. *Oraison funèbre de très-auguste princess Marie-Thérèse d'Austriche, reine de France et de Navarre, épouse du Roy Louis Le Grand XIV du Nom.* Paris: Guillaume de Luyne, 1783.

Gallot, Jacques. "La comète." D-Lem ms.II.6.14 f.52v–54r. Leipzig Städtische Bibliotheken, Musikbibliothek.

Gallot, Jacques. *Pièces de luth composées sur differens modes.* Paris: Bonneuil, n.d.

Gallot, Jacques, Alexandre Gallot, and Pierre Gallot. *Oeuvres des Gallot.* Edited and transcribed by Monique Rollin. Paris: Editions du Centre national de la recherche scientifique, 1987.

Godart de Beauchamps, Pierre-François. *Recherches sur les théâtres de France: Depuis l'année onze cens soixante et un; jusques à present.* 3 vols. Paris: Prault, 1735.

Grandmi, Jacques. *Le cours de la comète qui a paru sur la fin de l'année 1664, avec un traité de sa nature, de son movement et de ses effets.* Paris: Sébastien Cramoisy & Sébastien Mabre-Cramoisy, 1665.

Grandmi, Jacques. *Le Parallèle des deux comètes qui ont paru les années 1664 & 1665.* Paris: Sébastien Cramoisy & Sébastien Mabre-Cramoisy, 1665.

Hevelius, Johannes. *Cometographia, totam naturam cometarum . . .* Gdansk: Auctoris typis, & sumptibus, imprimebat Simon Reiniger, 1668.

Hevelius, Johannes. *Descriptio cometae anno Aerae Christ: 1665 exorti, cum genuinis observationibus, tam nudis, quam enodatis, mense Aprili habitis Gedani; Cui addita est mantissa prodromi cometici, observationes omnes prioris cometae 1664, ex iisque genuinum motum accuratè deductum, cum notis, & animadversionibus exhibens.* Gdansk: Auctoris typis, & sumptibus, imprimebat Simon Reiniger, 1666.

Hevelius, Johannes. *Prodromus cometicus, quo historia, cometae anno 1664: Exorti cursum, faciesq[ue] diversas capitis ac caudae accurate delineatas complectens nec non dissertatio, de cometarum omnium motu, generatione, variisque phaenomenis, exhibetur. Ad . . . Dn. J. Bapt. Colbert, Regis Christianissimi a Sanctioribus Consiliis.* Gdansk: Auctoris typis, & sumptibus, imprimebat Simon Reiniger, 1665.

Histoire de l'imprimerie et de la librairie: Où l'on voit son origine et son progrés jusqu'en 1689. 2 vols. Paris: Jean de la Caille, , 1689.

Huet, Pierre Daniel. *Traité de l'origine des romans.* In *Zayde histoire espagnole par Monsieur de Segrais: Avec un traité de l'origine des romans par Monsieur Huet,* vol. 1, by Marie-Madeleine de Lafayette. Paris: Claude Barbin, 1670.

Huygens, Christiaan, Johan Adriaan Vollgraff, and Hollandsche Maatschappij der Wetenschappen. *Oeuvres complètes de Christiaan Huygens.* 22 vols. La Haye: M. Nijhoff, 1888–1950.

Jombert, Charles-Antoine. *Catalogue raisonné de l'oeuvre de Sébastien Le Clerc, chevalier romain, dessinateur et graveur du cabinet du Roi.* Part 1. Paris: Charles-Antoine Jombert, 1773.

Jouvency, Joseph. H. Ferté. *De la manière d'apprendre et d'enseigner* [*De ratione discendi et docendi*]. Translated by Henri Ferté. Paris: Hachette, 1892.

La Fontaine, Jean de. *Les amours de Psyché et de Cupidon: Par monsieur de La Fontaine. Edition nouvelle, plus correcte que la précedente.* La Haye: Adrien Moetjens, marchand libraire, 1707.

Le Cat, Claude Nicolas. *Éloge de Monsieur de Fontenelle des trois académies de Paris, secrétaire de celles des sciences, &c. Par M. Le Cat, secrétaire perpétuel de l'Académie de Rouen, &c.* Rouen: J. N. Besogne, 1759.

Leclerc, Sébastien, Antoine Coypel, and Charles-Louis Simonneau. *Médailles sur les principaux événements du règne de Louis Le Grand, avec des explications historiques: Par l'Académie royale des médailles et des inscriptions.* Paris: Imprimerie Royale, 1702.

Lefèvre d'Ormesson, Olivier. *Journal d'Olivier Lefèvre d'Ormesson, et extraits des mémoires d'André Lefèvre d'Ormesson*, vol. 2, edited by Adolphe Chéruel. Paris: Imprimerie Impériale, 1861.

[La Grange]. *Registre de La Grange (1658–1685).* Edited by Edouard Thierry. Paris: J. Claye, 1876.

La Gravette de Mayolas et al. *Les continuateurs de Loret: Lettres en vers de La Gravette de Mayolas [et al.] 1665–1689.* Edited by James Armand de Rothschild. Paris: D Morgand et Ch. Fatout, 1881.

Le Large, François. *Explication des figures qui sont sur le globe terrestre de Marly.* Manuscript BnF.

Le Large, François. *Recueil des inscriptions des remarques historiques, et géographiques qui sont sur le globe terrestre de Marly, executé par le P. Coronelli.* Manuscript BnF.

Le Maire, Charles. *Paris ancient et nouveau, ouvrage très curieux où l'on voit la fondation, les accroissemens, le nombre des habitans et des maisons de cette grande ville, avec une description de ce qu'il y a de plus remarquable.* Vol. 3. Paris: Michel Vaugon, 1685.

Léris, Antoine de. *Dictionnaire portative historique et littéraire des théâtres: Contenant l'origine des differns théâtres de Paris, le nom de toutes les pièces qui y ont été représentées . . . et des pièces jouées en province ou qui ont . . . paru . . . depuis plus de trois siècles, avec des anecdotes et des remarques . . .* Paris: C. A. Jombert, 1754.

Lettres portugaises traduites en françois. Paris: Claude Barbin, 1669.

Loret, Jean. *La Muze historique, ou recueil des lettres en vers contentant les nouvelles du temp: Écrites à son altesse Mademoiselle de Longueville* par le Sr Loret. 2 vols. Paris: Charles Chenault, 1656–59.

Loret, Jean. *La Muze historique, ou recueil des lettres en vers contentant les nouvelles du temp: Écrites à son altesse Mademoiselle de Longueville, depuis duchesse de Nemours.* 5 vols. Edited by Charles-Louis Livet et al. Paris: P. Jannet, P. Daffis, and H. Champion, 1857–91.

Louis XIV. *Mémoires et réflexions.* Paris: Communication et tradition, 1995.

Lubieniecki, Stanisław. *Theatrum cometicum.* Amsterdam: Typis Daniel Baccamude apud Francis Cuyper, 1666–68.

Lubieniecki, Stanisław. *Theatrum cometicum.* Leiden: Pieter van der Meersch, 1681.

Magalotti, Lorenzo. *Saggi di naturali esperienze fatte nell'Accademia del Cimento sotto la protezione del serenissimo Principe Leopoldo di Toscano e descritte dal segretario dessa accademia.* Florence: Giuseppe Cocchini, 1666.

Mallement de Messange, Claude. *Dissertation sur les comètes à Monsieur le Procureur Général du Grand Conseil.* Paris: Jean Cusson, 1681.

Maupoint. *Bibliothèque des Théâtres.* Paris: Pierre Prault, 1733.

Ménestrier, Claude François. *Des ballets anciens et modernes selon les règles du théâtre.* Paris: René Guignard, 1682.

[Ménestrier, Claude François]. *Lettre d'un gentil-homme de province à une dame de qualité sur le sujet de LA COMÈTE*. Paris: Estienne Michallett, 1681.

Montpensier, Anne-Marie Louise d'Orléans. *Mémoires*. In *Oeuvres completes*, 2 vols. Edited by Jean Garapon. Paris: Honoré Champion, 2020.

Moréri, Louis et al. *Le grand dictionnaire historique, ou le mélange curieux de l'histoire sacrée et profane. Nouvelle edition, dans laquelle on a refondu les supplémens de M. l'abbé Goujet, le tout revu, corrigé et augmenté par M. Drouet*. Paris: Les libraires associés, 1759.

Newton, Isaac. *Principes thématiques de la philosophie naturelle, par feue Madame du Chastellet*. 2 vols. Translated and edited by Émilie du Châtellet. Paris: Desaint & Saillant, Lambert, 1759.

Newton, Isaac. *Sir Isaac Newton's Mathematical Principles of Natural Philosophy and His System of the World*. Translated by Andrew Motte. Edited by Florian Cajori. Berkeley: University of California Press, 1966.

Parfaict, François, and Claude Parfaict. *Dictionnaire des théâtres de Paris contenant toutes les pièces qui on été représentées jusqu'à present sur les différens théâtres François*. 7 vols. Paris: Lambert, 1756.

Pascal, Blaise. *Pensées de M. Pacal sur la religion et sur quelques autres sujets*. Paris: Guillaume Desprez, 1669.

Perrault, Charles. *Mémoires de ma vie*. Edited by Paul Bonnefon. Paris: Renouard, 1909.

Perrault, Charles, and Claude Perrault. "Pourquoi et comment l'observatoire a esté basty." In *Fortifications, sciences, lettres, beaux-arts, bâtiments*. Vol. 5 of *Lettres, instructions, et memoires de Colbert*. Edited by Pierre Clément, 515–16. Paris: de l'Imprimerie impériale, 1861.

Perrault, Claude. *Les dix livres d'architecture de Vitruve, corrigés et traduits nouvellement en François avec des notes et des figures*. Paris: Jean-Baptiste Coignard, 1673.

Perrault, Claude. *Mémoires pour servir a l'histoire naturelle des animaux*. Paris: Imprimerie Royale, 1671.

Pérelle, [Gabriel, Nicolas, Adam?]. *Veues des belles maisons de France*. Paris: N. Langlois, ca. 1680.

Petit, Pierre. *Dissertation sur la nature des comètes, au Roy: Avec un discours sur les prognostiques des eclipses et autres matières curieuses*. Par P. Petit, Intendant des Fortifications. Paris: Louis Billaine. 1665.

Picard, Jean. *Voyage d'Uranibourg, ou Observations astronomiques faites en Danemarck par Monsieur Picard de l'Académie royale des sciences*. Paris: Imprimerie Royale, 1680.

Pingré, Alexandre-Gui. *Annales célestes du dix-septième siècle: Ouvrage publié sous les auspices de l'Académie des sciences, par M. G. Bigourdan*. Edited by Guillaume Bigourdan. Paris: Gauthier-Villars, 1901.

Pingré, Alexandre-Gui. *Cométographie ou Traité historique et théorique des comètes*. 2 vols. Paris: Imprimerie Royale, 1783.

Renaudot, Théophraste. *La gazette de France*. Paris: Bureau d'adresse, 1653.

Rothschild, James, baron de, ed. *Les continuateurs de Loret: Lettres en vers de La Gravette de Mayolas, Robinet, Boursault, Perdou de Subligny, Laurent et autres (1665–1689)*. Paris: Damascène Morgand et Charles Fatout, 1881.

Saint-Simon, Louis de Rouvroy, duc du. *Mémoires complètes et authentiques du duc de Saint-Simon sur le siècle de Louis XIV et la régence.* 20 vols. Edited by Adolphe Chéruel. Paris: L. Hachette, 1856–58.

Sévigné, Marie de Rabutin Chantal. *Correspondance.* Edited by Roger Duchêne. Paris: Gallimard, 1972.

Sevin de Quincy, Charles. *Histoire militaire du règne de Louis-le-Grand, Roy de France enrichie des plans necessaires.* Paris: Denis Mariette, Jean-Baptiste Delespine, Jean-Baptiste Coignard, 1726.

Silvestre, Israël. *Les plaisirs de l'isle enchantée, ou les festes et divertissements du Roi à Versailles, divisez en sept journées et commencez le 7me jour de may l'année de 1664.* [Paris]: [Imprimerie Royale], [1673].

Sorbière, Samuel. *Discours de Mr de Sorbière sur la comète.* Paris: s.n., January 26, 1665.

Thevenot, Melchisedec. *Bibliotheca Thevenotiana, sive catalogus impressorum et manuscriptorum librorum bibliothecae viri clarissimi D. Melchisedicis Thevenot.* Paris: Florentinum & Petrum Delaulne, 1694.

Trousset de Valincour, Jean-Baptiste. *Lettres à Mme la marquise *** sur le sujet de la Princesse de Clèves.* Edited by Jacques Chupeau et al. Tours: l'Université de Tours, 1972. First published 1678 by Sébastien Marbre-Cramoisy (Paris).

Trublet, Nicolas-Charles-Joseph. *Mémoires pour servir à l'histoire de la vie et des ouvrages de Mr. de Fontenelle tires du 'Mercure de France' 1756, 1757, & 1758 par Mr. L'abbé Trublet.* 2d rev. ed. Amsterdam: Marc Michel Rey, 1759.

Villedieu, Marie-Catherine. *Les amours des grands hommes.* Edited by Nathalie Grande. Paris: Société des textes françaises modernes, 2015.

Villedieu, Marie-Christine. *Le favori, tragi-comédie (1665).* Edition established, edited, and annotated by Delphine Amstutz. Bibliothèque des littératures classiques. Paris: Hermann, 2017.

Voltaire, François-Marie Arouet. *The Age of Louis XIV: Translated from the Last Geneva Edition of M. de Voltaire . . . by R. Griffith, Esq.* 2 vols. London Fielding and Walker, 1779–80.

Voltaire, François-Marie Arouet. *Le siècle de Louis XIV, édition de 1751 préparée et présentée par Ulla Kölving.* Centre international d'étude du XVIIIe siècle, 2018. http://c18.net/vo/vo_textes_siecle.php?div1=29.

Modern Sources

Aït-Touati, Frédérique. *Fictions of the Cosmos: Science and Literature in the Seventeenth Century.* Translated by Susan Emanuel. Chicago: University of Chicago Press, 2011.

Albertini, Alexandra Willaume. "Rhétorique du discours contre la superstition dans les 'Pensées diverses sur la comète' de Pierre Bayle: Astronomie et littérature au service de la raison." *Studi Francesi* 174, no. 58.3 (2014): 455–66. http://journals.openedition.org/studifrancesi/1031.

Anderson, Benedict. *Imagined Communities. Reflections on the Origins and Spread of Nationalism.* Rev. ed. New York: Verso, 1998.

Apostolidès, Jean-Marie. *Le prince sacrifié: Théâtre et politique au temps de Louis XIV.* Paris: Les Éditions de Minuit, 1985.

Apostolidès, Jean-Marie. *Le roi-machine: Spectacle et politique au temps de Louis XIV*. Paris: Les Éditions de Minuit, 1981.

Ariew, Roger. "Theory of Comets at Paris during the 17th Century." *Journal of the History of Ideas* 53 (1992): 355–72.

Aubin, David. "The Fading Star of the Paris Observatory in the Nineteenth Century: Astronomers' Urban Culture of Circulation and Observation." In "Science and the City." Special issue, *Osiris*, 2d ser., 18, no. 1 (2003): 79–100.

Aubin, David, Charlotte Bigg, and H. Otto Sibum, eds. *The Heavens on Earth. Observatories and Astronomy in Nineteenth-Century Science and Culture*. Durham, NC: Duke University Press, 2010.

Ayala, Lucía. "Cosmology after Copernicus: Decentralisation of the Sun and the Plurality of Worlds in French Engravings." In *The Making of Copernicus: Early Modern Transformations of a Scientist and His Science*, edited by Wolfgang Neuber, Claus Zittel, and Thomas Rahn, 201–26. Leiden: Brill, 2015.

Ballon, Hilary. "Constructions of the Bourbon State: Classical Architecture in Seventeenth-Century France." In *Cultural Differentiation and Cultural Identity in the Visual Arts*, edited by Susan Barnes and Walter Melion, 135–48. Studies in the History of Art 27. Washington, DC: National Gallery of Art, 1989.

Barthes, Roland. *Le Plaisir du texte*. Paris: Seuil, 1982.

Beasley, Faith. *Revising Memory: Women's Fiction and Memoirs in Seventeenth-Century France*. New Brunswick, NJ: Rutgers University Press, 1990.

Beasley, Faith. *Salons, History, and the Creation of Seventeenth-Century France: Mastering Memory*. Aldershot: Ashgate, 2006.

Beasley, Faith. *Versailles Meets the Taj Mahal: François Bernier, Marguerite de La Sablière, and Enlightening Conversations in Seventeenth-Century France*. Toronto: University of Toronto Press, 2018.

Bellanger, Claude, et al. *Des origins à 1814*. Vol. 1 of *Histoire générale de la presse française*. Paris: Presses Universitaires de France, 1969.

Berger, Robert. *In the Garden of the Sun King: Studies on the Park of Versailles under Louis XIV*. Washington, DC: Dumbarton Oaks, 1985.

Berger, Robert. *A Royal Passion: Louis XIV as Patron of Architecture*. Cambridge: Cambridge University Press, 1994.

Betz, Louis-Paul. *Pierre Bayle und die "Nouvelles de la République des Lettres" (Este populärwissenschaftliche Zeitschrift) (1684–1687)*. Geneva: Slatkine Reprints, 1970. First published 1896 (Zurich).

Biagioli, Mario. *Galileo Courtier: The Practice of Science in the Culture of Absolutism*. Chicago: University of Chicago Press, 2018.

Biagioli, Mario. *Galileo's Instruments of Credit: Telescopes, Instruments, Secrecy*. Chicago: University of Chicago Press, 2006.

Biet, Christian. *Les miroirs du soleil: Le roi Louis XIV et ses artistes*. Paris: Gallimard, 2000.

Bjørnstad, Hall. *The Dream of Absolutism: Louis XIV and the Logic of Modernity*. Chicago: University of Chicago Press, 2021.

Blanchard, Jean-Vincent. *L'optique du discours au XVIIe siècle: De la rhétorique des jésuites au style de la raison modern*. Québec: Presses de l'Université Laval, 2005.

Blanchard, Jean-Vincent, and Hélène Visentin, eds. *L'invraisemblance du pouvoir. Mises en scene de la souveraineté au XVIIe siècle*. Paris: Schena Editore/Presses de l'Université de Paris-Sorbonne, 2005.

Blocker, Déborah, and Anne Piéjus. "Auctorialité, voix et publics dans le *Mercure galant*." *Dix-septième siècle* 270, no. 1 (January–March 2016): 3–8.

Blocker, Déborah, and Anne Piéjus, eds. "Nouveaux outils, nouveaux enjeux pour l'histoire de la culture et de la langue dans la presse française d'Ancien Régime: Le cas du 'Mercure galant.'" Special issue, *Dix-septième siècle* 270, no. 1 (January–March 2016).

Bloom, Gina. *Gaming the Stage: Playable Media and the Rise of English Commercial Theater*. Ann Arbor: University of Michigan Press, 2018.

Boch, Julie. *Approches de la pensée des Lumières*. Reims: Éditions et presses universitaires de Reims, 2012.

Bolduc, Benoît. *La fête imprimée: Cérémonies et spectacles politiques (1549–1662)*. Paris: Garnier, 2016.

Bollème, Geneviève. *Les almanachs populaires aux xviie et xviiie siècles: Essai d'histoire sociale*. Paris: Mouton & Co., 1969.

Bombart, Matilde. *Le savoir des clefs: Écritures et lectures à clef en France au XVIIe siècle*. Forthcoming.

Bonnefon, Paul. Introduction to *Mémoires de ma vie*, by Charles Perrault. Paris: Renouard, 1909.

Boquet, Félix. "L'Observatoire de Paris: A propos de deux gravures relatives à sa fondation." *L'astronomie* 25 (1911): 49–56.

Boschiero, Luciano. *Experiment and Natural Philosophy in Seventeenth-Century Tuscany: The History of the Accademia del Cimento*. Australasian Studies in History and Philosophy of Science 21. Dordrecht: Springer Verlag, 2007.

Boschiero, Luciano. "Giovanni Borelli and the Comets of 1664–1665." *Journal for the History of Astronomy* 40 (2009): 11–30.

Bost, Hubert. *Pierre Bayle*. Paris: Fayard, 2006.

Bost, Hubert. "Pierre Bayle et ses frères: les enseignements de la correspondence." In *Le rayonnement de Bayle*, edited by Philippe de Robert with Claudine Pailhès and Hubert Bost, 97–114. Oxford: Voltaire Foundation, 2010.

Bost, Hubert. *Un intellectual avant la lettre: Le journaliste Pierre Bayle, 1647–1706; L'actualité religieuse dans les Nouvelles de la République des Lettres, 1684–1687*. Amsterdam: APA-Holland University Press, 1994.

Bost, Joyce, and Hubert Bost. Introduction to *Pensées diverses sur la comète*, by Pierre Bayle. Paris: GF Flammarion, 2007.

Bots, Hans, and Françoise Waquet. *La République des Lettres*. Paris: Belin, 1997.

Boyce, Douglas. Partita no. 3, "La comète," for solo guitar, 2023.

Boysse, Ernest. *Le théâtre des jésuites*. Paris: Henri Vaton, 1880.

Brétéché, Marion. *Les compagnons de Mercure: Journalisme et politique dans l'Europe de Louis XIV*. Ceyzérieu: Champ Vallon, 2015.

Brétéché, Marion, and Dinah Ribard. "Qu'est-ce que les mercures au temps du *Mercure galant*?" *Dix-septième siècle* 270, no. 1 (2016): 9–22.

Briggs, Robin. "The Académie Royale des Sciences and the Pursuit of Utility." *Past and Present* 131 (May 1991): 38–88.

Brown, Harcourt. *Scientific Organizations in Seventeenth Century France (1620–1680)*. New York: Russell and Russell, 1967.

Burden, Michael. "A Spectacle for the King." In *Ballet de la nuit*, edited by Michael Burden and Jennifer Thorp, 3–8. Hillsdale, NY: Pendragon Press, 2009.

Burden, Michael, and Jennifer Thorpe, eds. *Ballet de la Nuit*. Hillsdale, NY: Pendragon Press, 2009.

Cajori, Florian. "Four Old Astronomical Observatory Buildings." *Scientific Monthly* 26, no. 4 (April 1928): 372–76.

Call, Michael. "Fortuna Goes to the Theater: Lottery Comedies in Seventeenth-Century France." *French Forum* 40, no. 1 (2015): 1–15.

Call, Michael. "Money for Nothing: Molière's Miser and the Risky World of Seventeenth-Century France." *Quidditas* 32 (2011): 8–29.

Callacciani, Domenico, and Alberto Frigo, eds. "Les academies avant l'Académie: L'essor des sociétés savants en France avant la foundation de l'Academie des sciences." *Dix-septième siècle* no. 292 (July 2021).

Campbell, Mary Baine. *Wonder and Science: Imagining Worlds in Early Modern Europe*. Ithaca, NY: Cornell University Press, 1999.

Canales, Jimena. "The Single Eye: Re-evaluating *Ancien Régime* Science." *History of Science* 39, no. 1 (March 2001): 70–94.

Canova-Green, Marie-Claude. "Scènes jésuites du marriage de Louis XIV et de l'infante Marie-Thérèse d'Autriche." In *Plaire et instruire*, edited by Anne Piéjus, 278–300. Rennes: Presses universitaire de Rennes, 2007.

Carlson, William R. "'Pensées' and 'Pensées Diverses': The Art of Persuasion in Pascal and Bayle." *French Forum* 4, no. 2 (1979): 137–46.

Carolus-Barré Louis. "Le collège Saint-Nicolas-aux-Clercs de Soissons au Moyen Age. Histoire et archéologie." *Bulletin de la Société Nationale des Antiquaires de France* (1973, 1975): 28–29.

Chaganti, Seeta. *Strange Footing: Poetic Form and Dance in the Late Middle Ages*. Chicago: University of Chicago Press, 2018.

Champion, Justin. "Bayle in the English Enlightenment." In *Le philosophe de Rotterdam: Philosophy, Religion and Reception. Selected Papers of the Tercentenary Conference held at Rotterdam, 7–8 December 2006*, edited by Wiep van Bunge and Hans Bots, 175–96. Leiden: Brill, 2008.

Cherbuliez, Juliette. *The Place of Exile. Leisure Literature and the Limits of Absolutism*. Lewisburg, PA: Bucknell University Press, 2005.

Cherbuliez, Juliette. "On Letting Sleeping Blonds Lie: Gender, Leisure Literature, and the Imagination in Fontenelle." *Romanic Review* 102, no. 1/2 (January–March 2011): 145–68.

Cherbuliez, Juliette. *In the Wake of Medea: Neoclassical Theater and the Art of Destruction*. New York: Fordham University Press, 2020.

Cherbuliez, Juliette. "Yesterday's Exiles." In *Self and State in the Early Modern World*, 24–25 Cambridge: University of Cambridge, 2021.

Chéreau, Achille. "Essai sur les origines du journalisme médical français." Paris: Aux Bureaux de l'Union Médicale, 1867.

Christout, Marie-Françoise. *Le ballet de cour au XVIIe siècle: The Ballet de Cour in the 17th Century*. Geneva: Minkoff, 1987.

Clark, Stuart. *Vanities of the Eye: Vision in Early Modern European Culture*. Oxford: Oxford University Press, 2007.

Clarke, Jan. "*La Devineresse* and the 'Affaire des Poisons." *Seventeenth Century French Studies* 28 (2006): 221–34.

Clarke, Jan. *The Guénégaud Theatre in Paris (1673–1680)*. 3 vols. Lampeter, Wales: Edwin Mellen Press, 1998.

Cohen, Sarah R. *Art, Dance, and the Body in French Culture of the Ancien Régime*. Cambridge: Cambridge University Press, 2000.

Conley, Tom. *The Self-Made Map: Cartographic Writing in Early Modern France*. Minneapolis: University of Minnesota Press, 2010.

Coquery, Natacha. *Tenir boutique à Paris au XVIIIe siècle: Luxe et demi-luxe*. Paris: CTHS Histoire, 2011.

Cornuaille, Philippe. "L'écran de *La devineresse*, ou Le miroir déformant de l'affaire des poisons." *Nouvelles de l'estampe* 258 (2017): 4–22.

Cosgrove, Dennis. *Apollo's Eye: A Cartographic Genealogy of the Earth in the Western Imagination*. Baltimore: Johns Hopkins University Press, 2001.

Cowart, Georgia. *Triumph of Pleasure: Louis XIV and the Politics of Spectacle*. Chicago: University of Chicago Press, 2008.

Cowdrick, Ruth Elizabeth. "The Early Reading of Pierre Bayle." PhD dissertation, Columbia University, New York, 1939.

Cuénin, Micheline. *Roman et Société sous Louis XIV: Madame de Villedieu*. 2 vols. Paris: Champion, 1979.

Daston, Lorraine, and Peter Gallison. *Objectivity*. New York: Zone Books, 2010.

Daston, Lorraine, and Katharine Park. *Wonders and the Order of Nature, 1150–1750*. New York: Zone Books, 2001.

Dear, Peter. *Discipline and Experience: The Mathematical Way in the Scientific Revolution*. Chicago: University of Chicago Press, 1995.

Dear, Peter. "Narratives, Anecdotes, and Experiments: Turning Experience into Science in the Seventeenth Century." In *The Literary Structure of Scientific Argument*, edited by Peter Dear, 135–63. Philadelphia: University of Pennsylvania Press, 1991.

DeJean, Joan. *The Age of Comfort: When Paris Discovered Casual and the Modern Home Began*. New York: Bloomsbury, 2009.

DeJean, Joan. *Ancients against Moderns: Culture Wars and the Making of a Fin de Siècle*. Chicago: University of Chicago Press, 1997.

DeJean, Joan. *How Paris Became Paris: The Invention of the Modern City*. New York: Bloomsbury, 2014.

DeJean, Joan. *Tender Geographies: Women and the Origins of the Novel in France*. New York: Columbia University Press, 1993.

Denise, Louis. *Bibliographie historique et iconographique du Jardin des plantes, Jardin royal des plantes médicinales, et Muséum d'histoire naturelle*. Paris: H. Daragon, 1903.

Desgraves, Louis. *Répertoire des programmes des pieces de théâtre jouées dans les Collèges en France (1601–1700)*. Geneva: Droz, 1986.

Djuth, Marianne. "Veiled and Unveiled Beauty: The Role of the Imagination in Augustine's Esthetics." *Theological Studies* 68 (2007): 77–91.

Dobre, Mihnea. "The Scientific Journals of the Seventeenth-Century: Cartesianism in Journal des Sçavans and Philosophical Transactions, 1665–1670." In *Branching Off: The Early Moderns in Quest for the Unity of Knowledge*, edited by Vlad Alexandrescu, 333–58. Bucharest: Zeta, 2009.

Doody, Margaret Anne. *The True Story of the Novel.* Camden, NJ: Rutgers University Press, 1996.

Dorvane, Jeannine. "Ballet de Collège." In *International Encyclopedia of Dance*, edited by Selma Jeanne Cohen, 1:282–85. Oxford: Oxford University Press, 1998.

Dufour-Maître, Myriam. *Les précieuses: naissance des femmes de lettres en France au XVIIe siècle.* Revised edition. Paris: Honoré Champion, 2008.

Eisenstein, Elizabeth. *The Printing Press as an Agent of Change: Communications and Cultural Transformations in Early-Modern Europe.* 2 vols. Cambridge: Cambridge University Press, 1979.

Febvre, Lucien, and Henri-Jean Martin. *The Coming of the Book: The Impact of Printing 1450–1800.* Translated by David Gerard. London: Verso, 1997.

Festa, Lynn. *Fiction without Humanity: Person, Animal, Thing in Early Enlightenment Literature and Culture.* Philadelphia: University of Pennsylvania Press, 2019.

Feuerbach, Ludwig. *Pierre Bayle, nach seinen für die Geschichte der Philosophie und Menschheit interessantesten Momenten.* Ansbach: C. Brügel, 1838.

Franko, Mark. *Dance as Text: Ideologies of the Baroque Body.* Cambridge: Cambridge University Press, 1993.

Franko, Mark, ed. *The Oxford Handbook of Dance and Reenactment.* Oxford: Oxford University Press, 2018.

Freedberg, David. *The Eye of the Lynx: Galileo, His Friends and the Beginnings of Modern Natural History.* Chicago: University of Chicago Press, 2002.

Frisch, Andrea. *Inventing the Eyewitness: Witnessing and Testimony in Early Modern France.* Chapel Hill: University of North Carolina Press, 2004.

Fumaroli, Marc. *Le poète et le roi: Jean de La Fontaine et son siècle.* Paris: Fallois, 1997.

Gal, Ofer, and Raz Chen-Morris. *Baroque Science.* Chicago: University of Chicago Press, 2013.

Gallagher, Catherine. "The Rise of Fictionality." In *The Novel*, edited by Franco Moretti, 1:337–63. Princeton, NJ: Princeton University Press, 2006.

Gallagher, Catherine. *Telling It Like It Wasn't: The Counterfactual Imagination.* Chicago: University of Chicago Press, 2018.

Garber, Daniel. "On the Frontlines of the Scientific Revolution: How Mersenne Learned to Love Galileo." *Perspectives on Science* 12, no. 2 (2004): 135–63.

Gauja, Pierre. *Institut de France, Académie des Sciences, Troisième Centenaire, 1666–1966. Quelques sourvenirs.* With Marie-Thérèse Delom. Observatoire de Paris 7–12 Juin 1966. Exhibition Catalog from 1966. Paris: Imprimerie Tournon et cie, 1966.

Gofflot, L.-V. *Le théâtre au college: Du moyen âge à nos jours*. Paris: Honoré Champion, 1907.

Goodman, Dena. *The Republic of Letters: A Cultural History of the French Enlightenment*. Ithaca, NY: Cornell University Press, 1994.

Gottlieb, Anthony. *The Dream of Enlightenment: The Rise of Modern Philosophy*. New York: Liveright, 2016.

Gottlieb, Anthony. "The Tolerant Philosopher: Why Pierre Bayle Is the Forgotten Figure of the Enlightenment." *New Statesman*, August 13, 2016. https://www.newstatesman.com/culture/books/2016/08/tolerant-philosopher-why-pierre-bayle-forgotten-figure-enlightenment.

Grande, Nathalie, and Edvige Keller-Rahbé, eds. "Madame de Villedieu, ou Les audaces du roman." Special issue. *Littératures classiques* no. 61 (Spring 2007).

Greenblatt, Stephen. *The Swerve: How the World Became Modern*. New York: Norton, 2012.

Gualandi, Andrea. *Teorie sulle Comete da Galileo a Newton*. Milan: Franco Angeli, 2009.

Guerrini, Anita. *The Courtiers' Anatomists: Animals and Humans in Louis XIV's Paris*. Chicago: University of Chicago Press, 2015.

Hahn, Roger. *The Anatomy of a Scientific Institution: The Paris Academy of Sciences, 1666–1803*. Berkeley and Los Angeles: University of California Press, 1971.

Hahn, Roger. "La foundation de l'Académie royale des sciences." In *Sciences et curiosités à la cour de Versailles*, edited by Béatrix Saule and Catherine Arminjon, 32–41. Paris and Versailles: Réunion des musées nationaux/Château de Versailles, 2011.

Hahn, Roger. "Les Observatoires en France." In *Enseignement et diffusion des sciences en France au XVIIIe siècle*, edited by René Taton, 653–58. Paris: École pratique des hautes études, 1964.

Hallays, André. *Les Perrault*. Paris: Perrin et Cie, 1926.

Hammond, Nicholas. *The Powers of Sound and Song in Early Modern Paris*. State College: Penn State University Press, 2019.

Harth, Erica. *Cartesian Women: Versions and Subversions of Rational Discourse in the Old Régime*. Ithaca, NY: Cornell University Press, 1992.

Harvey, Sara. "Les fins de l'obscurité dans les énigmes du *Mercure galant*." In *L'obscurité: Langage et herméneutique sous l'ancien régime*, edited by Delphine Denis, 171–82. Louvain-la-Neuve: Bruylant-Académia, 2007.

Harvey, Sara. "'L'Impartialité sera le premier de nos devoirs': Le critique journalistique dans les périodiques mondains (1650–1721)." In "Figures de lecteurs dans la presse et la correspondence en Europe (XVIe-XVIIIe siècle)." Special issue, *Mémoires du livre/Studies in Book Culture* 12, no. 1 (Spring 2021). https://doi.org/10.7202/1077803ar.

Hatin, Eugène. *Les gazettes de Hollande et la presse clandestine au 17e et 18e siècles*. Paris: René Pincebourde, 1865.

Hatin, Eugène. *Histoire politique et littéraire de la presse française*. 6 vols. Paris: Poulet-Malassis et de Broise, 1859.

Hazard, Paul. *The Crisis of the European Mind 1680–1715*. Translated by J. Lewis May. New York: New York Review Books, 2013.

Herrmann, Wolfgang. *The Theory of Claude Perrault*. London: Frances Schram, 1973.

Himelfarb, Hélène. "Versailles, Functions and Legends." In *The State*, vol. 1 of *Rethinking France: Les lieux de mémoire*, edited by Pierre Nora, translated by David P. Jordan and Mary Seidman Trouille, 269–333. Chicago: University of Chicago Press, 2001.

Hofmann, Catherine. "'Incomparable Globes' or 'Old Devices as Cumbersome as They Are Useless'? The Vicissitudes of the Great Globes of Coronelli 1683–1915." *Globe Studies* 53/54 (2005/2006): 24–41.

Hoffmann, Kathryn. *Society of Pleasures: Interdisciplinary Readings in Pleasure and Power during the Reign of Louis XIV*. New York: St. Martin's Press, 1997.

Hogg, Chloé. *Absolutist Attachments: Emotion, Media and Absolutism in Seventeenth Century France*. Evanston, IL: Northwestern University Press, 2019.

Hogg, Chloé. "Pour une esthétique des tablettes: Clélie et les tablettes à écrire au XVIIe siècle," *Seventeenth-Century French Studies* 28 (2006): 117–33.

Hogg, Chloé. "Some Observations on Seventeenth-Century Object Poems." Manuscript.

Howard, Skiles. *The Politics of Courtly Dancing in Early Modern England*. Amherst: University of Massachusetts Press, 1998.

Hughes, D. W. "The *Principia* and Comets." In "Newton's 'Principia' and Its Legacy." Special issue, *Notes and Records of the Royal Society of London* 42, no. 1, (January 1988): 53–74.

Hutchison, Keith. "Towards a Political Iconology of the Copernican Revolution." In *Astrology, Science and Society*, edited by Patrick Currey, 95–112. Woodbridge: Boydell & Brewer, 1987.

Ibbett, Catherine. *Compassion's Edge: Fellow-Feeling and Its Limits in Early Modern France*. Philadelphia: University of Pennsylvania Press, 2017.

Israel, Jonathan. *Enlightenment Contested: Philosophy, Modernity and the Emancipation of Man, 1670–1752*. Oxford: Oxford University Press, 2006.

Israel, Jonathan. "Pierre Bayle's Correspondence and Its Significance for the History of Ideas." *Journal of the History of Ideas* 80, no. 3 (2019): 479–500.

Jensen, Katherine Ann. *Writing Love: Letters, Women, and the Novel in France, 1605–1776*. Carbondale and Edwardsville: Southern Illinois University Press, 1995.

Johns, Adrian. "Miscellaneous Methods: Authors, Societies and Journals in Early Modern England." *British Journal for the History of Science* 33 (2000): 159–86.

Jorink, Eric. "Comets in Context: Some Thoughts on Bayle's *Pensées diverses*." In *Le philosophe de Rotterdam: Philosophy, Religion and Reception; Selected Papers of the Tercentenary Conference held at Rotterdam, 7–8 December 2006*, edited by Wiep van Bunge and Hans Bots, 51–68. Leiden: Brill, 2008.

Jouhaud, Christian. *Mazarinades: La Fronde des mots*. Paris: Aubier, 2009.

Kenley, McDowell E. "Mad Fools and the Praise of Folly: Matassins and the Ballets of Lully, Destouches and Campra (1660–1718). *Early Music* 45, no. 2 (August 2017): 445–57.

Klamt, Johann-Christian. "Astronomical Observatories in the Seventeenth and Eighteenth Centuries." In *Public Buildings in Early Modern Europe*, edited by Konrad Ottenhym, Krista De Jonge, and Monique Chatenet, 373–82. Architectectura Moderna 9. Turnhout: Brepolis, 2010.

Klestinec, Cynthia. "Practical Experience in Anatomy." In *The Body as Object and Instrument of Knowledge: Embodied Empiricism in Early Modern Science*, edited by Ofer Gal and Charles T. Wolfe, 33–58. Sydney: Springer, 2010.

Kronk, Gary W. *Ancient–1700*. Vol. 1 of *Cometography: A Catalogue of Comets*. Cambridge: Cambridge University Press, 1999.

Kronk, Gary W. *Comets: A Descriptive Catalog*. Hillside, NJ: Enslow, 1984.

Labrousse, Élisabeth. *Inventaire critique de la correspondance de Pierre Bayle*. Paris, Vrin, 1961.

Labrousse, Élisabeth. *Pierre Bayle: Du pays de foix à la cité D'Erasme*. La Haye: Nijhoff, 1963.

Labrousse, Élisabeth. *Pierre Bayle: Heterodoxie et rigorisme*. La Haye: Nijhoff, 1964.

Labrousse, Élisabeth. *Notes sur Bayle*. Paris: Vrin, 1987.

Lacoste, Edmond. *Bayle nouvelliste et critique littéraire*. Brussels: Maurice Lamertin, 1929.

Lallemand, Paul. *Histoire de l'éducation dans l'ancien oratoire de France*. Paris: Ernest Thorin, 1888.

Lankhorst, Otto. *Reinier Leers, uitgever en boekverkoper te Rotterdam (1654–1714): Een Europees "libraire" en zijn fonds*. Amsterdam and Maarssen: APA-Holland University Press, 1983.

LaPorta, Kathrina. *Performative Polemic: Anti-Absolutist Pamphlets in Late Seventeenth-Century France*. Newark: University of Delaware Press, 2021.

Laugaa, Maurice. *Lectures de Mme de Lafayette*. Paris: Armand Colin, 1971.

Lecomte, Nathalie. "Les interprètes des ballets dansés au collège Louis-le-Grand de 1684–1699." In *Plaire et instruire*, edited by Anne Piéjus, 114–37. Rennes: Presses universitaires de Rennes, 2007.

Lee, Natasha. "Planetary Perspectives in Enlightenment Fiction and Science." In *French Global: A New Approach to Literary History*, edited by Christie McDonald and Susan Rubin Suleiman, 94–109. New York: Columbia University Press, 2010.

Le Lutin d'Ecouves. "L'Esprit Français 12," November 23, 2015. Blog post. Accessed May 18, 2021. http://lelutindecouves.blogspot.com/2015/11/lesprit-francais-12.html.

Lennon, Thomas M. *Reading Bayle*. Toronto: University of Toronto Press, 1999.

Lewis, John. *Galileo in France: French Reactions to the Theories and Trial of Galileo*. New York: Peter Lang, 2006.

Lougee, Carolyn. *Le Paradis des femmes: Women, Salons and Social Stratification in Seventeenth-Century France*. Princeton, NJ: Princeton University Press, 1977.

Lux, David S. "Colbert's plan for the Grand Académie." *Seventeenth-Century French Studies* 12 (1990): 177–88.

Lyons, John D. *Before Imagination: Embodied Thought from Montaigne to Rousseau*. Stanford, CA: Stanford University Press, 2005.

Lyons, John D. *Kingdom of Disorder: Theory of Tragedy in Classical France*. West Lafayette, IN: Purdue Studies in Romance Literatures, 1999.

Lyons, John D. *The Phantom of Chance: From Fortune to Randomness in Seventeenth-Century French Literature*. Edinburgh: Edinburgh University Press, 2012.

MacLean, Ian. *Woman Triumphant: Women in French Literature. 1610–1652*. Oxford: Oxford University Press, 1977.

Mansel, Philip. *King of the World: The Life of Louis XIV*. Chicago: University of Chicago Press, 2020.

Mariage, Thierry. *The World of André Le Nôtre*. Translated by Graham Larkin. Philadelphia: University of Pennsylvania Press, 2010.

Marin, Louis. *Portrait of the King*. Translated and foreword by Tom Conley. Minneapolis: University of Minnesota Press, 1988.

Martin, Henri-Jean. *The French Book: Religion, Absolutism and Readership, 1585–1715*. Translated by Paul Saenger and Nadine Saenger. Baltimore: Johns Hopkins University Press, 1996.

Martin, Meredith, and Gillian Weiss. *The Sun King at Sea. Maritime Art and Galley Slavery in Louis XIV's France*. Los Angeles: Getty Research Institute, 2022.

Massip, Catherine. "Recherches Biographiques." In *Oeuvres des Gallot*, edited and transcribed by Monique Rollin, xv–xxiv. Paris: Éditions du Centre national de la recherche scientifique, 1987.

McClure, Ellen. *The Logic of Idolatry in Seventeenth-Century French Literature*. Rochester, NY: Boydell and Brewer, 2020.

McClure, Ellen. *Sunspots and the Sun King: Sovereignty and Mediation in Seventeenth-Century France*. Urbana and Chicago: University of Illinois Press, 2006.

McGowan, Margaret. "Dance in Sixteenth-and Early Seventeenth-Century France." In *Dance, Spectacle, and the Body Politick, 1250–1750*, edited by Jennifer Nevile, 94–116. Bloomington: Indiana University Press, 2008.

McKenna, Antony. *Études sur Pierre Bayle*. Paris: Honoré Champion, 2015.

McKenna, Antony, and Annie Leroux. "L'édition électronique de la correspondence de Pierre Bayle et l'étude de ses réseaux relationnels." In *Le rayonnement de Bayle*, edited by Philippe de Robert with Claudine Pailhès and Hubert Bost, 61–96. Oxford: Voltaire Foundation, 2010.

McLeod, Jane. *Licensing Loyalty: Printers, Patrons, and the State in Early Modern France*. State College: Penn State University Press, 2011.

Mélèse, Pierre. *Un homme de lettres au temps du grand roi: Donneau de Visé, fondateur du "Mercure galant."* Paris: Droz, 1936.

Mélèse, Pierre. *Répertoire analytique des documents contemporains d'information et de critique concernant le théâtre à Paris sous Louis XIV 1659–1715*. Paris: Droz, 1934.

Meli, Domenico Bertoloni. *Thinking with Objects: The Transformation of Mechanics in the Seventeenth Century*. Baltimore: Johns Hopkins University Press, 2006.

Melzer, Sara. *Colonizer or Colonized: The Hidden Story of Early Modern French Culture*. Philadelphia: University of Pennsylvania Press, 2012.

Melzer, Sara, and Katheryn Norberg, eds. *From the Royal to the Republican Body. Incorporating the Political in Seventeenth- and Eighteenth-Century France*. Berkeley: University of California Press, 1998.

Merlin, Hélène. *Public et littérature en France au XVIIe siècle*. Paris: Belles Lettres, 1994.

Merrick, Jeffrey. "The Body Politics of French Absolutism." In *From the Royal to the Republican Body: Incorporating the Political in Seventeenth- and Eighteenth-Century France*, edited by Sara Melzer and Katheryn Norberg, 11–31. Berkeley: University of California Press, 1998.

Middleton, William. *The Experimenters: A Study of the Accademia del Cimento*. Baltimore: Johns Hopkins University Press, 1971.

Mirimonde, M. P. de. Preface to *Le Ballet de Cour en France, 1581–1671*. Exhibition catalog. Aix en Provence: Cabinet des estampes de la Bibliothèque Nationale, 1971.

Mori, Gianluca. *Bayle philosophe*. Paris: Honoré Champion, 1999.

Mosely, Adam. "Past Portents Predict: Cometary *Historiae* and Catalogues in the Sixteenth and Seventeenth Centuries." In *Celestial Novelties on the Eve of the Scientific Revolution: 1540–1630*, edited by Dario Tessicini and Patrick Boner, 1–32. Florence: Leo S. Olschki Editore/Museo Galileo and Istituto e Museo di Storia della Scienza. 2013.

Mukerji, Chandra. *Territorial Ambitions and the Gardens of Versailles*. Cambridge: Cambridge University Press, 1997.

Naudeix, Laura. "Tragédie et ballet: Enjeux d'un spectacle mixte." In *Plaire et instruire*, edited by Anne Piéjus, 55–69. Rennes: Presses universitaire de Rennes, 2007.

Ndiaye, Noémie. *Scripts of Blackness: Early Modern Performance Culture and the Making of Race*. Philadelphia: University of Pennsylvania Press, 2022.

Needham, Maureen. "Louis XIV and the Académie Royale de Danse, 1664: A Commentary and Translation." *Dance Chronicle* 20, no. 2 (1997): 173–90.

Nevile, Jennifer, ed. *Dance, Spectacle, and the Body Politick, 1250–1750*. Bloomington: Indiana University Press, 2008.

Norman, Larry F. *The Shock of the Ancient: Literature and History in Early Modern France*. Chicago: University of Chicago Press, 2011.

Orlemanski, Julie. "Who Has Fictions? Modernity, Fictionality, and the Middle Ages." *New Literary History* 51, no. 1 (Winter 2020): 145–70.

Paige, Nicholas. *Before Fiction: The Ancien Régime of the Novel*. Philadelphia: University of Pennsylvania Press, 2011.

Pantin, Isabelle. "New Philosophy and Old Prejudices: Aspects of Reception of Copernicanism in a Divided Europe." *Studies in the History and Philosophy of Science* 30, no. 2 (1999): 237–62.

Parrott, David. "Art, Ceremony and Performance: Cardinal Mazarin and Cultural Patronage at the Court of Louis XIV." In *Ballet de la Nuit*, edited by Michael Burden and Jennifer Thorp, 9–18. Hillsdale, NY: Pendragon Press, 2009.

Pedersen, Olaf. "Some Early European Observatories." *Vistas in Astronomy* 20, no. 1 (1976): 17–28.

Pelletier, Monique. "Les globes de Louis XIV: Les sources françaises de l'oeuvre de Coronelli." *Imago Mundi* 34 (1982): 72–89.

Pelletier, Monique. "Scientific Connections, Ambitions and Financial Difficulties of the d'Estrées Family." *Globe Studies* no. 55/56 (2009): 153–60.

Pérez-Gómez, Alberto. *Architecture and the Crisis of Modern Science*. Cambridge, MA: MIT Press, 1983.

Pérez-Gómez, Alberto. Introduction to *Ordonnance for the Five Kinds of Columns after the Method of the Ancients*, by Claude Perrault. Translated by Indra Kagis McEwen. Los Angeles: Getty Publications, 1993.

Peters, Jeffrey N. *Mapping Discord: Allegoric Cartography in Early Modern French Writing*. Newark: University of Delaware Press, 2004.

Peters, Jeffrey N. *The Written World*. Space, Literature, and the Chorological Imagination in Early Modern France. Evanston: Northwestern University Press, 2018.

Petzet, Michael. "Claude Perrault als Architekt des Pariser Observatoriums." *Zeitschrift für Kunstgeschichte* 30, no. 1 (1967): 1–54.

Petzet, Michael. *Claude Perrault und die Architektur des Sonnenkönigs: Der Louvre König Ludwigs XIV und das Werk Claude Perraults*. Munich and Berlin: Deutscher Kustverlag, 2000.

Picon, Antoine. *Claude Perrault, 1613–1688, ou La Curiosité d'un classique*. Paris: Picard, 1988.

Piéjus, Anne. "Un spectacle collectif." *Dix-septième siècle* 238, no. 1 (2008): 9–17.

Piéjus, Anne, ed. *Plaire et instruire. Le spectacle dans les collèges de l'Ancien Régime*. Rennes: Presses universitaires de Rennes, 2007.

Poche, Jean. *Quelques addresses de libraires, imprimeurs, relieurs, marchands, etc. du 17e siècle*. Paris: Chamerot et Renouard, 1899.

Pognon, Edmond. "Les vicissitudes des globes de Marly." XVIIe Congrès international d'histoire des sciences. In *Actes, Histoire des instruments scientifiques* 10 (1971): 85–91.

Pont, Graham. "Plato's Philosophy of Dance." In *Dance, Spectacle, and the Body Politick, 1250–1750*, edited by Jennifer Nevile, 267–81. Bloomington: Indiana University Press, 2008.

Popkin, Richard. *The History of Scepticism from Savonarola to Bayle*. Oxford: Oxford University Press, 2003.

Prest, Julia. "Dancing King: Louis XIV's Roles in Molière's *Comédies-ballets*, from Court to Town." *Seventeenth Century* 16, no. 2 (2001): 282–98.

Prest, Julia. "The Politics of Ballet in the Court of Louis XIV." In *Dance, Spectacle, and the Body Politick, 1250–1750*, edited by Jennifer Nevile, 229–40. Bloomington: Indiana University Press, 2008.

Pruiksma, Rose. "Embodying Cosmological Order and Motion: Celestial Bodies, Royalty, and Mythology in French Court Ballets." Renaissance Society of America, Annual Meeting, Boston, MA, 2016.

Pruiksma, Rose. "Generational Conflict and the Foundation of the *Académie Royale de la Danse*: A Reexamination." *Dance Chronicle* 26, no. 2 (2003): 169–87.

Rabinovitch, Oded. "Chameleons between Science and Literature: Observation, Writing, and the Early Parisian Academy of Sciences in the Literary Field." *History of Science* 51, no. 1 (2013): 33–62.

Rabinovitch, Oded. *The Perraults: A Family of Letters in Early Modern Science*. Ithaca, NY: Cornell University Press, 2018

Racevskis, Roland. *Time and the Ways of Knowing under Louis XIV: Molière, Sévigné, Lafayette*. Lewisburg, PA: Bucknell University Press, 2003.

Reeves, Eileen. *Evening News: Optics, Astronomy, and Journalism in Early Modern Europe*. Philadelphia: University of Pennsylvania Press, 2014.

Ribard, Dinah. "La philosophie mise en recueils: les 'pièces fugitives.' In *GRIHL, De la publication: Entre Renaissance et Lumièeres*, 61–75. Paris: Fayard, 2002.

Richard, Hélène. *Les Globes de Coronelli*. Paris: Seuil/Bibliothèque nationale, 2006.

Riffaud, Alain. *L'aventure éditoriale du théâtre français au XVIIe siècle*. Paris: Presses de l'Université Paris-Sorbonne, 2018.

Riffaud, Alain. *Le libraire de Molière*. Arles: Les Sources Portaparole, 2022.

Roberts, William. "Perelle's Topographical Albums: Problems and Solutions." In "Relations and Relationships in Seventeenth-Century French Literature," Special issue, *Biblio* 17, no. 166 (2006): 163–76.

Rock, Judith. "The Jesuit College Ballets: What We Know and What's Next." *Journal of Jesuit Studies* 4 (2017): 431–52.

Rock, Judith. *Terpsichore at Louis-le-Grand: Baroque Dance on the Jesuit Stage in Paris*. Boston: Institute of Jesuit Sources, 1996.

Rollin, Monique. "La musique pour le luth des Gallot." In *Oeuvres des Gallot*, edited and transcribed by Monique Rollin, xxix–xxliv. Paris: Éditions du Centre national de la recherche scientifique, 1987.

Roman D'Elia, Una. "Classicism and the Problem of National Style: The Illustrations of Claude Perrault's Vitruvius." In *Coming About: A Festschrift for John Shearman*, edited by Lars Jones and Louisa Matthews, 333–39. Cambridge, MA: Harvard University Art Museums, 2001.

Roux, Sophie. "An Empire Divided: French Natural Philosophy (1670–1690)." In *The Mechanization of Natural Philosophy*. Boston Studies in the Philosophy of Science 282, edited by Daniel Garber and Sophie Roux, 55–95. Dordrecht: Springer, 2013

Roux, Sophie. "The Two Comets of 1664–1665: A Dispersive Prism for French Natural Philosophical Principles." In *The Idea of Principles in Early Modern Thought. Interdisciplinary Perspectives*, edited by Peter R. Anstey, 98–146. New York: Routledge, 2019.

Row, Jennifer Eun-Jung. *Queer Velocities: Time, Sex, and Biopower on the Early Modern Stage*. Evanston, IL: Northwestern University Press, 2022.

Rowell, Diana. *Paris: The "New Rome" of Napoleon I*. New York: Bloomsbury, 2012.

Russell, Nicolas, and Hélène Visentin, eds. *French Ceremonial Entries in the Sixteenth Century. Event, Image, Text*. Toronto: Center for Reformation and Renaissance Studies, 2007.

Rybczynski, Witold. *Home: A Short History of an Idea*. New York: Viking, 1986.

Sabatier, Gérard. *Versailles ou La disgrâce d'Apollon*. Rennes: Presses universitaires de Rennes, 2017.

Saunders, E. Stewart. "Politics and Scholarship in Seventeenth-Century France: The Library of Nicolas Fouquet and the College Royal." *Journal of Library History* 20, no. 1 (1985): 1–24.

Schechner Genuth, Sara. *Comets, Popular Culture, and the Birth of Modern Cosmology*. Princeton, NJ: Princeton University Press, 1997.

Schuwey, Christophe. *Un entrepreneur des lettres au XVIIè siècle: Donneau de Visé, de Molière au Mercure galant*. Paris: Classiques Garnier, 2020.

Schuwey, Christophe. "Naissance d'une figure déviante: la genèse des nouvellistes." In *L'Errance au xviie siècle*, edited by Lucie Desjardins, Marie-Christine Pioffet, and Roxanne Roy, 303–19. Papers on Seventeenth Century Literature. Tübingen: Gunter-Narr, 2016.

Sedley, David L. "La Fin de La princesse de Clèves comme fait divers." *Etudes epistémè* (2020): n.p.

Sgard, Jean. *Dictionnaire des journalistes, 1600–1789*. Voltaire Foundation. https://dictionnaire-journalistes.gazettes18e.fr.

Sgard, Jean. *Dictionnaire des journaux, 1600–1789*. Voltaire Foundation. https://dictionnaire-journaux.gazettes18e.fr.

Shank, J. B. *The Newton Wars and the Beginning of the French Enlightenment*. Chicago: University of Chicago Press, 2008.

Shapin, Steven. *Never Pure: Historical Studies of Science as If It Was Produced by People with Bodies Situated in Time, Space, Culture, and Society, and Struggling for Credibility and Authority*. Baltimore: Johns Hopkins University Press, 2010.

Shapin, Steven, and Simon Schaffer. *Leviathan and the Air Pump: Hobbes, Boyle and the Experimental Life*. Princeton, NJ: Princeton University Press, 1985.

Snyder, Laura. *Eye of the Beholder: Johannes Vermeer, Antoni van Leeuwenhoek, and the Reinvention of Seeing*. New York: Norton, 2016.

Söderlund, Inga Elmqvist. "Taking Possession of Astronomy: Frontispieces and Illustrated Title Pages in 17th Century Books on Astronomy." PhD dissertation. Stockholm University, Department of Art History, Stockholm, Sweden, 2010.

Soll, Jacob. *The Information Master: Jean-Baptiste Colbert's Secret State Intelligence System*. Ann Arbor: University of Michigan Press, 2009.

Stafford, Barbara Maria. *Artful Science. Enlightenment Entertainment and the Eclipse of Visual Education*. Cambridge, MA: MIT Press, 1994.

Stedman, Allison. *Rococo Fiction in France, 1660–1715: Seditious Frivolity*. Lewisburg, PA: Bucknell University Press, 2012.

Steinberger, Deborah. "*Le Mercure Galant* and Its Student Body: Donneau de Visé's Inclusive Pedagogy." *Cahiers du dix-septième* 17 (2016): 41–56.

Stone, Harriet. *Crowning Glories: Netherlandish Realism and the French Imagination during the Reign of Louis XIV*. Toronto: University of Toronto Press, 2019.

Stone, Harriet. *Tables of Knowledge: Descartes in Vermeer's Studio*. Ithaca, NY: Cornell University Press, 2006.

Storer, Mary Elizabeth. "Information Furnished by the *Mercure Galant* on the French Provincial Academies in the Seventeenth Century." *PMLA* 50, no. 2 (1935): 444–68.

Strong, Roy. *Art and Power: Renaissance Festivals 1450–1650*. Berkeley: University of California Press, 1984.

Stroup, Alice. *Botany, Patronage, and Community at the Seventeenth-Century Parisian Royal Academy of Sciences*. Berkeley: University of California Press, 1990.

Sturdy, David. *Science and Social Status: The Members of the Académie des Sciences 1666–1750*. Woodbridge, UK: Boydell Press, 1995.

Surgers, Anne. "Une image unifiée, un regard captif: Le décor du théâtre jésuite dans les traités du F. Andrea Pozzo S.J. (1642–1709)." In *Plaire et instruire*, edited by Anne Piéjus, 85–98. Rennes: Presses universitaires de Rennes, 2007.

Taton, R. "Les origines et les débuts de l'Observatoire de Paris." *Vistas in Astronomy* 20, no. 1 (1976): 65–71.

Tavernier, Eugène. *Du journalisme, son histoire, son rôle politique et religieux*. Paris: H. Oudin, 1902.

Tellier, Pierre Jean. *Aventurier médical au XVIIe siècle: Nicolas de Blégny*. Paris: Librairie Louis Arnette, 1932.

Tucker, Holly. *Blood Work: A Tale of Medicine and Murder in the Scientific Revolution*. New York: Norton, 2012.

Tucker, Holly. *City of Light, City of Poison: Murder, Magic and the First Police Chief of Paris*. New York: Norton, 2017.

Turnovsky, Geoffrey. "Les lecteurs du *Mercure Galant*: Trois aperçus." *Dix-septième siècle* 270 (January–March 2016): 65–80.

van Helden, Albert. "Essay Review: Galileo Galilei: *Le Messager des Étoiles*, transl. by Fernand Hallyn, and Galileo Galilei: *Sidereus Nuncius*, transl. by Isabelle Pantin," *Journal for the History of Astronomy* (February 1994): 56–58.

van Nouhuys, Tabitta. *The Age of the Two-Faced Janus: The Comets of 1577 and 1618 and the Decline of the Aristotelian World View in the Netherlands*. Leiden: Brill, 1998.

Veret-Basty, Elsa. "'Être quelque chose comme un auteur:' Le statut paradoxal de l'énigmatiste dans le *Mercure galant*." *Dix-septième siècle* 270, no. 1 (2016): 97–114.

Viala, Alain. *La France galante*. Paris: Presses universitaires de France, 2008.

Viala, Alain, ed. *L'esthétique galante*. Toulouse: Société des littératures classiques, 1989.

Viala, Alain. *Naissance de l'écrivain: Sociologie de la littérature à l'âge classique*. Paris: Minuit, 1985.

Vincent, Monique. *Anthologie des nouvelles du 'Mercure galant,' 1672–1710*. Paris: Société des textes français modernes, 1996

Vincent, Monique. *Donneau de Visé et le "Mercure galant."* 2 vols. Paris: Aux Amateurs de Livres, 1987.

Vittu, Jean-Pierre. "Diffusion et reception du *Journal des savants* de 1665 à 1714." In *La Diffusion et la lecture des journaux de langue française sous l'ancien régime / Circulation and Reception of Periodicals in the French*

Langue During the 17th and 18th Centuries, edited by Hans Bot, 167–75. Amsterdam and Maarssen: APA-Holland University Press, 1988.

Vittu, Jean-Pierre. "La formation d'une institution scientifique: *Le journal des savants* de 1665 à 1714." *Journal des savants* no. 1 (2002): 179–203.

Vittu, Jean-Pierre. "'Le people est fort curieux de nouvelles' l'information périodique dans la France des années 1690." *SVEC* 320 (1994): 105–44.

Wallis, Helen. "Coronelli's *Libro dei Globi*." *Der Globusfreund* no. 18/20 (1969–71), 124–29.

Welch, Ellen. "Fictions of the Courtly Self: French Ballet in the Age of Louis XIV." *Early Modern French Studies* 39, no. 1 (2017): 17–30.

Welch, Ellen. "State Truths, Private Letters, and Images of Public Opinion in the *Ancien Régime*: Sévigné on Trials." *French Studies* 67, no. 2 (April 2013): 170–83.

Welch, Ellen. *A Theater of Diplomacy: International Relations and the Performing Arts in Early Modern France*. Philadelphia: University of Pennsylvania Press, 2017.

Wellington, Robert. *Antiquarianism and the Visual Histories of Louis XIV*. Farnham, UK: Ashgate, 2015.

Williard, Ashley. *Engendering Islands: Sexuality, Reproduction, and Violence in the Early French Caribbean*. Lincoln: University of Nebraska Press.

Wolf, Charles. *Histoire de l'Observatoire de Paris de sa foundation à 1793*. Paris: Gauthier-Villars, 1902.

Yeomans, Donald. *Comets: A Chronological History of Observation, Science, Myth and Folklore*. New York: Wiley, 1991.

Zanger, Abby. "Le nouvelliste et son public." In *Order et contestation au temps des classiques*, edited by Roger Duchene and Pierre Ronzeaud, 2:197–209. N.p.: Papers on Seventeenth Century Literature, 1992.

Zanger, Abby. *Scenes from the Marriage of Louis XIV: Nuptial Fictions and the Making of Absolutist Power*. Stanford, CA: Stanford University Press, 1997.

Zuber, Roger. *Les "belles infidèles" et la formation du goût classique*. New rev. ed. Paris: Albin Michel, 1995.

INDEX

Numbers in italics refer to illustrations.